Transforming Nuclear Safeguards Culture

The Belfer Center Studies in International Security book series is edited at the Belfer Center for Science and International Affairs at the Harvard Kennedy School and is published by the MIT Press. The series publishes books on contemporary issues in international security policy, as well as their conceptual and historical foundations. Topics of particular interest to the series include the spread of weapons of mass destruction, internal conflict, the international effects of democracy and democratization, and U.S. defense policy.

A complete list of Belfer Center Studies in International Security books appears at the back of this volume.

Transforming Nuclear Safeguards Culture

The IAEA, Iraq, and the Future of Non-Proliferation

Trevor Findlay

Belfer Center Studies in International Security
The MIT Press
Cambridge, Massachusetts
London, England

This book was set in Minion Pro and Trajan by Jen Jackowitz.
Printed and bound in the United States of America.

Library of Congress Cataloging-in-Publication Data

Names: Findlay, Trevor, 1951– author. | Project on Managing the Atom (Harvard University), sponsoring body.
Title: Transforming nuclear safeguards culture : the IAEA, Iraq, and the future of non-proliferation / Trevor Findlay.
Other titles: Belfer Center studies in international security.
Description: Cambridge, Massachusetts : The MIT Press, 2022. | Series: Belfer Center studies in international security | This study is a product of the Project on Managing the Atom (MTA) at the Belfer Center for Science and International Affairs at Harvard University. The research was funded by the Carnegie Corporation of New York—Preface. | Includes bibliographical references and index.
Identifiers: LCCN 2021046104 | ISBN 9780262543736 (paperback)
Subjects: LCSH: International Atomic Energy Agency. | Nuclear nonproliferation—Iraq. | Nuclear energy—Security measures—Iraq. | Nuclear industry—Security measures—Iraq. | Nuclear industry—Iraq. | Nuclear weapons—Iraq.
Classification: LCC JZ5675 .F56 2022 | DDC 327.1/747—dc23/eng/20211210
LC record available at https://lccn.loc.gov/2021046104

10   9   8   7   6   5   4   3   2   1

Contents

List of Illustrations

List of Abbreviations

ABACC	Argentine-Brazilian Agency for Accounting and Control
ACDA	U.S. Arms Control and Disarmament Agency
C&S	Containment and Surveillance
CSA	Comprehensive Safeguards Agreement
CTBT	Comprehensive Nuclear Test Ban Treaty
CTBTO	Comprehensive Nuclear Test Ban Treaty Organization
CWC	Chemical Weapons Convention
DDG	IAEA Deputy Director General
DPRK	Democratic People's Republic of Korea (North Korea)
EAEC	European Atomic Energy Community (also EURATOM)
ECOSOC	UN Economic and Social Council
EMIS	Electro-Magnetic Isotope Separation
ESARDA	European Safeguards Research and Development Association
EURATOM	European Atomic Energy Community (also EAEC)
ExPo	IAEA Office of External Relations and Policy Coordination
FAO	UN Food and Agriculture Organization
GAO	U.S. General Accountability (formerly Accounting) Office
HEU	Highly-enriched uranium
IAEC	Iraq Atomic Energy Commission
IAEA	International Atomic Energy Agency
ICAO	International Civil Aviation Organization
ICNND	International Commission on Nuclear Non-Proliferation and Disarmament
IMO	International Maritime Organization
INFCE	International Fuel Cycle Evaluation
INFCIRC	IAEA Information Circular
INMM	Institute of Nuclear Materials Management
INPO	Institute of Nuclear Power Operations
INVO	Iraq Nuclear Verification Office
ISO	International Organization for Standardization
ITU	International Telecommunication Union
JCPOA	Joint Comprehensive Plan of Action
JIU	UN Joint Inspection Unit
kW	Kilowatt/s

kW(th)	Kilowatt/s (thermal)
LEU	Low-enriched uranium
MW	Megawatt/s
MW(th)	Megawatt/s (thermal)
NAM	Non-Aligned Movement
NASA	U.S. National Aeronautics and Space Administration
NEA	OECD Nuclear Energy Agency
NGO	Non-Governmental Organization
NNWS	Non–Nuclear Weapon State/s
NPT	Nuclear Non-Proliferation Treaty
NRC	U.S. Nuclear Regulatory Commission
NTM	National Technical Means
OECD	Organization for Economic Cooperation and Development
OMV	Ongoing Monitoring and Verification (Iraq)
OPCW	Organization for the Prohibition of Chemical Weapons
OTA	U.S. Office of Technical Assessment
QMS	Quality Management System
RSAC	Regional System/s of Accounting and Control
SAC	IAEA Scientific Advisory Committee
SAGSI	IAEA Standing Advisory Group on Safeguards Implementation
SAL	IAEA Safeguards Analytical Laboratory
SEERT	State-Level Effectiveness Evaluation Review Team
SEG	State Evaluation Group
SIPRI	Stockholm International Peace Research Institute
SIR	Safeguards Implementation Report
SLA	State-Level Approach
SLC	State-Level Concept
SQP	Small Quantities Protocol
SSAC	State System/s of Accounting for and Control of Nuclear Material
TC	Technical Cooperation (IAEA)
UNAEC	UN Atomic Energy Commission
UNESCO	UN Educational, Scientific and Cultural Organization
UNMOVIC	UN Monitoring, Verification and Inspection Commission
UNSC	UN Security Council
UNSCOM	UN Special Commission
UNSCR	UN Security Council Resolution
VCDNP	Vienna Center for Disarmament and Non-Proliferation
VERTIC	Verification Research, Training and Information Center

VIC	Vienna International Center
WiN	Women in Nuclear
WINS	World Institute for Nuclear Security
WMD	Weapon/s of Mass Destruction

Preface

THIS STUDY IS a product of the Project on Managing the Atom (MTA) at the Belfer Center for Science and International Affairs at Harvard University. The research was funded by the Carnegie Corporation of New York. The project was stimulated by the growing international interest in nuclear safety and security cultures and the likelihood that similar cultural concepts could also be applied to nuclear safeguards. It was also prompted by reports that the International Atomic Energy Agency (IAEA) had sought to change its safeguards culture after the Iraqi non-compliance case of the early 1990s revealed the shortcomings of its safeguards system.

I began research for this project by reviewing the general literature on organizational culture and consulting management theory experts. Professor John Carroll, Morris A. Adelman Professor of Management at the Massachusetts Institute of Technology's Sloan School of Management, was of great assistance in helping me understand the finer points of organizational culture theory. I then conducted archival and current documentary research on the origins of IAEA safeguards culture and its evolution since the Iraq case. One challenge in pursuing such research is that the IAEA Secretariat does not publicly report to its member states the details of its organizational, management, or personnel changes.[1] This secrecy extends even to its archives. Unlike other organizations in the UN system, the IAEA will only release documents after thirty years, rather than the standard twenty years elsewhere at the United Nations, and even then the Agency insists that certain documents remain classified indefinitely.[2] The analysis in this book is necessarily hampered by such constraints.

I have addressed the lack of official documentation on organizational change within the IAEA through interviews and email correspondence with current and former IAEA staff, including inspectors and safeguards experts; officials of IAEA member states; and academic and government researchers. Official interviews with current staff were difficult to arrange at IAEA headquarters. Instead, it was necessary to conduct them off-site, at conferences or in informal meetings. In addition, I drew on the invaluable oral history interviews of IAEA safeguards staff conducted by the Pacific Northwest Nuclear Laboratory (PNNL) from 2005 to 2009 and by Elisabeth Roehrlich for the IAEA History Project at the University of Vienna from 2015 to 2016. I also scoured published materials—personal reminiscences and memoirs—of former Directors General and other IAEA officials for cultural stories and references. My research was further supplemented by a workshop on safeguards culture convened by MTA at Harvard in May 2014, which involved former IAEA officials and academics.

I presented a preliminary version of the study at the IAEA's International Safeguards Symposium in Vienna in October 2014. I received comments both during the formal session and afterward. Furthermore, I attended and interviewed participants at workshops and conferences held by the IAEA and the Institute of Nuclear Materials Management (INMM). Such events, in addition to being devoted to policy and technical issues, are also jamborees for the nuclear community and are thus culturally revealing. I also had access to papers presented at workshops on safeguards culture convened by the INMM and the European Safeguards Research and Development Association (ESARDA) in 2005 in Santa Fe, New Mexico and by the INMM at Texas A&M University in April 2016.

One of the ways in which organizational culture can be investigated—often used by management consultants—is to conduct a formal survey and systematic interviews of an organization's personnel. This method was not possible at the IAEA due to the then-management's constraints on transparency and openness, an unfortunate hallmark of the organization's culture. In any case, formal surveys do not necessarily reveal the deepest aspects of culture, as respondents are not always willing to be frank about their true beliefs.

Due to such constraints, much of the evidence for this study is therefore unavoidably anecdotal rather than based on systematic collection of data. Moreover, given the sensitivities associated with revealing information about organizational culture, most of those who were interviewed or provided comments for this book prefer anonymity. The 2014 Harvard workshop was run according to Chatham House rules, according to which comments may not be attributed to individuals without permission. I am grateful to all my interlocutors for their frank insights and willingness to humor me with my probing and sometimes uncomfortable questions.

I am thankful for the enthusiastic assistance of the staff of the IAEA Archives, especially director Leopold Kammerhofer and Marta Riess, despite the constraints under which they operate. In addition, I am grateful to those who read the manuscript, in whole or in part, for their insightful comments and suggestions; among them were Matthew Bunn, John Carlson, Jim Casterton, Vilmos Cserveny, Olli Heinonen, Shirley Johnson, Robert Kelley, Martin

Malin, Steven Miller, Tariq Rauf, Therese Renis, Laura Rockwood, and John Tilemann.

I would also like to thank Raoul Awad, Hans Blix, Patrick Burton, Colin Carroll, Jill Cooley, Jean-Maurice Crete, Ken Desson, Greg Dupuy, Craig Everton, Robert Floyd, Sarah Frazar, Kurt Kessler, John Lepingwell, Steve Mladineo, Herman Nackaerts, Dimitri Perricos, Joe Pilat, Steve Pullinger, Jenni Rissanen, Carl Robichaud, Elisabeth Roehrlich, Mark Schanfein, Tom Shea, Elena Sokova, Anne Starz, Tero Varjoranta, and Kevin Veal for their comments and suggestions at various stages of the project.

This book would not have been possible without the enthusiastic support and assistance of the faculty and staff of the Belfer Center for Science and International Affairs. Among them were Graham Allison, Matthew Bunn, Sean Lynn-Jones, Steven Miller, and William Tobey. I owe a special debt of gratitude to the late Martin Malin, MTA's former Executive Director, who was an unflagging guide and mentor during the many challenges I faced during the completion of this study.

MTA Research Assistant Brett Cox conducted a valuable literature search on safeguards culture, while Katie Miller, Josh Anderson, and Casey Campbell provided critical production support. My partner Christopher Reberger assisted at crucial moments with IT and graphics. Beth Clevenger and Anthony Zannino at the MIT Press were model collaborators in ushering the book into print. Above all, I am thankful to my editors at the Belfer Center, Morgan Kaplan and Karen Motley, for their moral and editorial support in getting a book to publication during a global pandemic. I am eternally grateful to Karen for her superb editorial skills and Herculean efforts in the final sprint to publication.

Naturally, notwithstanding the assistance I have received, I am completely responsible for the content of this book.

This study is dedicated to my friend, the late Dimitri Perricos (1936–2016), a legendary IAEA inspector who, as a cultural skeptic, introduced me to the term "candy concept."

Trevor Findlay
Melbourne, January 2022

Introduction

IN MAY 1991, following Iraq's defeat in the Gulf War, inspectors from the International Atomic Energy Agency (IAEA) entered Iraq to conduct inspections of the country's nuclear facilities. They were astonished to discover that for years, Iraq had been violating its international commitment not to acquire nuclear weapons. The Iraqis had assembled a substantial nuclear weapons infrastructure "over the berm" from the peaceful nuclear facilities and materials that had been declared to the Agency and which had been subject to IAEA nuclear safeguards. Not only did Iraq have a substantial secret nuclear weapons program hidden beyond the view of IAEA inspectors, but it had also successfully misled inspectors about the purposes of its declared nuclear activities.[1] Saddam Hussein's regime had come closer to acquiring nuclear weapons than anyone, including the IAEA itself, had imagined.

Iraq was among the first countries to join the IAEA, and one of the first to sign and ratify the 1968 Nuclear Non-Proliferation Treaty (NPT) and a comprehensive safeguards agreement (CSA). Iraq also became the first country to confront the Agency with a major safeguards violation.[2] The apparent ease with which it fooled the Agency meant that the credibility of the entire safeguards system was at stake. The then–Director General of the IAEA, Hans Blix, said that the revelations "shook the world."[3] Critics accused the IAEA Secretariat of being "complacent and unobservant."[4] Veteran IAEA official David Fischer lamented at the time that "the IAEA was seen by many as having failed its (presumably) first diversion detection test; it has patently been unable to detect a large and longstanding undeclared programme. Without the Gulf War, the IAEA may not have discovered the programme until the Iraqi Government openly demonstrated that it had acquired the bomb."[5] He records in his official history of the Agency that the Iraq case "demonstrated serious defects in the system."[6]

How did this happen? How did the international organization charged with detecting non-compliance with the NPT fail to discover Iraq's clandestine weapons program, which had been planned since 1972, pursued in research and development mode until 1988, and then visibly ramped up into large-scale facility construction for three years prior to 1991? With almost half a century of experience with safeguards, how was it that the IAEA failed to detect a nuclear weapons program located in plain sight of its inspectors? Why had they chosen not to look and inquire? Wasn't the whole point of nuclear safeguards that violations be detected early enough for the international community to act? Long before the Iraq case, safeguards expert Paul Szasz explained that:

> It hardly seems conceivable that even a large State could maintain entirely separate two massive nuclear cycles, one open and controlled and the other secret—keeping in mind that the frequent presence of inspectors in the country and their contacts with many persons dealing with nuclear matters would mean that even in the most repressed society rumors and news of the construction of large "secret" facilities would reach the Agency long before these became operational.[7]

Mohamed ElBaradei, IAEA Director General from 1997 to 2009, himself asked: "Why had the IAEA not challenged the Iraqis on the incompleteness of their declarations? Why had there been no calls for special inspections? How could the IAEA have 'missed' Iraq's broader nuclear ambitions?"[8]

This book seeks to answer such questions by considering the possible role of the IAEA's organizational culture, especially its safeguards culture. While there is no agreed definition of an ideal IAEA nuclear safeguards culture, as there is for both nuclear safety and nuclear security, I draw on those models to propose such a definition as: "That assembly of characteristics, attitudes, and behavior, exhibited by the Agency's personnel, which supports and enhances effective and efficient nuclear safeguards as a vital contribution to the non-proliferation of nuclear weapons and international peace and security."[9]

In the decades after the inception of safeguards in the late 1950s, safeguards culture had become cautious, process-driven rather than goal-oriented, and complacent about potential non-compliance. It had drifted away from the original goal of deterring states' clandestine acquisition of nuclear weapons by detecting any illicit acquisition of nuclear materials. Rather, it became entirely focused on nuclear material and facilities declared by states, while ignoring the possibility that they might fail to declare everything that they possessed.

Organizational theorists, especially those concerned with investigating major accidents, have increasingly come to use three broad "lenses" or perspectives to analyze organizational success and failure: the institutional, the political, and the cultural.[10] Each lens magnifies a stereotype that is never found in pure form in real life, but which nonetheless provides useful analytical insights into what drives organizations. As John Carroll argues, "By using all the lenses to analyze an organization or problem, we gain new insights and a richer understanding of organizational dynamics and human behavior within organizations."[11]

First, there is the institutional perspective, which Carroll calls the strategic design lens; it portrays organizations as "a kind of machine" that can be rationally designed for optimal performance. As Carroll notes, "People, money, equipment, and information are assigned using logical principles of efficiency and effectiveness to achieve organizational goals."[12] In this case, Carroll says, "action comes about through planning."[13]

This model, applied to the IAEA, would assume that the original negotiators of the IAEA's safeguards system strove for maximum effectiveness and efficiency and that the organizational design reflects these aspirations. As implementation of the system proceeded, the Board of Governors and the Secretariat would make decisions designed to maximize the operation of the system and ensure that it evolved optimally. Problems with safeguards can, this perspective holds, be fixed with better planning, management, administration, and resources.

One hears this view, naturally, from IAEA managers, who explain that if only they are given the right tools, they could do the job more effectively. Management consultants and some critics contend that the IAEA is unable to meet the challenge of implementing safeguards principally because of its management and administrative flaws. Both characteristics, in this view, derive from the Agency being a United Nations–type institution, rather than from any inherent deficit in safeguards as negotiated. This perspective, while clearly of some merit, is nonetheless inadequate in entirely explaining organizational behavior. Organizational theorists such as John Child bemoan the fact that "most of the literature on organizational design treats it as a purely technical matter, a question of adjusting structure to suit prevailing contingencies."[14]

Second, there is the political lens. In contrast to the institutional or strategic design lens, the political lens assumes that "everything is political," or at least that most important decisions are driven by politics. It is the quest for power, rather than rational analysis, that has the greatest effect on organizational performance. As Carroll, citing Jeffrey Pfeffer, puts it, "The political lens views the organization as a contested struggle for power (the ability to get things done) among stakeholders with different goals and underlying interests."[15] Action in this case comes through jockeying for power and negotiations between various interest groups. One hears this perspective on the IAEA most vividly, often expressed cynically, from diplomats accredited to the Agency. But it also dominates the views of some Agency employees and some external observers, especially the media.

The governance of international organizations such as the IAEA is arguably particularly prone to political machinations. Unlike entities such as corporations, which are driven principally by profit-maximization goals that management, boards of directors, and shareholders have in common, international organizations have as their principal stakeholders their member states, which often have widely differing opinions on their goals for and expectations of the organization. Unlike the shareholders of corporations, the stakeholders of international organizations themselves provide the management, funding, and staffing, creating multiple opportunities for political maneuvering and intrusion. Member state involvement can range from strong support for an organization's declared goals and mission at one end of the spectrum (such as Western support for safeguards) to efforts to hobble the organization's ambitions at the other (as India, Iran, Iraq, Pakistan, and Russia have all attempted to do in the case of IAEA safeguards).

Explanations by political scientists and international relations scholars of the behavior of international organizations thus emphasize the pivotal role of international politics.[16] Oran Young writes of the United Nations that "it does not operate above the hurley-burley of power politics. . . . The states which engage in power politics are its constituent units, and the problems generated by power politics constitute its reason for being."[17] Global affairs theorists highlight an eternal struggle between states' desire for international organizations to assist them in meeting collective challenges that they cannot meet individually, on the one hand, and their desire to preserve their national sovereignty, power, and prerogatives on the other. As Ian Hurd notes, "The tension between state obligations and state sovereignty provides the fuel that drives world politics in and around international organizations."[18] This struggle affects the IAEA's ability to detect and deal with violations of safeguards by limiting the capabilities of its verification system and constraining its continuing operation and enhancement. Problems faced by the safeguards system, the political perspective assumes, can only be solved by negotiations between states.

Third, there is culture theory, which, in contrast to the other two approaches, contends that the key to organizational dysfunction is often found in the values, norms, perceptions, attitudes, and resulting behavior of those who work in the organization, rather than in flaws in the formal structures and processes or in political machinations. Organizational behavior, viewed through this lens, often reflects implicit or even unconscious understandings of "the way we do things around here," rather than formal job descriptions or the official line about how things are done. The cultural approach focuses on how organizations find solutions to the challenges they face by institutionalizing "rules, rituals and values of the group."[19] From a cultural perspective, action comes through habit and routine.[20] In the formation of culture, political and institutional arrangements develop routine and unmindful behavior going beyond the original intent. In investigating culture, the most revealing comments often come from rank-and-file employees when explaining how they do their jobs day-to-day—in contrast to their job description and how management envisages those jobs being done pursuant to political or strategic planning decisions. A key insight of the cultural perspective is that decisions made for political and institutional reasons may be absorbed, shaped, distorted, ignored, or resisted by an organization's culture. Despite the best efforts of political and institutional forces, culture may prevail. A cultural perspective presumes that there is something ineffable in organizational behavior that is not reducible to factors such as material interests or power. As renowned organizational theorist Peter Drucker reportedly famously said, "culture eats strategy for breakfast."[21]

This study does not assume that culture explains all or even most organizational behavior. Rather, it accepts that culture is a useful lens through which organizations may be viewed to gain greater insights into why they operate the way they do. Exploration of the IAEA's organizational culture, especially its safeguards culture, will be used to supplement the standard explanations of the Iraqi non-compliance case—the political and the institutional—and to address nagging suspicions that they do not tell the whole story.

How could it be, for instance, that no one at the IAEA, whose main business is to worry about nuclear weapons proliferation, was sufficiently concerned to blow the whistle on Iraq? Why were disquieting media reports and the concerns of some member states not taken more seriously, prompting a review of the Agency's information? After all, Iraq was not a standard case; its nuclear research reactor, which it had claimed was for peaceful purposes, had been bombed by Israel in 1981 on the suspicion that it was part of a nuclear weapons program. There were clearly some aspects of Iraq's declared nuclear activities that did not make sense in terms of a standard peaceful research agenda, an issue that bothered both French and Italian technology suppliers. Was the organizational culture of the IAEA Secretariat, which had had almost half a century to gestate, blind to the type of non-compliance pathway taken by Iraq? Did IAEA member states, which are responsible for the Agency's governance, imbibe the same safeguards culture as the Agency's Secretariat, or did they help to create it? Even after the IAEA gained unprecedented access to Iraq in 1991 and began to uncover evidence of Iraq's non-compliance, why was the Agency reluctant to accept what it was finding? And why did the Agency, after Iraq's transgressions were revealed, initially begin reapplying its standard safeguards approach when it was clear that something more ambitious was required?

Was this case like the 1986 *Challenger* space shuttle disaster, which was allegedly spawned by the National Aeronautical and Space Administration's cultural "normalization of deviance"?[22] Or was it more like a "normal accident," such as the 2011 Fukushima nuclear power plant disaster, involving systems so complex that they are bound to fail sooner or later because human beings cannot conceptualize all the potential pathways to failure?[23] In the Iraq case, the IAEA safeguards system had some of the qualities of "an accident waiting to happen," because it dealt with the proliferation of complex technologies and failed to take account of all potential pathways to non-compliance. The question for this study is how much of this may be attributed to culture.

The principal motivation for this book is to help ensure that the cultural perspective is taken into account as continued efforts are made to strengthen IAEA nuclear safeguards. Understanding of and attention to organizational cultures in the fields of nuclear safety (prevention of nuclear accidents) and nuclear security (prevention of nuclear terrorism) are well advanced compared to that of nuclear safeguards culture, partly because the chain of events that produces the greatest threats are more imaginable.[24] This study addresses this lacuna; I consider these differences in more detail in Chapter 3. This book does not consider the national safeguards cultures of IAEA member states, except where they have contributed significantly to the Agency's safeguards culture (which is true of the United States and some other Western states). National safeguards culture has been analyzed extensively by others.[25] In fact, when most safeguards experts discuss safeguards culture, they are invariably referring to national safeguards cultures, not that of the IAEA. Furthermore, this volume does not examine in detail the safeguards culture of the two regional safeguards organizations, the European Atomic Energy Community (EAEC), also known as EURATOM, and the Argentine-Brazilian Agency for Accounting and Control (ABACC), except where they may have contributed to IAEA safeguards culture. This does not imply that these are unimportant regimes, but they pale in comparison to the size and complexity of the IAEA system and its foundational and transformative role in safeguards culture globally.

Unlike nuclear safety and nuclear security, where the state is held to be principally responsible, it is the IAEA that is primarily accountable for applying nuclear safeguards to states. It is the Agency alone that negotiates safeguards agreements with each state, establishes the appropriate safeguards regimes, conducts inspections, assesses a state's compliance with its obligations, and reports serious non-compliance to the UN Security Council. If safeguards fail, it is the Agency, rightly or wrongly, that is blamed. In fact, one of the assumptions of the IAEA's safeguards culture is that the credibility, professionalism, and technical competence of the IAEA's safeguards performance should be the gold standard for states and regional organizations to emulate in fulfilling their respective safeguards obligations.

In light of these considerations, the purpose of this study is to:

- discern what the IAEA's organizational culture, particularly its safeguards culture, can tell us about the failure of safeguards in the Iraq case, in a way that usefully supplements the traditional explanations;
- explore the challenges that the Iraq case posed to the then-prevailing IAEA's safeguards culture;
- examine whether a new safeguards culture has embedded itself at the Agency since the Iraq case along with the considerable strengthening of the safeguards system overall; and
- suggest how the Agency might deal with safeguards culture in the future.

In the first chapter of this volume, I analyze the IAEA nuclear safeguards system, the Iraqi nuclear program, and the way in which safeguards were applied to that program before 1991. In Chapter 2, I discuss the standard political and organizational explanations for why the IAEA missed evidence of illicit Iraqi nuclear weapons activity. In Chapter 3, I investigate the explanatory power of the cultural approach to the Iraq safeguards puzzle and explain how it complements the standard approaches. In the fourth chapter, I probe the culture shock experienced by the IAEA and its safeguards system as a result of its failure in the Iraqi case. In Chapter 5, I consider the state of IAEA safeguards today and measures that should be taken to cultivate an optimal safeguards culture. In my conclusion in Chapter 6, I argue that, alongside the standard political and organizational factors, culture played a discernible role in the IAEA's failure to detect Iraqi non-compliance in the several years prior to 1991. I conclude that as a result of the culture shock experienced by the Agency, a new safeguards culture has arisen since the Iraq case, though this was not due to a deliberate, systematic strategy of cultural change. Safeguards culture remains largely unexamined and, among some IAEA officials, dismissed as irrelevant. I recommend that the IAEA regularly assess the state of its safeguards culture using widely accepted techniques, and incorporate its findings into its strategic planning and organizational change strategies.

One

IAEA Nuclear Safeguards and Iraq's Nuclear Program

NUCLEAR SAFEGUARDS ARE a form of verification designed to detect the diversion of nuclear material from peaceful to military purposes. This chapter explores the main features of the safeguards system of the International Atomic Energy Agency (IAEA) and its standing prior to the discovery of Iraq's non-compliance in 1991, followed by an account of Iraq's declared nuclear activities and the application of safeguards to that program that had taken place up to that time.

The IAEA Safeguards System

Established in 1957, the IAEA is an international organization based in Vienna, currently with 173 member states. It has a permanent Secretariat headed by a Director General, with approximately 2,600 international civil servants (professional and support staff). It is governed by a General Conference of all member states and a 35-member Board of Governors comprising semi-permanent and elected representatives. The IAEA Statute mandates it to "seek to accelerate and enlarge the contribution of atomic energy to peace, health and prosperity throughout the world."[1] It is required to do this without furthering "any military purpose," notably by means of a monitoring and verification system known as "safeguards" that is intended to detect the misuse of peaceful applications of nuclear energy. In addition to safeguards, the IAEA also has programs on nuclear safety and security, nuclear technology, and technical cooperation for development. In 2021, its regular budget was €389.7 million, supplemented by an additional €204 million in voluntary contribution from member states.[2] Safeguards is its largest program, its most labor-intensive and its most expensive, consuming 39 percent (€151.1 million) of the regular budget in 2021. In 2020, €26.8 million in voluntary contributions from member states was spent on safeguards.[3]

The IAEA began developing its nuclear safeguards system from the outset of its establishment in Vienna in 1957. Initially, safeguards were aimed at ensuring that any assistance the Agency provided to its member states did not help them acquire nuclear weapons. This model followed bilateral U.S. agreements with other states that were designed to prevent recipients of U.S. assistance from using it to develop nuclear weapons. The Agency could also apply safeguards to bilateral or multilateral arrangements between states at their request. This system, which developed into what became known as INFCIRC/66 safeguards, was applied only to designated amounts of nuclear material and facilities.[4] U.S. bilateral safeguards agreements were eventually superseded by IAEA agreements.

To flesh out the practicalities of safeguards inspections, the Board of Governors approved a brief, three-page document, "The Agency's Inspectors" (attached to an explanatory note by the Director General that was confusingly titled "The Agency's Inspectorate"), in 1961.[5] Known colloquially as the "Inspectors' Document," it established that "the visits and activities of the Agency's inspectors shall be so arranged as to ensure on the one hand the effective discharge of their functions and on the other hand the minimum inconvenience to the State and disturbance to the facilities inspected."[6] Although states were generally to be given one week's notice of an inspection, such notice "need not exceed 24 hours for any inspection to investigate any incident requiring a 'special inspection.'"[7] Even more remarkably in view of what happened in Iraq, inspectors were to "have access at all times to all places and data and to any person, to the extent provided for in Article XIII.A.6 of the [IAEA] Statute."[8] That article has identical language about access rights, and it specifies that they apply to "materials, equipment and facilities which are required by this Statute to be safeguarded, as necessary to account for source and fissionable materials supplied and fissionable products and to determine whether there is compliance with the undertaking against use in furtherance of any military purpose."[9]

The first Agency safeguards on nuclear material were applied in 1959 to three tons of natural uranium supplied by Canada to Japan.[10] In 1961, the Board approved the first safeguards system for a type of nuclear facility—small research reactors. The IAEA made its first inspection of such a facility in 1962 to verify the design of a Norwegian reactor.[11] Early safeguards efforts were described as "rudimentary" and "technically amateurish," which is understandable because there were no multilateral models or precedents to follow.[12] Even among the IAEA's member states, only the United States had experience with safeguards. Symbolizing their Cinderella status in the early years, safeguards personnel were originally banished to the attic of the Grand Hotel, the IAEA's first home, with no elevator access.[13] Carlos Büchler records that "we were regarded by the rest of the staff, particularly the group that dealt with technical assistance, as a disruptive element and responsible for delaying action by the Board on the truly important functions of the Agency," the peaceful uses of nuclear energy.[14]

In 1968, the Nuclear Non-Proliferation Treaty (NPT) delegated to the IAEA the task of verifying compliance by its non–nuclear weapon state parties (NNWS) with their non-proliferation obligations.[15] Designed to stop the proliferation of nuclear weapons beyond the five states that already had them—China, France, the Soviet Union, the United Kingdom, and the United States—the treaty made safeguards mandatory for all NNWS parties. In vastly expanding the scope and nature of safeguards, the treaty, which entered into force in 1970, boosted the IAEA's status and reputation and gave it its true *raison d'être.*

The purpose of NPT safeguards, as set out in IAEA Document INF-CIRC/153, is "the timely detection of significant quantities of nuclear material from peaceful nuclear activities to the manufacture of nuclear weapons or of other nuclear explosive devices or for purposes unknown, and deterrence of such diversion by the risk of early detection."[16] This system came to be known as comprehensive or full-scope because it was intended to cover all peaceful nuclear activities in a state, whether indigenous or based on imported technology and materials.[17] The details were negotiated by a Committee of the Whole, open to all IAEA member states, which was established by the Board of Governors in 1970. Safeguards techniques would include nuclear material accountancy, seals on equipment and materials, remote monitoring of facilities and, most important, a variety of on-site inspections. To ensure the smooth implementation of safeguards, each state was obliged to sign a bilateral safeguards agreement and subsidiary arrangements with the Agency. Standard texts of such agreements were devised by the Secretariat.[18]

Nuclear Material Accountancy

Accounting for nuclear material was deemed to be "of fundamental importance" to comprehensive safeguards, with containment and surveillance (C&S) as "important complementary measures."[19] Nuclear material accountancy involved establishing accounting areas; record-keeping; nuclear material measurement; preparation and submission of accounting reports; and verifying the correctness of the data.[20] Such activity would be carried out in the first instance by the state itself, with inspections to confirm the results. The IAEA was not limited to verifying the state's findings, however; it could carry out independent measurements and observations. The role of inspectors was as follows:

> On the basis of inventory data submitted by the safeguarded state, inspectors look for changes in the inventory since the last inspection, based on receipt and transfer notices. This book inventory then becomes the basis for inspection. To verify the inventory, IAEA inspectors may count, weigh, measure the material, take samples for independent analysis, perform non-destructive tests, inspect seals, and make comparisons with accounting records.[21]

Containment and surveillance would include placing tamper-proof seals on equipment and installing remote monitoring devices. Such instruments would be used instead of human inspectors wherever feasible.[22]

Additional new concepts were later introduced that were not present in the INFCIRC/66 safeguards, including dividing a state's nuclear industry into "material balance areas" bounded by "strategic points," at which verification measures would be targeted. These activities were designed to make safeguards more cost-effective and efficient, lower the risk of exposing proprietary information, and keep procedures as non-intrusive and invisible as possible.[23] An important principle was that information obtained by the Agency through safeguards would be treated as "safeguards confidential." Data obtained during inspections or from the state itself would thus be classified and not released to unauthorized persons, other states, or the public.[24]

Unlike INFCIRC/66 safeguards, NPT safeguards were not, in theory, to be applied to facilities but only to materials. Yet, because most nuclear material is located at facilities, in practice this was a distinction without a difference. In any event, states were obliged to provide design information for facilities, which would be subject to on-site verification to confirm the details. The IAEA was to be notified about construction of new facilities as soon as possible, but no later than 180 days before the introduction of nuclear material.[25]

The new system recognized that reprocessing and enrichment facilities, not just nuclear reactors, should be subject to safeguards, because they could be used to produce bulk material for both peaceful and military purposes. Safeguards would also be applied to significant amounts of nuclear material at locations outside facilities (LOF). The IAEA was to be informed of international transfers of nuclear material subject to safeguards no later than the time they left their destination or arrived at it.[26] Procedures were envisaged for withdrawing nuclear material from safeguards for non-explosive military purposes, such as submarine propulsion, but the details awaited future elaboration.

Technical Goals

Although INFCIRC/153 set out the broad goals of verification, the Agency still needed to determine exactly what this meant technically, because it was not feasible to keep track of every gram of a state's nuclear material in real time. On the advice of the Standing Advisory Group on Safeguards Implementation (SAGSI), which had been established in 1975 to provide advice to the Director General, the Secretariat set goals for "timely detection" and "significant quantity." Timely detection—the estimated time needed to convert diverted material into a form suitable for a nuclear explosive device—was set at 7–10 days for plutonium or highly-enriched uranium (HEU) in metallic form; 1–3 months for plutonium in irradiated fuel; and about a year for natural or low-enriched uranium (LEU).[27] Significant quantity was defined as roughly the amount of material a state would need to make its first nuclear explosive device, taking into account waste and other losses: 8 kilograms of plutonium; 25 kilograms of uranium-235 contained in uranium enriched to 20 percent or more; or 8 kilograms of uranium-233.[28]

As for the risk of detection, David Fischer and Paul Szasz pointed out in one of the first critical appraisals of the safeguards system that this "lies in the eye of the potential diverter" and cannot be quantified.[29] Nonetheless, the Secretariat aimed to achieve, through material accountancy, a 90–95 percent likelihood of detecting a diversion and a less than 5 percent likelihood of sounding a false alarm.[30] The Secretariat asserted, somewhat enigmatically, that these were all goals rather than strict requirements against which safeguards performance should be judged.[31]

Inspections

INFCIRC/153 described in detail the purposes of on-site inspections and the activities that inspectors could and could not undertake.[32] New rules were set out for designating inspectors to conduct inspections in each state, including a state's right to refuse an inspector without specifying grounds. Repeated refusals would, however, be viewed as impeding effective safeguards implementation and be subject to action by the Board of Governors.[33]

"Routine" inspections would form the bulk of on-site inspections and would be carried out on a pre-determined schedule. They would be limited to locations within a declared facility or other sites where nuclear material was located or through which nuclear material was expected to flow. Some of these could be carried out as "unannounced" inspections through random sampling.[34] Such inspections would become the "core" of safeguards at enrichment plants, but rarely used elsewhere.[35]

Ad hoc (non-routine) inspections would be used chiefly to verify a state's initial report on its holdings of nuclear material and to confirm changes thereafter, but also to verify imports and exports of nuclear material.[36] Such activities were hardly the "surprise" inspections of popular imagination. They required the Secretariat to apply for visas for inspectors; alert national authorities at least 24 hours in advance and obtain their cooperation; arrange air travel and accommodation; and transport inspectors to inspection sites. Inspectors could be accompanied by state officials while conducting their duties.

"Special" inspections, already provided for under INFCIRC/66, were possible if they were "urgent and essential to verify non-diversion."[37] They could now also be conducted at "additional locations" to those declared by the state, or when substantial amounts of safeguarded nuclear material were transferred outside a state's jurisdiction.[38] Such inspections were further permitted to verify information in "special reports" from a state about significant changes in its nuclear materials inventory, or if the IAEA considered the information provided by the state or obtained from routine inspections inadequate. The only catch was that the Agency had to obtain the state's permission to conduct a special inspection.

The Limitations of Safeguards

Comprehensive NPT-type safeguards, in seeking to encompass all of a state's peaceful nuclear activities, were clearly an advance on the previous material- or item-specific agreements. But there were significant limitations, most of them obvious and widely acknowledged at the time.

While these safeguards were called "comprehensive," not all nuclear activities in the fuel cycle were covered. Uranium (or thorium) mining and processing activities were exempt, although imports and exports of "source materials" such as uranium and thorium were to be reported to the IAEA as "a form of safeguards."[39] Limits were also placed on the amounts of natural uranium, depleted uranium, and thorium in a state's possession, below which safeguards did not have to be applied.[40] The real starting point of safeguards at that time was when nuclear material was prepared for fuel fabrication.[41] Moreover, at a state's request, research laboratories were not safeguarded if they used small amounts of special fissionable material (below one kilogram in total for the state). Furthermore, the Agency had no right to obtain information about a new plant as soon as it was envisaged, planned, or under construction, but only "as early as possible" before nuclear material was introduced, usually specified in Subsidiary Arrangements as 180 days.[42] In addition, export controls were applied only to exports to non-NPT states as a way of encouraging them to seek the benefits of NPT membership. An entire reprocessing or enrichment plant could be exported to an NPT state party without any notification to the IAEA.[43]

Under "comprehensive" safeguards, the Agency's own powers and authorities were also far from comprehensive. Despite provisions in the IAEA Statute that essentially gave the Agency virtually "anytime, anywhere" inspection powers, such inspections were no longer envisaged.[44] As Hans Blix has lamented, IAEA inspectors "had no right to roam around a country looking for undeclared installations or activities."[45] Such activity would, in any case, not have been "meaningful" without intelligence from member states, he says, and "no channels had been established to provide such intelligence." Inspected states had the right to reject individual inspectors, "and many made use of that right."[46] Aerial overflights or other types of remote surveillance that had been foreshadowed in early international safeguards schemes promoted in the mid-1940s were no longer envisioned.[47] Satellite reconnaissance capabilities were restricted to only the most advanced states until the 1990s, and were certainly not contemplated for the IAEA.[48] Blix also says the system "had few teeth," presumably meaning that the IAEA had few tools to induce or compel states to comply, short of the drastic step of reporting them to the UN Security Council.[49]

One of the biggest flaws in the system, obvious at the time but little remarked upon, was the overwhelming emphasis of safeguards on detecting the diversion of declared material, rather than considering the possibility that a state might retain or acquire undeclared material or facilities or an entire undeclared nuclear fuel cycle. Moreover, safeguards were fixated on the nature and quantity of declared nuclear material at each declared location, rather than on the territory of the whole state.[50] The INFCIRC/153 negotiating record "reemphasizes the Agency's right and obligation to apply safeguards on all nuclear material in all peaceful nuclear applications." Negotiators rejected an alternative proposal which would have limited safeguards to nuclear material reported by the state.[51] In practice, however, safeguards focused on what was declared and could be observed and counted, ignoring what, hypothetically, might be undeclared, untraceable, and unverifiable.

Allan McKnight, the first IAEA Inspector General (the equivalent of today's Deputy Director General for Safeguards), along with many other early observers, acknowledged that no one believed safeguards were a "panacea" for proliferation.[52] Even the IAEA Statute recognized that the Agency's goal was to avoid nuclear energy being used for any military purpose only "so far as it is able."[53] Given the weaknesses in the system, Fischer and Szasz warned as early as 1975 that the IAEA "cannot entirely discount the possibility of a breach of its undertakings by any state in which it applies safeguards."[54] The two renowned experts described safeguards as:

> first and foremost a means of promoting greater confidence between nations: the extent to which they actually deter states from breaking their word is secondary to their role of building confidence. . . . Since governments accept the NPT and its safeguards by their own free decision (although sometimes as a condition for continued nuclear supply) there is a strong presumption that they will normally abide by their peaceful, non-explosive use pledges and that breaches of safeguards agreements are unlikely.[55]

The Status of Safeguards by 1990

By the end of 1990, five months before Iraq's non-compliance was revealed, safeguards agreements pursuant to the NPT were in force for 86 states.[56] All NNWS with significant nuclear holdings had been brought under the safeguards regime. The number of inspections performed in 1990 was 2,188, involving 10,381 person-hours.[57] Research reactors and critical assemblies under safeguards numbered 170.[58] The annual regular budget for safeguards was approximately $60 million.[59] The Safeguards Department staff totaled 1,749.

The Secretariat had continued developing the system as it gained experience in implementing it.[60] One incentive was the need to apply safeguards to Germany and Japan. Both of those countries were sensitive about discrimination between IAEA member states, concerned that safeguards might compromise their state and commercial secrets, and determined that safeguards should be as non-intrusive and cost-effective as possible. Myron Kratzer notes that further improvements to safeguards were catalyzed by India's nuclear test in 1974, which, "in the perverse way in which important events often unfold, tended to discredit the effectiveness of safeguards, even though none was involved."[61] The Board usually accepted or acquiesced to changes in safeguards—when it was informed about them. Member states more broadly, but not always, welcomed them and apparently complied without protest.

NPT states parties seemed satisfied with the performance of the system. In 1990, the Fourth NPT Review Conference's Committee II agreed on language for a final document reaffirming participants' conviction that:

> IAEA safeguards provide assurance that States are complying with their undertakings and assist States in demonstrating this compliance. . . . The Conference notes with satisfaction that since the last Review Conference, the IAEA has again, in carrying out its safeguards activities for the States Parties, not detected any diversion of a significant amount of safeguarded material to the production of nuclear weapons or other nuclear explosive devices or to purposes unknown.[62]

The IAEA Secretariat itself also seemed satisfied with the performance of its system. At the 1990 General Conference, its annual report, which was endorsed by member states, recorded that:

> In 1989, as in previous years, the Secretariat, in carrying out the safeguards obligations of the Agency, did not detect any event which would indicate the diversion of a significant amount of safeguarded nuclear material, or the misuse of facilities, equipment, or nonnuclear material subject to safeguards for the manufacture of any nuclear weapon, for any other military purpose, for the manufacture of any other nuclear explosive device, or for purposes unknown. It is considered reasonable to conclude that the nuclear material under Agency safeguards in 1989 remained in use for peaceful nuclear activities or was otherwise adequately accounted for.[63]

In carrying out its verification activities, the Secretariat had found "a number of discrepancies and anomalies," but these were "mostly minor and were satisfactorily explained upon subsequent appraisal or investigation."[64] No further details were given due to safeguards confidentiality.

Iraq's Nuclear Program and Safeguards

In 1956, Iraq became one of the first developing countries to launch a peaceful nuclear program, although it was initially limited to basic research, which progressed slowly, and was indifferently managed.[65] Iraq was also one of the first countries to sign and ratify the NPT and to negotiate a comprehensive safeguards agreement with the IAEA. With the rise to power of then–Vice President Saddam Hussein after a 1968 coup, Iraq's nascent nuclear program received greater official attention, and its scientists were accorded a "privileged position."[66] Intimations of nuclear weapons ambitions at its nuclear research establishment began to surface, although there was no formal decision at this time to begin a weapons program.[67] International controversy over Israel's destruction of Iraq's Tamuz-1 research reactor in June 1981 aroused doubts in the West about Iraqi intentions, as well as the effectiveness of IAEA safeguards.[68] Yet, after the controversy faded, the Iraqi program appeared to atrophy, and safeguards continued apparently uneventfully and effectively. We now know, however, that by 1979, when Saddam formally became president, Iraq had initiated a secret nuclear weapons program, which was beyond the reach of IAEA inspectors, alongside its declared program. The following section explores Iraq's nuclear program, the IAEA's application of safeguards, and how much the IAEA knew about the extent of Iraq's program prior to 1991.

Iraq's Declared Nuclear Program

Iraq's nuclear program began with the creation of the Iraq Atomic Energy Commission (IAEC) in the 1950s.[69] Amer Al-Sa'idi, a key player in Iraq's military industrialization, claimed that its original goal was research into the peaceful uses of nuclear energy, not the acquisition of nuclear weapons.[70] The program developed lethargically, initially boosted by UK and U.S. assistance and later with aid from the Soviet Union and the IAEA.

By the late 1960s, Iraq had set up a modest Nuclear Research Center based at Al-Tuwaitha, twenty miles south of Baghdad. Its only notable facility was a small 2 MW(th) research reactor, the IRT-2000, which had been supplied by the Soviet Union. The reactor was fueled with about 3.2 kilograms of 80-percent-enriched uranium provided by the Soviets. A further 27 kilograms was in storage as fresh or irradiated fuel.[71] Iraq was obliged, in compliance with standard Soviet non-proliferation policy, to return the spent fuel to the Soviet Union after its use. In 1968, the year Iraq signed the NPT, the reactor was commissioned, along with several small facilities for radioisotope production. Prior to 1973, the Iraqi nuclear program focused on theoretical research, which was "exploratory and, for all practical purposes, aimless," as the IAEA itself observed in its dealings with the Iraqis.[72]

After the 1973 Arab-Israeli War and a substantial increase in its oil revenues, Iraq sought to embark on an ambitious expansion of its nuclear program.[73] It was, however, unsuccessful in persuading France to sell it a plutonium production reactor similar to the one used in France's nuclear weapons program, along with a reprocessing plant to recover the plutonium produced in the reactor. Concerned about Iraqi intentions, France eventually agreed in 1976 to provide only a 40 MW(th) light-water reactor, which was considered more proliferation-resistant, along with HEU for fuel. It was designated Tamuz-1 by the Iraqis.[74] France also provided a smaller zero-power 500 kW(th) reactor called Tamuz-2, mainly as a neutronic mock-up for Tamuz-1; a materials-testing hot laboratory (called LAMA); workshops; and a radioactive waste treatment station.[75] Khidhir Hamza, the former head of the Iraqi nuclear program, who styled himself as "Saddam's bombmaker," subsequently described Tamuz-1 as "totally inappropriate" for a peaceful nuclear energy program, but ideal for irradiation of target materials to produce plutonium-239—a material that could be used to create a bomb.[76] IAEA safeguards were apparently not seen by the Iraqis as an insurmountable barrier to using it for this purpose.[77] The French pressed them unsuccessfully to accept their so-called Caramel fuel, which was allegedly more proliferation-resistant.[78]

Iraq's Relationship with the IAEA

Iraq joined the IAEA in 1959, two years after the Agency's establishment. Hamza described Iraq's initial involvement with the Agency, under the allegedly "indifferent watch" of Iraqi ambassador Baquir Hasani, as "nonexistent."[79] This is not quite true. Iraq was elected to a two-year term on the Board of Governors from 1960 to 1962, and Hasani chaired the Board in 1961.[80] Iraq's program also received modest technical assistance from the Agency. Still, Iraq was not a major player in Vienna.[81] For instance, although any member state could participate in the committee that negotiated INFCIRC/153, Iraq chose not to.[82]

In September 1973, Iraq's relationship with the Agency expanded when Saddam allegedly ordered his officials to "get inside it and turn it to our purposes."[83] After an intensive and Machiavellian lobbying effort, Iraq succeeded in being elected to a second two-year term on the Board of Governors from 1974 to 1976.[84] From 1980 to 1991, Iraq was elected to four terms.[85] Iraqi officials would consequently have become intimately familiar with safeguards matters during Board debates about the evolving system.

Hamza claims that a special intelligence office was created at the Iraqi permanent mission in Vienna to further "penetrate" the IAEA's operations.[86] Suroor Mahmoud Mirza, a brother of Saddam's senior bodyguard, was appointed as the scientific attaché. In addition, the Iraqis managed to have an Iraqi nuclear physicist, Abdul-Wahid al-Saji, appointed as an IAEA inspector.

Thereafter, Hamza boasts, "classified information began to flow from our spy in the IAEA, proving invaluable in circumventing the IAEA's detection of Iraqi cheating."[87]

It is not clear how revealing this access would have been, because safeguards techniques are publicly known and information derived from safeguards about states' nuclear programs is treated as confidential and given appropriate protection in the Safeguards Department. Senior IAEA officials were aware that Iraqi staff at the Agency had "pledged allegiance to Saddam" and presumably treated them circumspectly.[88] Dieter Goethel, IAEA Director of Personnel from 1989 to 2002, has revealed that the policy toward staff with "potential conflicts of interest" was to move them to areas where they could do less harm, presumably where they had no national interests to protect.[89] Still, the Iraqi inspector's experiences in conducting inspections in other countries, as well as discussions among inspectors and safeguards managers, could have alerted him to potential diversion pathways and the efficacy of other non-compliance scenarios.

Safeguards in Iraq

Nuclear safeguards were first applied to Iraq soon after it became party to the NPT. It signed the treaty on July 1, 1968, the date it was opened for signature, and ratified it on October 29, 1969.[90] Article II committed it to the following agreement:

> not to receive the transfer from any transferor whatsoever of nuclear weapons or other nuclear explosive devices or of control over such weapons or explosive devices directly, or indirectly; not to manufacture or otherwise acquire nuclear weapons or other nuclear explosive devices; and not to seek or receive any assistance in the manufacture of nuclear weapons or other nuclear explosive devices.[91]

As for all NNWS parties, Article III required Iraq to negotiate a bilateral agreement with the IAEA to accept safeguards "for the exclusive purpose of verification of the fulfilment of its obligations assumed under this Treaty with a view to preventing diversion of nuclear energy from peaceful uses to nuclear weapons or other nuclear explosive devices."[92] Safeguards were to be applied "to source or special fissionable material whether it is being produced, processed or used in any principal nuclear facility or is outside any such facility" and "in all peaceful nuclear activities within the territory of such State, under its jurisdiction, or carried out under its control anywhere."[93]

Iraq was among the earliest NPT parties to have a comprehensive safeguards agreement. States parties were obliged to begin negotiations with the Agency within 180 days of entry into force of the NPT, which occurred for Iraq on March 5, 1970. Following successful talks, Iraq's agreement entered into force on February 29, 1972.[94] The text was supplemented in 1979 by an exchange of letters with France committing Iraq to accepting safeguards in perpetuity on French-supplied nuclear facilities and materials.[95]

While the INFCIRC/153 model for safeguards is subject to minor changes when a state's agreement is negotiated, each agreement is essentially the same.[96] Iraq's document, as with all such agreements, obliged it to declare to the Agency, in an Initial Declaration, all the nuclear material it possessed for peaceful purposes and where it was held, including in facilities and other locations. This declaration is the baseline for the initial application of safeguards and is required to be updated if the country acquires more material or facilities. Iraq was also obliged to establish a State System of Accounting and Control (SSAC) for its nuclear material, the output of which the IAEA would seek to verify. In addition, Iraq was expected to designate a government authority with which the Agency could deal on safeguards implementation matters. The IAEC was so assigned.[97] Like all states with significant amounts of nuclear material, Iraq was also obliged to negotiate subsidiary arrangements to flesh out the details of safeguards on its territory, as well as "facility attachments" with even more detail about safeguards procedures at its nuclear installations. These documents, as usual, were confidential.

Iraq's Initial Declaration and safeguards obligations would have been relatively straightforward, given how basic the country's nuclear establishment was at the time. The rules for facility inspections would have provided for the sole reactor to be inspected by IAEA representatives at least twice a year to check that Iraq's declaration remained correct and that no unreported changes had taken place. In 1978, the IRT-2000 reactor was upgraded to 5 MW(th) and renamed the IRT-5000, but this would not have significantly changed safeguards at the facility.[98]

While Iraq permitted inspections to take place as required, Fischer and Szasz reported as early as 1975 that the Agency "encountered difficulties" with Iraqi officials in arranging them.[99] It is not clear what these issues were, how long they persisted, or whether they were resolved to the Agency's satisfaction. In 1981, however, former IAEA inspector Roger Richter testified to the U.S. Senate Foreign Relations Committee that when the Iraqis finally agreed to inspections, "They do that under very strict security. The inspections have to be planned quite a long time in advance. Very long periods of notice are given, and the Iraqis pretty well determine exactly what time the inspectors can arrive."[100] He then partly exonerated the Iraqis, though, by saying that "part of this can be ascribed to their war situation."[101] What is known is that the Iraqis systematically rejected inspectors of various nationalities until eventually, from 1976 onward, apparently only Soviet and Hungarian IAEA inspectors were allowed to conduct inspections in Iraq.[102]

France's 1976 sale of the Tamuz-1 reactor and other facilities and material to Iraq were duly declared to the IAEA. The Secretariat began preparing the usual safeguards approach for the facility well in advance of the reactor going critical (achieving a sustainable fission chain reaction). Initial inspections were planned every two weeks until criticality was achieved, and then twice yearly thereafter. As added reassurance, France planned to have its technical staff work there for ten years on a joint research program.[103] In 1981, Deputy Director General (DDG) for Safeguards Hans Grümm, casting off the cloak of "safeguards confidential" because so much information had already been revealed, described in detail the IAEA's approach to the reactor.[104] He reported that in June 1980, inspectors were present when the first batch of fuel arrived from France and immediately inspected it. The fuel elements were counted, identified, and determined to amount to 12 kilograms of HEU.[105] The Agency decided to perform two to three inspections a year if no further fuel elements were shipped, but one every two weeks in the event of further shipments.

After an air attack by Iran in September 1980 caused minor damage at the site, but not to the reactor, the Iraqis temporarily declined to accept further inspections.[106] Eventually they agreed to a second inspection in January 1981. Fearing a further Iranian attack, the Iraqis restricted access to and imposed a blackout on the facility, so the inspection took place at night under "war conditions." Lighting was only "adequate," except in rooms where large windows were not blacked out. Flashlights were permitted. Despite these restrictions, the fuel elements, stored in a channel under water, were "satisfactorily identified and accounted for," as was the stored natural and depleted uranium.[107]

Meanwhile, Iraq added to its facilities in Tuwaitha. In 1979, it contracted with Italian firm SNIA-Techint for a pilot plutonium separation and handling facility, as well as a uranium refining and fuel-manufacturing plant. This occurred despite misgivings within the company, but apparently not in the Italian government, about the proliferation implications.[108] Iraq did submit design information for the fuel fabrication facility to the IAEA, but none of these facilities would be subject to safeguards until shortly before nuclear material was introduced into them. As Grümm points out, however, the plutonium-related facilities would have been useless without a source of plutonium, which could only have been produced in Tamuz-1, which was under safeguards.[109]

Between 1979 and 1982, Iraq also bought several hundred tons of uranium in various forms, including yellowcake, uranium dioxide, LEU, and depleted uranium from Brazil, Germany, Italy, Niger, and Portugal.[110] Some of the purchases were reported to the IAEA, as required by Iraq's safeguards agreement and those of the exporting states, but some were not. Brazil and Niger were not required to report their transactions because they were not party to the NPT at that time.

The Destruction of Tamuz-1: Safeguards Pilloried

Safeguards for the new Tamuz-1 reactor became irrelevant on June 7, 1981, when it was destroyed by an Israeli bombing raid before fuel could be loaded into it.[111] Israel claimed that U.S. intelligence based on "highly reliable sources" had revealed a secret tunnel under the reactor for processing plutonium and U-235 for nuclear warheads.[112] Iraq's attempt to delay IAEA inspections due to their fear of Iranian attack had only increased Israeli suspicions.[113]

The Israeli action was condemned almost universally, including by the United States and the UN Security Council. IAEA Director General Sigvard Eklund decried it as an attack on the safeguards system itself.[114] A major crisis for the Agency emerged when the non-aligned countries (known as the Group of 77) moved to expel Israel from the General Conference. When this succeeded after a contested vote, the United States temporarily withdrew from the Agency until the expulsion was reconsidered.[115]

Some Western politicians, experts, and media used the Israeli attack to question the efficacy of safeguards.[116] During hearings of the U.S. Senate Foreign Relations Committee in June 1981, Senator Alan Cranston alleged that IAEA inspectors would have been unaware of clandestine plutonium production because Iraq could unload and reload the reactor between inspections without being detected.[117] Richter, who said he had just resigned as an IAEA inspector due to his concerns about the weakness of safeguards, testified that Iraq's intention to develop nuclear weapons was "obvious."[118] He alleged that current safeguards were "totally incapable of detecting the production of plutonium," presumably meaning clandestine production.

The IAEA and France, in response, "put forward convincing arguments to show that any violation of the IAEA's safeguards agreement with Iraq . . . would have been detected."[119] French nuclear industry sources insisted that Iraq could not have produced plutonium clandestinely because of the nature of the reactor and the possibility of detection by on-site French technicians and IAEA inspectors.[120] They disputed the existence of a tunnel. Furthermore, because France would not deliver multiple fuel loadings but only transported each shipment as it was required, it was unclear where Iraq would have obtained additional HEU to refuel its reactor illicitly for plutonium production between inspections.

The IAEA also disputed Richter's testimony, reaffirming that since Tamuz-1 was not yet operating, the Agency was still determining its safeguards approach to the reactor.[121] Their plan would deal with what the Agency recognized as the possibility of illicit plutonium production in research reactors.[122] Additional measures may have included surveillance cameras and extra inspection activities. Warren Donnelly, a non-proliferation expert at the U.S. Congressional Research Service, concurred that IAEA inspectors would have detected non-compliance on the scale of Richter's hypothetical scenario.[123]

A second diversion scenario—removal of the reactor's fuel to build an HEU-based nuclear device—would, the Agency reaffirmed, be readily detected by inspectors. Even though the amount was, by itself, insufficient for a single weapon, it was apparently envisioned that it might be combined with other discrete amounts that Iraq had acquired. DDG for Safeguards Grümm asserted that "it is completely out of the question that such an overt act of diversion, which would have made it impossible for the reactor to operate, could have escaped the attention of [the] Agency's inspectors."[124] The Agency clarified that the Iraq safeguards agreement, along with all other NPT-type safeguards agreements, allowed for up to fifty person-days of inspections per year, including some unannounced, not just the twice- or thrice-yearly visits commonly presumed. In addition, *ad hoc* inspections were permitted for imports of HEU fuel from France. If such measures were insufficient, special inspections could be requested.

After the Israeli attack, Iraq permitted the IAEA to conduct an inspection from November 15 to 17, 1981. Unusually, the Agency revealed the names and nationalities of the two inspectors, G. Rabot of France and V. Seleznev of the Soviet Union (who presumably had previously carried out inspections in Iraq).[125] These inspectors confirmed that the fuel assemblies remained intact. They also inspected the stores of natural and depleted uranium, as well as yellowcake (which would become subject to safeguards when chemically purified). They found nothing amiss.

Safeguards after the Tamuz-1 Attack, 1981–1990

In the decade following the destruction of the Tamuz-1 reactor by Israel in 1981, the safeguards system operated on the basis that the Iraqi nuclear program had fallen into abeyance. Iraq made no attempt to reconstruct Tamuz-1, especially since the French ultimately declined to assist, and Tamuz-2 was repurposed for training, research, and neutronic radiography.[126] Safeguards remained in place on the old Soviet reactor, as well as on the Tamuz-1 fuel, Tamuz-2, the Experimental Research Laboratory for Fuel Fabrication, and a storage facility.[127]

From 1981 through 1990, the IAEA Secretariat is not known to have recorded any significant anomalies for Iraq; no non-compliance issue is known to have been raised by inspectors with the Director General or by the Director General with the Board; and certainly no special inspection was sought. Iraq was listed without comment in the Agency's annual Safeguards Implementation Report (SIR) among all of the other states that were complying with their comprehensive safeguards agreements. As late as November 1990, the Agency's report on Iraq contained the standard safeguards conclusion that "all nuclear material under safeguards was accounted for."[128] In early 1991, DDG for Safeguards Jon Jennekens even praised Iraq's cooperation as exemplary.[129] Unfortunately, apart from these bare facts, there is little publicly available official information on safeguards activities and findings in Iraq at this time.

Anecdotally, though, Kratzer says that inspectors apparently "simply walked past" undeclared activities and buildings, of which there were many at Tuwaitha.[130] Australian safeguards expert John Carlson says that inspectors ignored facilities that they should have been curious about. They did not inquire about them because they did not see that as their role.[131] While most of the undeclared structures were physically separated from declared buildings and activities, often with sand berms between them, some were intermingled. Matthew Fuhrmann claims that an undeclared facility was located "within yards" of one of the safeguarded reactors.[132] David Albright, Corey Gay, and Hamza report that the Iraqi authorities developed procedures to limit access to the undeclared areas by inspectors, who only had the right to inspect the safeguarded facilities.[133] On days when inspectors were scheduled to visit, only the fuel fabrication rooms were open to them. Employees were usually told to take the day off or to stay in their offices so that inspectors did not see an unusually large number of people.

Inspectors apparently did not ask their Iraqi minders about what went on in the other buildings or what was located on the other side of the berms. Nor did they ask permission to visit locations beyond those Iraq had declared. They could not, of course, do this on a whim, but would have needed credible evidence to make such requests. Back at IAEA headquarters, some safeguards personnel obviously had qualms about the Iraqi program, but those concerns are not known to have gone beyond internal discussion in the Safeguards Department, much less pursued with the Iraqi authorities.

Iraq's Nuclear Weapons Program

As we now know, no sooner had Iraq signed its safeguards agreement in 1972 than Saddam Hussein approved a 50-page plan for an ambitious and "carefully designed" civilian nuclear energy program ultimately aimed at creating nuclear weapons.[134] It has been described as "an experiment in how far a nation could proceed with the acquisition of nuclear weapons under cover of a civilian nuclear program and in violation of its sworn adherence to [an] international treaty, without being called to account for this behaviour."[135] The Tamuz-1 reactor was a key part of the plan.

If its nuclear program had been ambiguous, slow-moving, and disorganized before the Israeli attack, the assault motivated Iraq to attempt a secret, accelerated, and multi-pronged nuclear weapons program thereafter.[136] Al-Sa'idi claims that the IAEC took an abrupt turn toward a weapons program after the destruction of Tamuz-1.[137] Remarkably, planning and preparations proceeded during Iraq's bitter war with Iran from 1980 to 1988. But it was not until the last year of that war that the Iraqi government began to take serious steps toward developing the necessary infrastructure. Incredibly, despite its seemingly ever-changing priorities and often rudderless management, many of the program's major facilities were constructed in just three years, from 1988 to 1991.

It is beyond the scope of this volume to describe the clandestine Iraqi program in detail, especially as others have done it so masterfully.[138] Thanks to reports by the IAEA and UN Special Commission (UNSCOM) inspectors after they gained unprecedented access to Iraq in 1991, as well as investigations by the U.S.-led Iraq Survey Group after the coalition invasion in 2003, we now know a great deal about the state of the program in 1990.[139] It shows the extent of the activity that IAEA inspectors missed (see Figures 1.1 and 1.2 below).

Infrastructure was built not just at the sprawling Tuwaitha site that had been visited by IAEA inspectors, but also at several other locations deliberately scattered around the country. Significant clandestine facilities were erected and an astonishing range of research, development, and testing activities were conducted alongside Iraq's declared peaceful program. In its endeavor to produce enriched weapons-grade uranium, Iraq attempted "virtually every feasible uranium enrichment process."[140] Their efforts included electromagnetic

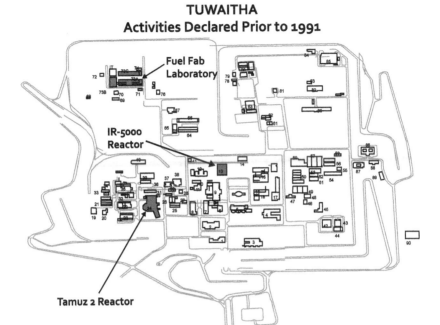

Figure 1.1

Tuwaitha: Activities Declared Prior to 1991.

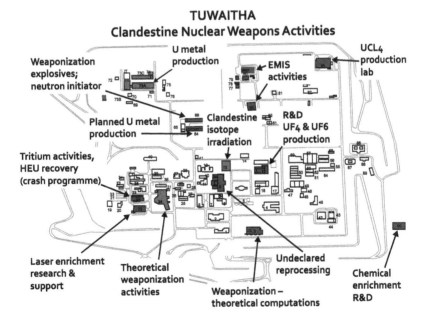

Figure 1.2

Tuwaitha: Clandestine Nuclear Weapons Activities.

isotope separation (EMIS), a Cold War technology dating from the Manhattan Project that used so-called calutrons. Considered too slow and expensive by other countries, EMIS had long been abandoned in favor of gaseous diffusion. Undeterred, the Iraqis constructed a large EMIS plant at Tarmiya, directly north of Baghdad, along with facilities to fabricate equipment and prepare uranium feedstock for separation. Additionally, the Iraqis tried gas centrifuges, chemical enrichment, gaseous diffusion, and laser isotope separation. Weaponization studies and activities were conducted, notably at a research, development, and production complex at Al-Athir, southwest of Tuwaitha. By 1988, the number of scientists, engineers, and technicians in the IAEC exceeded 6,000, including the "brightest and best qualified" in Iraq.[141]

This is not to say that Iraq came close to acquiring a usable nuclear weapon. Experts from France, Russia, the United Kingdom, and the United States who examined documents collected by IAEA inspectors in Iraq between 1991 and 1992 concluded that despite the extent of its program, the country "faced such significant bottlenecks that [it] was at least three years and possibly more from acquiring [its] first crude nuclear device."[142] Jacques Hymans graphically presents the mismanagement, corruption, incompetence, and political machinations that characterized what he calls "Iraq's fruitless search for nuclear weapons."[143] In her pointedly titled book, *Unclear Physics*, Målfrid Braut-Hegghammer portrays Iraq as lacking the institutional capacity to plan and oversee a nuclear weapons program, which was adrift "without clear objectives and guidelines."[144] There was hence no immediate danger that Iraq would have obtained nuclear weapons despite its vast clandestine efforts.

Shortly after the Iraqi invasion of Kuwait in August 1990, however, General Hussein Kamel, desperate to show Saddam some results from the exhaustive research so far, ordered a "crash" program to produce a single nuclear weapon using materials to be secretly withdrawn from IAEA safeguards.[145] The intention was to extract and further enrich the HEU in the safeguarded Tamuz-1 fuel, which had survived the Israeli attack, and combine it with the material in the old Soviet reactor. It was assumed that by the end of April 1991, this would amount to enough for a single low-yield nuclear device.[146] Iraq intended to build a 50-machine centrifuge cascade to further enrich the material, although there is no evidence that this construction ever began.[147] Also planned were fabrication of the implosion package, preparation of a nuclear explosion test site, and studies of a delivery system. Iraqi scientists estimated that assembly of a device for testing or deployment would be possible after the end of 1992.[148]

Hymans reports that leading Iraqi scientists opposed the crash program, realizing that "breaking the IAEA seals on the nuclear fuel would represent the definitive end to their long-term ambitions," presumably because they would be caught red-handed by the IAEA.[149] Most analysts consider the crash program completely unfeasible, not least because Iraq would have had to resort to the black market; it didn't have enough time to manufacture everything required to meet these "crash" goals.[150] Legitimate suppliers of sensitive equipment would soon have become suspicious about the sudden urgency, volume, and nature of the imports being sought.[151] Ultimately, all of Iraq's proliferation pursuits were ended during the Gulf War by the destruction of the Tuwaitha site by coalition aircraft in January 1991. The Gulf War had been launched to drive Iraqi forces from Kuwait, which it had invaded in August 1990, but the ceasefire that ended it had the unintended consequence of revealing—and destroying—what was left of the nascent Iraqi nuclear weapons program.

What the IAEA Knew and When

Although the Agency was not officially permitted to gather information on a state's nuclear activities beyond what it acquired through safeguards, there were significant indicators in the public domain that Iraq's nuclear program was not benign. Certainly, there had been continuing media speculation that Iraq harbored ambitions to develop nuclear weapons and may have been violating its safeguards agreement to pursue them.[152] Albright, Gay, and Hamza describe "rampant" press speculation about Iraq's intentions.[153] The Secretariat would have known about such media reports, as well as those from other open sources, such as academics and commercial companies. Academic papers, for example, alleged an Iraqi nuclear weapons program, including attempts to acquire nuclear technology abroad, both openly and secretly.[154] There were reports that Iraq had imported components of a gas centrifuge uranium enrichment plant, notably from Germany.[155] There was also a report in the early months of 1991 that Iraq was trying to use EMIS to enrich uranium, although no one dreamed that anyone would revive such outdated technology.[156] Most observers believed that Iraq would find it difficult, if not impossible, to indigenously develop centrifuge technology, which is notoriously difficult to master.

The media, however, reported two startling incidents that on their own should have raised alarms. First, in mid-August 1989, a huge explosion occurred at the Al-Qa Qaa General Establishment, which was involved in researching high explosives. The detonation was heard in Baghdad and Tuwaitha, leading to foreign press speculation that it could have been a nuclear accident.[157] Second, in March 1990, a sting operation at Heathrow Airport caught five Iraqis trying to smuggle high-speed detonators called krytrons to Iraq. This again drew international media attention to the Iraqi program.[158] In addition, and most alarming of all, Saddam had not been shy about his nuclear intentions. Before flying to France to close the Tamuz-1 sale in September 1975, he had told a Beirut-based Arab-language news magazine that his country was engaged in "the first Arab attempt at nuclear arming."[159] The difficulty for the Agency was that it was not permitted to use open sources in compiling dossiers on states, a situation that would change after the Iraq case.

The IAEA obviously knew through the application of safeguards that Iraq had enough HEU for a crude nuclear weapon and significant amounts of imported uranium of other types. It also knew about Iraq's failed attempts to discover and exploit uranium on its territory, because assistance was provided in this endeavor by the IAEA's Technical Cooperation (TC) program. The report by the post-war Iraq Survey Group, the Duelfer Report, states that the existence of the Iraqi centrifuge enrichment program was known before 1991, presumably by the IAEA, but that "the Iraqis did not fully declare its extent and maintained that it was only a limited research and development activity located at Tuwaitha."[160] The IAEA also knew about the Iraqis' interest in plutonium, because the Agency gave them permission to exempt from safeguards some of the fuel pins from the IRT-5000 reactor for use in plutonium separation experiments.[161] At the time, the IAEA did not have the capability for wide-area sampling, but it did have techniques to detect enrichment if this had been suspected, and if the Iraqis had permitted inspectors to take samples.[162] As veteran inspector Shirley Johnson notes, "In Iraq they had incredible, huge filters on their nuclear facilities, but atoms do get out!"[163] Enriched uranium particles were indeed discovered on the clothes of Western hostages who had been kept near nuclear sites to deter airstrikes.[164]

As for the IAEA's member states, Israel had certainly made everyone aware that it suspected Iraq of pursuing a nuclear weapons program and distrusted the assurances provided by IAEA safeguards. Israel had the best intelligence sources in the region, but its poor relationship with the Agency precluded it from handing over credible information. Braut-Hegghammer reports that in 1989–1990, other states also had suspicions about Iraq's intentions, although these may only have been shared with other governments and not also with the IAEA.[165] Blix says that "although suspicions against Iraq existed, at the time no government had any concrete evidence of Iraq's large, secret uranium enrichment and weapons construction facilities. They were neither known to the IAEA nor, it appears, to any national intelligence service."[166]

The French clearly had concerns when they decided not to help Iraq reconstruct Tamuz-1 or provide new facilities. Italian suppliers, if not the Italian government, also had qualms. It is not known whether these governments ever made their views known to the IAEA. The Agency was certainly not provided with evidence from so-called National Technical Means—U.S. and Soviet satellites that could have taken photographic images of Iraq at the time.[167] Judging by the blatant way in which Iraq constructed large facilities at Tuwaitha and elsewhere, it must have assumed, correctly, that U.S. intelligence agencies, which had the most sophisticated technology, would not share their fruits with the Secretariat. The Agency was thus ignorant even of the extensive building activity at the Tuwaitha site and at other locations, which would have been revealed by satellites. At that time, no member states were prepared to provide intelligence information of any type to the Agency, due both to concerns about revealing "sources and methods" and because they believed an international organization was ill-equipped to handle such information sensitively.

In any case, Under Secretary of State Walter J. Stoessel Jr. said in public testimony before the U.S. House Foreign Affairs Committee in June 1981 that the U.S. intelligence community had itself not concluded that Iraq was planning to develop a nuclear weapon.[168] Several committee members responded that they had heard a conflicting assessment from a Central Intelligence Agency officer in a private session the day before. Stoessel would not be drawn out further, but told the committee that the administrations of both Jimmy Carter and Ronald Reagan had been concerned about the purpose of the Iraqi program, and both had pressed the French and the Italians not to export technology that could enable Iraq to make a bomb. It is not clear, though, whether these concerns were expressed to the IAEA. The difficulty for the Agency was that without credible evidence to present to the Iraqis, there was little point in confronting them.

The Agency itself had, however, assisted Iraq's nuclear research efforts since the late 1950s through its TC program, and would have obtained some information about its progress as a result. For example, under the program, Poland supplied a dense plasma focus device to Iraq.[169] Braut-Hegghammer has revealed that IAEA experts submitted confidential assessments to the Agency concerning the Iraqi state's ability to absorb technical assistance.[170] Such reports should have given the Secretariat some insights into Iraq's capacity to mount a nuclear weapons program. Paradoxically, though, they probably would have reinforced the notion that such a technologically backward and poorly governed state such as Iraq could never have achieved its ambitions.

At the time, moreover, the Safeguards Department was insulated from other parts of the Agency, especially TC, and thus apparently ignorant of technology transfers and other forms of technical assistance provided.[171] Carlson mentions instances in which TC was helping states acquire conversion plants, but neither the states nor TC informed the Safeguards Department. When "safeguards finally caught up with them states were technically in non-compliance."[172] Johnson says, "I don't remember anyone suggesting 'oh, look at TC in that country.' Shame on us. Why? So strange. I didn't know what TC did. It was our fault. We know what we are doing so don't bother us."[173] This was typical of the IAEA's legendary and much criticized programmatic "stove-piping."[174]

In any event, secrecy surrounding Iraqi nuclear activities was tightened by Saddam after 1973, limiting IAEA assistance to a few areas and restricting access that TC projects might have provided. Information about the Iraqi nuclear program and certain lines of research were "compartmentalized, even within the nuclear establishment."[175] In a letter to the Agency, the IAEC's Secretary-General, Moyassar al-Mallah, informed it that "unforeseen circumstances and current developments have dictated on us [sic] the revision of our entire programme, especially the priority of committed projects."[176] Despite its vagueness, this missive itself should have raised a red flag at the Agency, but it was sent to the head of TC. It is not known whether it was copied to the Safeguards Department, but it has been characteristic of the IAEA that each department is an informational silo, reporting up the chain of command (and sometimes not even then) but not horizontally across departments.

One signal of a state's intent to begin a nuclear weapons program can be the retreat of key nuclear scientists from international academic circles. To reduce suspicion, Iraq cleverly continued its non-weapons research activities, international publishing, and conference participation, completely separating these from the military program. Braut-Hegghammer reports that Iraq's Nuclear Research Center continued to carry out basic research in chemistry, physics, biology, agriculture, medical diagnosis, and radioisotopes, and reported this work in scholarly publications and at academic conferences.[177] This would have helped conceal the fact that Iraq was training hundreds of additional nuclear technicians; four hundred were sent abroad for such training.[178]

While the Agency could have begun an informal process to assemble a more complete picture of Iraq's activities from these and other open-source nuggets of information, it was not authorized or organized to do so. This, too, would change after the Iraq case. In answering his own question of whether the Secretariat could have done more, Blix subsequently suggested that "it could have systematically scanned media for information and found a few suspicious items regarding Iraqi imports."[179] The implication is that he would have been happy for the Secretariat to have done so, which begs the question of why it did not.

Paradoxically, even if the Secretariat had attempted to compile a coherent picture of Iraq's activities, it would have been impossible given that there was no coherence to be discerned. Braut-Hegghammer makes a convincing case that Iraqi scientists, officials, and leaders were themselves perennially divided over the goals of the program, which changed frequently. Even Saddam himself was at times apparently not particularly interested in the nuclear program and never made it the country's number one priority. Moreover, as the Duelfer Report suggests, "Saddam did not lead by espousing detailed goals and objectives. He tended to allow ideas to float up and he would consider them—often never pronouncing on them one way or the other. This meant that much guidance to the government was implicit rather than explicit. For investigators, a consequence is that forensic evidence of Presidential direction may not exist."[180] Public signals of Iraq's intent were thus inherently ambiguous. This echoes the confusion among Western intelligence agencies about Iraq's extant weapons of mass destruction (WMD) capabilities prior to the 2003 invasion. Still, there were enough publicly aired inconsistencies to raise questions in the Secretariat.

As for the crash program, Hymans argues that "there was zero doubt that the IAEA inspectors were going to discover the Iraqi smash-and-grab operation in their regular, semi-annual inspection."[181] In 1995, Blix revealed that in order to delay detection of the diversion of safeguarded fuel by the Agency for as long as possible, the Iraqis had planned the extraction operation to start immediately after what they assumed would be a twice-yearly safeguards inspection in November 1990 and before the next scheduled visit six months later.[182] Iraq would no doubt have tried to delay the latter inspection for as long as possible, presumably using the excuse that threatened Allied air attacks would endanger IAEA inspectors. This was the same ploy that it had used prior to the Israeli attack in 1981. In this case, the IAEA would have had grounds for seeking a special inspection. If that had been refused, the Board may ultimately have reported Iraq to the UN Security Council. The Agency would not have had to prove a diversion, but only that it was unable to verify the whereabouts of all of Iraq's declared material.[183]

Doubtless, the IAEA cannot be blamed for failing to detect Iraq's "crash" diversion plan because it had not yet been put into operation—and safeguards famously do not consider states' intentions. As Fischer and Szasz insisted at the time, "international safeguards cannot, except very indirectly, give any indication about the nuclear intentions of states. Safeguards can only deal with measurable and observable facts—past and present—and not future intentions."[184] DDG Grümm declared openly that "clearly it is not the task of the Inspectorate to speculate about the intentions of States which possess safeguarded nuclear material or facilities."[185] Rather, "the IAEA has to assume and take into account the possibility of diversion in *all* States and in *all* applications of safeguards." He compared inspectors to airport security guards who must consider all passengers as possible carriers of weapons, and "cannot afford to exempt well-dressed gentlemen."[186] So, Grümm says, "We did not exempt Iraq and considered all technical possibilities for diversion of the nuclear material present and to be expected in Iraq."[187]

Divining Iraqi intentions would have required intelligence information about internal Iraqi discussions that only some states (most likely Israel, the United Kingdom, and the United States) might have possessed, none of which was shared with the IAEA.[188] Western intelligence agencies, with all their resources, failed to detect either Iraq's initial decade-long effort to acquire nuclear weapons or the crash program. Despite the vast range of their technical assets, U.S. intelligence agencies apparently had little success in penetrating Saddam's regime; even the United Kingdom, which historically had more success using low-technology methods, failed to do so as well. For such agencies, the challenge was not only that Saddam's intentions were often indecipherable because they were implicit rather than explicit, but also that Western cultural assumptions about how a nuclear weapons program should logically be initiated, organized, and managed were misleading.[189] Iraq did not undertake its nuclear weapons program in the same way as the Western nuclear weapon states, the United States and the United Kingdom, that were most influential in shaping IAEA safeguards.

Beyond the minutiae of what the IAEA may or may not have known at the time, the important question is whether there are systemic explanations for why the Agency was not better prepared and equipped to detect a safeguards violation such as the Iraqi nuclear weapons program, which in some respects was hidden in plain sight. It is to these possible systemic explanations that we now turn.

Two

Missing Iraq: Political and Organizational Explanations

T HE MOST COMMON systemic explanation for the failure of IAEA safeguards to detect Iraq's non-compliance is international politics. According to this argument, in setting up the safeguards system, states conspired to
ensure that it was not comprehensive, overly intrusive, or expensive. In over-
seeing implementation of the system, states continued, for political reasons,
to hobble it. Multiple factors conspired to render the system weaker than
originally intended: political differences; concerns about sovereignty, confidentiality, and non-discrimination; budgetary calculations; and the resulting
compromises necessary to reach agreement. A second major explanation is
organizational. Although there were acknowledged shortcomings in the safeguards regime as negotiated, it was argued by some member states, outside
observers, and elements of the media that it was the organizational shortcomings of the IAEA itself that allowed Iraq's behavior to escape notice.[1] The managerial, technical, and personnel challenges involved in implementing such a
ground-breaking multilateral verification enterprise reputedly conspired to
render it less effective than its negotiators had envisaged. This chapter will
examine how far these explanations take us.

International Politics as the Culprit

Political scientists, governments, non-governmental organizations, and IAEA
officials often explain the weaknesses of the safeguards system as largely the
result of politics. From this perspective, politics determined the bargains
reached by states in establishing the initial safeguards system, in negotiating the Nuclear Non-Proliferation Treaty (NPT) and INFCIRC/153, in
implementing comprehensive safeguards, and in pursuing subsequent new
iterations of safeguards.[2] A declassified 1984 report by the U.S. Arms Control and Disarmament Agency (ACDA) on the negotiations on the original
INFCIRC/66 safeguards concludes that "for the most part, these instruments
reflect compromise formulations between more extensive and more restric-
tive safeguards provisions for the IAEA, which in turn reflect the position of

countries which actively participated in the development of the basic safe-guards document or with whom the Agency negotiated particular safeguards agreements."[3] The report goes on to say that efforts by the IAEA to improve safeguards implementation have been opposed "on the basis *inter alia* that explicit provision is not in the relevant agreement."[4] Former IAEA Director General Hans Blix explains that although the Agency's safeguards "represented a dramatic leap forward simply by being the first global on-site inspection system," it "eventually proved the difficulty of designing an inspections regime capable of satisfying all states parties while simultaneously fulfilling its mandate."[5]

States insisted that safeguards operate with respect for their sovereignty, strict non-discrimination between them, and preservation of the confidentiality of safeguards-derived information.[6] Most states were more concerned with preserving the secrets of their commercial nuclear enterprises and avoiding espionage by international inspectors than with instigating a more intrusive and effective system. The main proponent of stronger safeguards was the United States, along with two other nuclear weapon states, the Soviet Union and the United Kingdom, as well as Canada, the Netherlands, most of the Scandinavian countries, and Soviet allies in Eastern Europe.[7] Leading industrial states that were likely to experience the full impact of safeguards, such as the then–Federal Republic of Germany, Italy, Japan, and to some extent Sweden, were reportedly suspicious of the two superpowers and of the IAEA Secretariat, which they felt was dominated by them.[8]

The non-NPT states that took an active part in the safeguards negotiations—Argentina, Brazil, India, Pakistan, South Africa, and, at that time, Australia—"generally supported proposals to limit or 'soften' safeguards."[9] Uranium suppliers such as Australia and South Africa successfully fought to prevent safeguards from being applied to mining activities.[10] The flaws of the system, however, were not just due to its opponents. Proponents "deliberately set out to make NPT safeguards less 'burdensome' to encourage acceptance of the NPT" and achieve consensus in the negotiations.[11] They did this chiefly by setting limits on the IAEA's authority and freedom of action. In the negotiations on INFCIRC/153, nine states spoke in favor of flexibility in the IAEA's methodologies, while the principle of non-discrimination had strong support from thirteen states.[12] Strangely, although the word "flexibility" does not appear in the text, there are provisions that "build in flexibility and differentiation in how safeguards can apply in different states."[13]

A significant economic driver in the negotiations was the determination of the major industrial states not to handicap their nuclear industries in ways that the same industries of unsafeguarded competitors, especially in the nuclear weapon states, were not hampered. In the rush for nuclear energy in the 1970s, no one wanted safeguards to get in the way of legitimate peaceful uses.[14] The NPT was deliberately drafted to provide reassurance that nothing in the treaty could be interpreted as "affecting the inalienable right" of its parties to the use of nuclear energy for peaceful purposes.[15] Certainly, no one wanted IAEA inspectors wandering around freely looking for infractions. Safeguards agreements gave states the right to object to receiving individual inspectors or classes of inspectors on essentially any grounds, including political ones, such as the inspector's nationality. This right was not unlimited, however, and it was up to the Board to decide whether such refusals were justified or not, whether they would likely impede inspections, and whether "appropriate action" should be taken.[16]

The Secretariat was expected to have a "technical" orientation and be apolitical rather than "subjective" in its compliance judgments, meaning that even if there were suspicions about a state's nuclear intentions, it could not be subjected to increased scrutiny by the safeguards system. Even after the controversy over the adequacy of safeguards following the destruction of Iraq's Tamuz-1 reactor, Blix records that "despite its questioning of the safeguards system's reliability, neither the U.S. nor any other government took the initiative to strengthen the system during the 1980s. There would at this time have been insuperable resistance to more intrusive inspections."[17]

While it is not quite true that there were no initiatives to strengthen safeguards in the 1980s (see the discussion on the Agency's technical capabilities below), Blix's remarks illustrate the "nightmarish dilemma about all attempts to control atomic weapons" that was originally identified by the 1962 Advisory Committee on U.S. Policy Toward the International Atomic Energy Agency (the Smyth Report): "The political acceptability of any system proposed," it noted, "seems to be inversely proportional to its effectiveness."[18] The United States judged at the time it was negotiated that INFCIRC/153 did achieve the objective of maintaining the effectiveness of the Agency's safeguards system, "although in the case of some provisions, particularly those relating to inspection effort, the margin was close."[19]

The operation of the new safeguards system, once negotiated, was also subject to political considerations, whether in the allocation of key posts in the Secretariat, the interpretation of safeguards provisions, the budgetary allocations for each major program, or new measures to strengthen safeguards. One IAEA official described the Agency as typical of an organization with a "political top and an operational bottom."[20] Slobodan Nakićenović, former Director of Safeguards Operations from 1966 to 1977, recounted numerous examples of U.S. and Soviet Cold War rivalry influencing decisions that affected the Safeguards Department—which he rather intemperately declared at the time "stinks of politics."[21] He claimed, for instance, that contrary to the advice of a 1974 Panel of Experts, the Director General had in 1977 split the Operations Division in two so that a Western ally, Japan, could get one post, and Czechoslovakia another (Nakićenović, who had been the sole operations director, apparently retired soon afterward).[22] In 1976–1977, the Soviets had even threatened to withdraw from the IAEA unless their nationals were given more positions.[23] Nakićenović conceded, however, that by 1981, the "dominance of the superpowers was starting to fade."[24] The important question is whether this had a notable impact on the effectiveness of safeguards at the time.

Aside from U.S.-Soviet Cold War rivalry, the politics of member states did interfere in the optimal operation of safeguards. Member states continued to insist that the Secretariat should not seek to estimate the likelihood that a state might seek to acquire nuclear weapons or the risk that it would divert material, as this could be viewed as "subjective" and "discriminatory." They also insisted that the Secretariat could not use non–safeguards derived information in compiling a non-compliance case against a state, lest the IAEA become an "intelligence agency." Although informally discussed within the Safeguards Department and at times even with member states, an open debate on this issue was officially taboo. Attempts by the Secretariat to collect information on unsafeguarded facilities in 1970 and 1973, Nakićenović claims, led to successful U.S. and Soviet pressure (prompted by him) to desist.[25]

What is more widely recognized as having an impact on safeguards effectiveness is that states would not give the IAEA the resources—financial, technical, or human—to mount an intrusive verification effort to keep track of all declared nuclear material, much less search for undeclared materials and facilities. As early as 1971, Allan McKnight, the first and only IAEA Inspector General, noted a "pattern of parsimony in relation to IAEA activities and particularly the safeguards function."[26] Even routine inspections could be cut back due to budgetary and staff shortages. In 1980, Tom Shea recalled, the Agency was only able to provide 42 percent of the inspection effort considered necessary.[27] That was the year that member states imposed zero real growth on the IAEA budget, along with similar constraints on other UN organizations (which remain in effect to this day). Blix laments that the Agency could have performed more inspections at Iraq's declared facilities, but in fact it did fewer, "in order to save resources."[28] Former Deputy Director General (DDG) Hans Grümm explained that during his 1978–1983 term as head of safeguards, it was "very hard to wrest from the Board an increase of the staff from 200 to 400 within five years."[29] In 1990, just before the Iraq case broke out into the open, DDG Jon Jennekens lamented that the IAEA's "capability to carry out effective safeguards operations" was being "seriously tested," not only by political developments and the increasing sophistication of nuclear fuel cycles, but also by financial limitations.[30]

As developing countries joined the IAEA in greater numbers, one emerging political factor that complicated these issues was their increasing demands for funding of technical assistance, later known euphemistically as Technical Cooperation (TC), as a *quid pro quo* for agreeing to increased safeguards budgets. Indian Ambassador Kamal Bakshi characterized this position in a debate with Blix in 1994: "In short what we are trying to say is, not that you spend 50 percent on safeguards and 50 percent on peaceful promotion of nuclear energy. No. We are saying, please, what has to be done, has to be done. But while doing so, please don't neglect the peaceful uses of nuclear energy."[31] This rather understates the persistent hard line taken by the Group of 77 developing countries in insisting on linkages between the safeguards budget and the voluntary funding of TC.

It is difficult to calculate the precise impact of the political machinations involved in negotiating and operating the safeguards regime on the specific case of Iraq. Blix says the system "was designed primarily with open, advanced industrial countries in mind. Clandestine facilities in closed societies," he stated, "were not what the system focused on."[32] Yet in implementing safeguards, the Secretariat never made such a distinction between member states, applying them without discrimination. Major developing countries certainly did not believe that safeguards were designed only for developed states; on the contrary, they argued that they suspected safeguards were aimed at suppressing their legitimate rights to the peaceful uses of nuclear energy. Grümm sees the focus on advanced industrial countries as an unintended consequence of the principle of non-discrimination:

The principle of the so-called "equality of misery" was strongly emphasized . . . by some States, which were very reluctant to accept safeguards. We had to apply safeguards without any consideration of particular situations in specific States. This led to undue concentration of inspection efforts in countries with open democratic societies where, I dare to say, at that time the press would have reported any diversion before the Agency inspectors had even arrived.[33]

By the 1980s, the IAEA had, due to its non-discriminatory approach, gained plenty of experience in safeguarding the unsophisticated, basic nuclear programs of developing countries and those of the closed societies of the Soviet bloc. Following U.S. President Dwight D. Eisenhower's 1953 Atoms for Peace proposal and the acceptance of the concept by the Soviet Union in 1954, both countries had begun providing assistance in the peaceful uses of energy, including nuclear research reactors and materials, to a large number of countries. The U.S. program focused largely on developing countries, while the Soviets largely assisted their Eastern European allies and China.[34] Most of these facilities and materials eventually came under IAEA safeguards. U.S. reactors, like Iraq's, invariably used highly-enriched uranium (HEU) as fuel, thereby increasing the proliferation risk—a danger of which the IAEA Secretariat was fully aware.[35] Because research reactors were the first type of nuclear facility to which safeguards were applied, the Agency had extensive experience with them long before comprehensive safeguards were introduced.

The fact that eventually all of the inspectors who dealt with Iraq were Soviet and Hungarian does not suggest that politics played a role in their performance. Even if they had acted more loyally to their states of origin rather than to the international organization to which they had pledged allegiance, the Soviets and their allies were by then strongly in favor of non-proliferation efforts, including safeguards.[36] They were especially concerned about having another nuclear weapon state not far from their borders in the volatile Middle East. While the Agency admitted that there were "differences" in the experience and training of individual inspectors, it "generally believed" that they were "motivated and [so] geographically mixed that any collusion with inspected states is extremely unlikely."[37] While an inspection team comprising only Soviet and Hungarian inspectors was hardly geographically mixed, they were obliged to report to the Middle East section of the Safeguards Department, which was staffed mostly by non–Eastern Europeans, including at least one American.[38]

Where the political constraints on the conceptualization and operation of safeguards did affect the Iraq case was regarding undeclared facilities. Despite explicit agreement that safeguards should cover all nuclear material in peaceful uses, the safeguards system was designed to focus narrowly on declared activities and, within those activities, to be concentrated on "strategic points" (although special inspections were not to be so constrained).[39] The

Secretariat judged that its use of risk assessments to identify potential proliferators, including utilizing open source or intelligence information, would be opposed by member states.[40] Furthermore, states repeatedly declined to provide the necessary funding, technology, and staff for a more robust safeguards system. All of these factors were rooted, at least in part, in the political positions of influential member states.

The Politics of Reporting Non-Compliance

An additional political explanation for the Iraq case focuses not on the politics that plagued the safeguards system itself, but on the Secretariat's supposed unwillingness to report a state suspected of non-compliance. Because international organizations are beholden to their member states for political, institutional, and financial support, their secretariats are, by their nature, presumed to be overly deferential toward governments and willing to challenge them only in the most extreme circumstances, if at all.[41]

In the case of the IAEA, it is the inspectors who have the initial responsibility to report a case of suspected or actual non-compliance to their section head and then up the chain of command through the head of the division and Deputy Director General for Safeguards to the Director General.[42] The Director General, in turn, has the right to seek further information from the member state, including, in extreme cases, requesting a special inspection. Ultimately, he or she may refer a case of non-compliance to the Board of Governors and prepare a report for the UN Security Council if the Board so decides.[43] In this respect, for an international civil servant, the Director General has considerable responsibility and influence.

After Israel's 1981 destruction of the Tamuz-1 reactor, U.S. congressional critics alleged a "lack of political spine on the part of inspectors or the agency to promptly sound the alarm if they are not able to affirm peaceful use or if they detect a serious anomaly that cannot be reconciled."[44] This implied that the inspectors or the Secretariat's leadership might be overly responsive to the concerns of the countries they inspect, or apprehensive of the broader political implications of reporting a member state.[45] As Lawrence Scheinman explains, "The possibility cannot be ignored that an anomalous situation laced with ambiguities and involving a less than cooperative state could lead the secretariat to defer, even indefinitely, informing the board."[46]

Nakićenović asserts that in his eleven years as head of safeguards operations under Director General Sigvard Eklund, inspectors were willing to report non-compliance, but the more "permissive bureaucracy" was "unwilling to undertake an adequate political response."[47] He claimed that "many inspectors argued strongly that Agency inaction in the face of noncompliance with a safe-guards agreement constituted an unhealthy *de facto* acceptance of such behav-ior."[48] In September 1981, three months before leaving office, Eklund did report two cases to the Board involving facility-specific safeguards applied to plants in India and Pakistan.[49] These cases, however, were not characterized as non-compliance, and were resolved by renegotiating the relevant agreements.[50]

Regarding alleged undeclared facilities and the possibility of special inspections to resolve anomalies, Eklund insisted that these were not to be pursued. Beginning in the late 1970s, the Safeguards Department repeatedly sought guidance from him about whether to pursue suspected undeclared activities or materials by means of a special inspection. The first case raised was Iraq's Tamuz-1 reactor. Eklund reportedly rebuffed all such approaches. Tom Shea, who served at the IAEA from March 1977 to July 1981 and helped craft the safeguards attachments for Tamuz-1, recalls that "there were no cases where we raised the point of these other issues [undeclared activities] where the answers didn't come down, from the DG typically. These matters [the DG noted] were the subject matter of [member states'] national technical means, but the Secretariat did not need to concern itself with this."[51]

There is no indication that any member state pressured the Director General to take this approach. Rather, it seems to have been part of Eklund's own attitude toward safeguards: they were concerned only with detecting and deterring the non-diversion of declared nuclear materials and facilities.[52] Any evidence of undeclared activities would come, he believed, not from IAEA investigations but from intelligence provided by member states, notwith-standing his reported episodic attempts to gather information on unspeci-fied unsafeguarded facilities.[53] Eklund would certainly have been sensitive to the political pressure likely to arise from member states if he was seen to be exceeding his mandate. Such considerations tend to produce caution in all heads of international organizations, as well as a proclivity toward self-censorship in anticipation of states' reactions. Eklund was also criticized for allegedly being too beholden to the two Cold War–era superpowers, which, as nuclear weapon states, were not subject to safeguards, but which certainly had major interests in the way the Agency operated.

At least prior to the Iraq case, Blix held similar views to Eklund's about the Agency's mandate when it came to undeclared materials, activities, or facili-ties. Blix believed that it was beyond the Agency's mandate for its inspectors to "roam around" looking for undeclared facilities, saw no "channels" for using intelligence information, and never asked a state to permit a special inspec-tion (although it is not clear whether inspectors ever pressed for one during his term).[54] Although he concedes that the relatively positive findings about Iraq in the Safeguards Implementation Report (SIR) as late as 1990 could have been formulated "with stronger reservations," he has always maintained that he was no softer on Iraq than on any other country—though this begs the question of whether he was too soft on all of them.[55] It is clear that before 1991, the Secretariat did not have credible evidence of Iraqi non-compliance to take to the Board, whether about undeclared or declared activities, and was taken completely by surprise by the nature and extent of Iraq's deception.

Blix was not, however, shy about reporting suspected non-compliance to the Board when there was credible evidence. Scheinman records that between 1981 and 1983, the Safeguards Department determined several times that it was unable to verify that no diversion had taken place at certain safeguarded facilities.[56] Instead of seeking findings of non-compliance, the Secretariat, with the support of the Board and with "some difficulty," renegotiated the pertinent arrangements. In another case, in 1984, a state failed to notify the Agency of the export of a quantity of depleted uranium, an anomaly detected by safeguards and subsequently judged to have been an accounting error.[57] This matter was also brought to the attention of the Board. It seems clear that Blix would not have hesitated to report credible evidence of Iraqi non-compliance to the Board if such evidence had emerged.

Far from states pressing Blix to either ignore or pursue suspicions about Iraq's nuclear program, he notes that during the Iran-Iraq War from 1980 to 1988, many were more concerned about fundamentalist Iran and "were probably not too keen to ask questions and possibly rock the Iraqi boat."[58] The United States had increasingly allied itself with Iraq out of fear that Iran might gain the upper hand, even providing targeting information to the Iraqis and, at least initially, turning a blind eye to Iraq's use of chemical weapons. It is unlikely, however, that the United States would have pressed Blix not to raise concerns about Iraq's nuclear program with the Board, especially as he is unlikely to have responded well to such an overture, as he demonstrated in his feisty dealings with the Bush administration over its claims that Iraq retained hidden nuclear capabilities prior to the U.S. invasion in 2003.[59] David Waller, an American who served as Deputy Director General for Management and Budget from 1993 to 2011, says he is proud that he never once felt pressured by the U.S. government in his twenty-five years at the Agency.[60]

Directors General must inevitably take political considerations into account in judging whether to raise a non-compliance issue. They must strike a balance between failing to report, on the one hand, and "crying wolf" on the other. In either case, they risk jeopardizing the credibility of the inspectors and the Secretariat as a whole. An additional factor in the Iraq case was that there was no precedent, at the time, for reporting a state to the Board for non-compliance with its safeguards agreement.[61] International secretariats are devoted to precedent and careful about setting a new one. A quite powerful case against Iraq would have been needed to set the precedent of reporting a state to the Board. Clearly there was no such case at the time. The option of special inspections, meanwhile, had become moot through non-use and, in any case, would have required credible information as a basis for requesting one. After the dam broke with the Iraq case, there were seven significant non-compliance reports (Egypt, Iran, Libya, North Korea, Romania, South Korea, and Syria) by the Secretariat to the Board.[62] Iraq had set the precedent.

Organizational Explanations

Notwithstanding the flaws in the design of safeguards and the political and financial constraints on the effective operation of the system, a second explanation for the Secretariat's failure to detect Iraq's non-compliance was that it was at best feckless and at worst incompetent in implementing safeguards. The assumption here is that, even with all the known lacunae in safeguards, it was the lack of safeguards implementation that enabled Iraq to blindside the Agency. After examining both the pre- and post-NPT safeguards models, the Stockholm International Peace Research Institute (SIPRI) predicted as early as 1971 that "both safeguards documents and the agreements that are based on them generally give the Agency the legal tools to achieve its objectives. Any flaws that may be found in the application of safeguards would be due to the way in which the documents are implemented rather than to any inherent inadequacy in them."[63] The confidential 1984 ACDA report on the negotiations that produced the INFCIRC/153 noted that:

> From the vantage point of the present, it is difficult to separate the issue of the adequacy of the document from that of the manner in which it has been implemented. In general, however, it appears that, to the extent that safeguards implementation in practice has fallen below the anticipated level, this reflects in some measure the fact that the rights and opportunities available to the Agency under the agreements . . . have not been fully and effectively utilized.[64]

There were several organizational critiques of IAEA safeguards at the time of the Iraq case. The first was the assumption that as a "typical" UN organization, the IAEA was, by its nature, incapable of operating a verification system as complex and intrusive as safeguards. A second critique focused on three specific aspects of the Safeguards Department: management and administration; technical capacity; and the inspectorate.

The IAEA as a "Typical" UN Organization

Generic criticism of the IAEA assumed that it was a typical UN-type organization, and therefore ill-equipped to verify state compliance with an international security agreement like the NPT.[65] A common prejudice among the public, diplomats, analysts, and even international civil servants themselves is that UN-type organizations are inherently ineffective and inefficient. With their multinational staff (some appointed according to political rather than competency criteria), rigid bureaucratic procedures, arcane financial rules and regulations, and outdated technology, UN-type organizations are presumed to be institutionally flawed. To assign to a traditional UN-type bureaucracy like the IAEA a global verification role, such as nuclear safeguards, is greeted by this school of thought with incredulity.

It is true that the IAEA reproduces, for better or for worse, many of the UN's organizational features as practiced at UN headquarters and in its specialized agencies. The Agency's diplomatic, managerial, administrative, and financial practices essentially replicate those of the UN system. This model was designed for collaborative multilateral bodies charged with delivering important but relatively benign public goods like economic and social development, food and agriculture, and global health (although the global COVID-19 pandemic has exponentially politicized the latter field). Specialized UN agencies such as the UN Food and Agriculture Organization (FAO) and the UN Economic, Social and Cultural Organization (UNESCO) are often regarded as among the least effective and efficient UN organizations.[66]

While it is deeply influenced by UN organizational structures and traditions, the IAEA is not a "typical" UN organization. Although the Agency is considered a member of the UN's organizational family, it has a degree of autonomy from the UN system. It is not classified as a specialized agency. It has its own founding statute and membership. Its unprecedented inspection powers on member states' sovereign territory and its direct access to the UN Security Council in the event of safeguards violations make it unique among international organizations.[67] The IAEA's authorities are much more like those of a national regulatory body that issues regulations, conducts inspections, and considers penalties for non-compliance.[68] John Tilemann notes that Agency staff saw the culture of national regulatory bodies as dominant in shaping safeguards culture.[69] The closest analogues in the international system are the International Civil Aviation Organization (ICAO), the International Telecommunications Union (ITU), and the International Maritime Organization (IMO). All of these organizations have quasi-regulatory functions, although none matches the IAEA's on-site inspection powers. Tom Shea calls it "somewhat magic" that the IAEA Secretariat was ever permitted by states to do what it does.[70]

Moreover, the IAEA has always been widely regarded as one of the most competent, effective, and efficient multilateral organizations, as attested by numerous independent commissions, UN reports, and U.S. congressional studies.[71] A report to the U.S. Secretary of State in 1972 concluded that "the IAEA is an effective, well-run organization by comparison with most international organizations."[72] The Agency's performance in fiscal management was "conservative and efficient." Successive U.S. administrations have singled out the IAEA as deserving special attention whenever the Congress or U.S. allies were contemplating across-the-board cuts to international organizations.[73] Neither internal nor external audits of the Agency have revealed major organizational shortcomings.[74] Shea and David Thompson judged that by the early 1980s, "to a very great extent the Agency succeeds in providing an amalgamation in which its international civil servants can work in a productive setting."[75]

Viewed through John Carroll's strategic planning lens, the IAEA, despite financial and personnel shortfalls, was appropriately organized for its safeguards role. A Safeguards Department—eventually the largest and best-funded department in the organization—reported directly to the Director General under a remarkably flat organizational structure that ensured safeguards received the highest possible level of attention. It drew on the necessary technical expertise of member states through expert committees, a Scientific Advisory Board (until 1988), and a Standing Advisory Group on Safeguards Implementation (SAGSI). The Secretariat itself "came to take the initiative for most of the IAEA's technical work."[76] By 1975, the department had its own Safeguards Analytical Laboratory (SAL) at Seibersdorf, near Vienna, and a Network of Analytical Laboratories in member states.[77]

The organizational or strategic planning lens, while focusing on organizational structure and capabilities, also naturally recognizes the role of leadership. It tends, however, to emphasize the managerial and technical competence of leadership rather than its role in shaping organizational culture, even though cultural markers constantly arise in any consideration of leadership. The drawbacks of such an approach are illustrated by the following discussion of IAEA leadership as viewed through a strategic planning lens. A full consideration of the cultural implications of IAEA leadership follows in Chapter 3 of this volume.

From the beginning, the IAEA was blessed with highly competent Directors General who, through its earliest decades, steered the Agency successfully through treacherous political and technical waters. Prior to the Iraq case, there had been three: Sterling Cole, from the United States (1957–1961); Sigvard Eklund of Sweden (1961–1981); and Hans Blix, also of Sweden (1981–1997). Berhanykun Andemicael and John Mathiason, former UN officials and management specialists, contend that all of the early Directors General were "quiet, technologically competent and politically adept in their relations with Member States."[78] The support of successive Directors General for the safeguards work of the Agency was, however, mixed.

Sterling Cole, formerly a Republican Congressman, surprised everyone with his strong support for IAEA safeguards in the face of the Eisenhower administration's unexpected favoritism toward the European Atomic Energy Community (EAEC or EURATOM), the nascent European safeguards agency.[79] This was due partly to the rise of EURATOM as a competing safeguards organization for European states, but also because the role envisaged for the Agency as a fuel bank and nuclear information clearinghouse was being undermined by bilateral arrangements between states, regardless of whether they were IAEA members or not. Cole apparently fought valiantly to bring the IAEA safeguards system into operation.[80]

Eklund, a distinguished scientist, was reportedly "a little bit paternalistic" in his management style. He knew all of the staff members and took an interest in them (the Agency was much smaller at that time). Perhaps more important, James Goodby credits Eklund and his senior IAEA staff with ensuring the Agency's survival in the early years, when it came close to becoming irrelevant and was "skating close to disaster" due to a lack of support from the Eisenhower administration and the absence of a clear purpose.[81] Although Eklund was criticized for being too close to the superpowers, this was probably unavoidable at the height of the Cold War, when all international organizations had to perform a delicate dance between the competing demands of the Soviet Union and the United States.[82] A more problematic factor was that Eklund's scientific background, while useful in grounding safeguards in a technical framework, may have blinded him to the political context of the Agency's work. Although during his tenure the NPT accorded the IAEA its most important mandate and vastly expanded the scope of safeguards, Eklund reportedly "didn't believe in nuclear safeguards or verification."[83]

Nonetheless, when Blix became Director General in late 1981, he felt that he was taking over an organization that was "in good shape."[84] When the Iraq case broke in 1991, he had been Director General for a decade, more than enough time to put his personal stamp on the Agency. A 2002 external consultants' report concluded that under Blix, "the Agency [had] consistently shown its determination to enhance its effectiveness and efficiency."[85] Blix was outspoken, unpretentious, had an "open door" policy, and sought to explain his decisions to his staff.[86] Unlike Eklund, Blix reportedly took a personal interest in safeguards and literally "walked the corridors" to see what was happening.[87] Blix attended meetings of SAGSI, his safeguards advisory body, and received briefings afterward.

Given that two Swedes were Directors General for most of the Agency's history prior to the Iraq crisis, it is tempting to see Nordic values of effectiveness and efficiency at work. Both Eklund and Blix suggested that such values made the Agency different from other members of the UN organizational "family." Goodby praises the IAEA as having "a remarkable set of managers" and as having "outperformed expectations," although at times these were apparently quite low.[88]

In summary, viewed through Carroll's strategic planning or institutional lens, it is not clear that any Agency-wide organizational or leadership deficit was responsible for the IAEA "missing" the safeguards violations in Iraq. Indeed, shortly after Iraq's non-compliance was revealed, the Secretariat successfully identified North Korea as being in violation of its recently concluded safeguards agreement and instigated the necessary processes leading to action by the Board of Governors and the UN Security Council.[89] The system redeemed itself. Blix noted at the time that:

> From the IAEA horizon, the DPRK [North Korea] case is straightforward: safeguards findings suggest that the DPRK has more plutonium than it has declared and when the country refuses to co-operate by allowing requested inspections of two non-declared sites and fails to give satisfactory explanations, the Agency turns to the Security Council. It is not really the effectiveness of safeguards as an alarm mechanism which is now being tested. Thanks to the use of new analytical techniques which are part of the strengthened safeguards, the system has produced warning signals.[90]

The North Korean case came, of course, after the shock of Iraq, when the Agency was on much higher alert for non-compliance cases and when reforms were being suggested and implemented.[91] As Blix noted, however, the North Korean case was also quite different from that of Iraq.[92]

Safeguards Management and Administration

If the Agency overall was in good organizational shape by 1990, what about the specific management of safeguards at the time of the Iraq case? Could mismanagement have hindered the proper application of safeguards to Iraq and the timely discovery of its non-compliance?

The IAEA Safeguards Department, which manages, develops, and implements safeguards, is headed by a Deputy Director General (DDG) for Safeguards.[93] Underneath the DDG, the structure of the department is divided between operational divisions that implement safeguards in different regions of the world and divisions dealing with information, development, and technical support. The department is supported in policy matters and in drafting and negotiating safeguards agreements by the management and legal departments. The Safeguards Department was originally a single division, but its operations were divided into two parts—A and B—in 1977. A confidential report on the Agency's management practices by the consulting firm Coopers and Lybrand in 1981, the year of transition between the twenty-year tenure of Sigvard Eklund and the new regime of Hans Blix, resulted in the creation of a third operational division, Operations C, and two supporting divisions.[94] Iraq was covered by the South and South East Asia Section of Operations B.[95]

The relatively flat organizational structure of the IAEA means that all department heads, all of whom are DDGs, report directly to the Director General, or at least to his executive office. This arrangement is particularly well suited for safeguards non-compliance matters, because the DDG for Safeguards has direct access, at least in theory, to the Director General or his executive office at any time. This is especially important in urgent situations involving safeguards violations. John Tilemann confirms that all of the DDGs for Safeguards that he observed kept in close contact with the responsible Director General's Office staff.[96]

The Safeguards Department has always been seen by outside observers, and by many within the Agency itself, to be a particularly professional and competent part of the IAEA, with a technical, apolitical orientation and a strong commitment to strengthening its performance. After a slow beginning, it gradually became the largest and best funded of the Agency's departments.[97] From the outset, it employed large numbers of physical and chemical engineers, mathematicians, statisticians, and accountants, including as inspectors. It ran its conferences and workshops along scientific lines. It also developed strong links with certain national technical organizations, especially in North America and Western Europe, and with professional nuclear organizations that emerged, such as the European Safeguards Research and Development Association (ESARDA) and the U.S-based Institute of Nuclear Materials Management (INNM). All of this burnished the department's reputation for professional and technical competence.

There is evidence, however, that the optimal management of safeguards was blighted by several organizational fissures that are common in bureaucracies in general but had features specific to the Agency. As Shea and Thompson note, "The problems inherent in any bureaucracy of establishing and maintaining a sense of purpose and vitality required to effectively discharge its mission exist in the IAEA. As with any service organization, performance cannot be gauged by any monetary measure such as profit or loss, but only in terms of its credibility—the degree to which it performs its intended functions."[98] The first opportunity for discord is the relationship between the Director General and the DDG for Safeguards. The DDG is regarded as having special authority and influence due to his role in helping manage the IAEA's relationship with member states on safeguards issues. His most sensitive task is deciding how far to pursue anomalies in a state's compliance with its safeguards agreement. This may start with the DDG communicating directly with the state party's authorities, at a suitably low, technical level, but may ultimately involve a decision to inform the Director General so that higher-level action may be initiated.

Potentially, the result could be a report to the Board and, in turn, to the UN Security Council. The Safeguards Department, while keen to demonstrate that it is acting professionally and legitimately, is at the same time conscious of a potential loss of credibility if it makes ill-supported allegations about a state. The DDG of Safeguards will also unavoidably be aware, through his dealings with the Director General and his experience with the Board, of what the "political traffic" might bear in each case. This is a high-stakes game.

Several DDGs for Safeguards have regarded themselves—and not the Director General, as the press often insists—as the "world's top nuclear inspector."[99] This has, on occasion, created discord between Directors General and their DDGs for Safeguards. Eklund replaced the title of "Inspector General" created by Cole for the head of safeguards with "Deputy Director General for Safeguards." This brought the post into line with other designations, but was also reportedly done because the incumbent, Rudolf Rometsch, had become too imperious.[100] Meanwhile, Nakićenović, the first Director of Safeguards Operations, was startlingly distrustful of his seniors, including

Eklund, although this probably had personal as well as organizational origins.[101] Blix, in contrast, seemed to have a much better relationship with the staff in general, including his DDGs for Safeguards.[102]

The post of DDG for Safeguards has traditionally been regarded as being "assigned" to the West (or at least Western neutrals), the result of a political deal at the inception of the Agency.[103] The Western states were determined to oversee the development of safeguards to ensure that the system was not derailed by the Soviets or other safeguards skeptics such as India. Equally important, they wanted to prevent the Secretariat from exceeding its carefully bounded safeguards remit. Finally, they sought to ensure that safeguards were effectively and efficiently managed, as they were largely paying for it.

Table 2.1. Heads of IAEA Safeguards, 1958–2021.

Head of Safeguards*	Nation of Origin	Term of Office
Roger Smith	Canada	1958–1960
Dragoslav Popovích	Yugoslavia	1960–1964
Allan McKnight	Australia	1964–1969
Rudolf Rometsch	Switzerland	1969–1978
Hans Grümm	Austria	1978–1983
Peter Tempus	Switzerland	1983–1987
Jon Jennekens	Canada	1987–1993
Bruno Pellaud	Switzerland	1993–1999
Pierre Goldschmidt	Belgium	1999–2005
Olli Heinonen	Finland	2005–2010
Herman Nackaerts	Belgium	2010–2013
Tero Varjoranta	Finland	2013–2018
Massimo Aparo	Italy	2018–present

Note: The Safeguards Department was not established until 1964. Prior to that it was a section. Massimo Aparo was still Deputy Director General for Safeguards as of this writing in 2021.
Sources: IAEA, *IAEA Bulletin*, March 1994, p. 13; Jeffrey Lewis, "Replacing Olli," Arms Control Wonk, July 3, 2010, http://lewis.armscontrolwonk.com/archive/2792/replacing-olli; and comments by Mark Hibbs, "Replacing Olli."

Successive DDGs of Safeguards have had a Western management style deriving from their experience in their national regulatory bodies, favoring top-down direction and strict procedures and processes. In a sense, they are the international nuclear regulator—in popular parlance, the "nuclear watchdog." As in the case of national regulators, this involves "deploying material resources to induce or compel actors to comply with existing rules."[104] The IAEA, being unable to impose fines or cancel licenses like national regulators, deploys its safeguards system to achieve "regulation by revelation," relying ultimately on "publicity of behavior to prompt remedial action."[105]

DDGs of Safeguards, as managers and planners of the enterprise, are concerned not just with the technical details of verification and compliance—a Western member state preoccupation—but also with achieving maximum effectiveness and efficiency, including the ways in which resources, budgets, and staff are utilized. They are naturally also concerned with the political and institutional reputation of the department, both within the Agency and externally. Such priorities have not necessarily been shared by the operational staff, including inspectors. A discussion at the 1994 IAEA Safeguards Symposium involving Rudolf Rometsch, Hans Grümm, Peter Tempus, and Jon Jennekens—all of the early DDGs who had progressively overseen the introduction and implementation of comprehensive safeguards—revealed that all of them had sought to improve the effectiveness and efficiency of safeguards, despite resistance from member states and, on occasion, from within the Agency and even within their own department.[106]

Although all IAEA departments, including Safeguards, have traditionally been reputed to operate as separate little principalities within the Agency, this does not mean that DDGs have been all-powerful. All have had to operate within the traditional political and management constraints of international secretariats, with their multiple political masters and multinational staff. As Mathiason notes, "Leadership in international organizations is not based on the ability to give orders but rather on the ability to convince member states to endorse proposals and to motivate staff to implement requests."[107] Accordingly, DDGs for Safeguards appear to have faced difficulties at times in establishing their authority, partly due to resistance from directors, section heads, and independently minded inspectors.

Writing in the early 1980s, Shea and Thompson contended that DDGs "operated in the European professorial style of management, attempting to control all subordinate functions, but on occasion have been frustrated by the reluctance of subordinates to conform."[108] On some occasions, directives were "ignored with impunity." They attributed this at least partly to the different nationalities of staff who may not share a "directive management philosophy"—which points to national cultural differences.

The potential for discord between the DDG and the department was compounded by the fact that DDGs were external appointments, not from within the department itself, and certainly not from the inspectorate (Olli Heinonen was a later significant exception). They would have been confounded to find themselves in a UN-type organization that ran on different principles from commercial industrial corporations or national regulators. Grümm, who became DDG for Safeguards in 1978, noted the organizational splits within the department: "I had the feeling when I came that I inherited not a real department but a federation of safeguards divisions. I spent my 4 years in this way, in consolidation. It was an attempt to unify the activities of the divisions so that action of the department as a whole would come out."[109]

As Andemicael and Mathiason note, in contrast to the DDG, most division heads and section chiefs are career staff "who have emerged through the ranks."[110] "This reflects," they say, "an understanding that in management of the inspection function an ability to navigate the complex political environment in which inspections take place is more important than technical skills, which are also essential but can be acquired more easily."[111]

In a scathing assessment for the U.S. Congressional Research Service in 1981, Nakićenović contended that the major weaknesses of safeguards were due to implementation by the "bureaucracy," as he called them, which he portrayed as "incompetent" and "unduly influenced by expediency."[112] Because he himself was the chief safeguards bureaucrat, it appears that he was referring to anyone at the Agency who was not involved in safeguards operations. Among other charges, he alleged that: "they did not sufficiently and efficiently use inspection rights available under safeguards agreements"; they relied too much on an "intelligence type approach" and too little on the use of sophisticated technology and the cooperation of safeguarded states; and information about them was over-classified "by a self-centered bureaucracy."[113] He also accused them of using safeguards confidentiality to mask their own failures and suppress criticism; being too beholden to the superpowers and other powerful nuclear states like Japan, as well as EURATOM; and providing insufficient information to the Board.[114] He even reproached the Secretariat for weakening safeguards by negotiating facility attachments that limited the Actual Routine Inspection Effort (ARIE) to a fraction of what was legally permissible. This originated, he charged, during the Secretariat's negotiations with Sweden, implying that Eklund, as a Swede, had gone easy on his compatriots and then allegedly extended this "novelty" subsequently to all other member states without consulting the Board.[115]

Unfortunately, because Eklund neither wrote memoirs nor publicly refuted these serious allegations, his side of the story is not available. While it is thus difficult to establish the veracity of these allegations, their mere existence illustrates the gulf that existed, at least during that era, between safeguards management and the inspectorate. In their study of the Agency for ACDA in 1981–1982, Shea and Thompson concluded that by the early 1980s, management problems at the IAEA had resulted in "forming non-productive work attitudes and habits" which had caused some inspectors to leave.[116] This sounds in part like a problem of organizational culture.

By the time Dieter Goethel became Director of Personnel in 1989, less than two years before the Iraq case emerged, there was, he says, a "breakdown in personnel management" at the Agency, due to three factors.[117] One was the large number of British staff, especially in information technology (IT), apparently due to poor job prospects in the United Kingdom at the time. In contrast to the Austrian "social partnership" approach to management, the British had an adversarial style, presumably deriving from their long trade union tradition. This too sounds like a cultural issue. A second problem was the floating U.S. dollar, which had essentially devalued IAEA salaries. The Board had to be asked for salary adjustments, which "took years to fix."[118] The third factor was the increasingly vociferous demands of developing countries for staff positions. Such organizational dysfunction undoubtedly produced organizational inefficiencies. Whether any of them affected the outcome in Iraq is considered further below in assessing the Agency's technical capacities, its use of technology, and the quality of its inspectorate in the decades before 1990.

Technical Capacity

Consistent with the IAEA's claim to be a technical organization, the Safeguards Department stressed that "to constitute an effective deterrent, safeguards must be technically capable (and be *seen* to be capable) of promptly detecting the diversion [of nuclear materials or misuse of facilities]."[119] The political value of safeguards, it noted, depends "significantly on the way in which its detection capability is perceived by those States which expect the IAEA to provide assurance as well as by any State which might contemplate diversion and wished to know the risk of being detected."[120]

Paradoxically, one way to enhance the deterrent effect of safeguards was for the Secretariat to stretch the concept of safeguards confidentiality beyond its original meaning, to cover not just information that governments wished to keep secret, but also the operational details of safeguards implemented by the Agency that might allow others to judge the Agency's own effectiveness and efficiency. Preserving opacity seemed designed to enhance the mystique of safeguards, strengthen the system's credibility, and fend off criticism that might weaken safeguards' deterrent effect. This was not what was intended by safeguards negotiators: the negotiating record makes clear that "INFCIRC/153, in common with corresponding provisions of INFCIRC/66, requires the Agency to protect information belonging to a state, and coming to the Agency's knowledge through the application of safeguards, but it does not require [emphasis in original] the Agency to withhold information on its own implementation activities."[121] Nakićenović accused the Secretariat of deliberately keeping even the Board in the dark about the state of safeguards in order to avoid criticism of its shortcomings.[122]

Despite its opacity, the Safeguards Department consistently sought to make technical improvements to safeguards wherever it could, while being acutely aware of the political sensitivities involved and the impracticalities of radical change. The advent of comprehensive safeguards in the 1970s significantly strengthened the hand of the department in seeking such improvements. From 1975, it was further assisted by the establishment of SAGSI, which, though it advised the Director General, not the Safeguards Department directly, and comprised mostly Western experts endorsed by their governments, made a series of invaluable suggestions to strengthen the system. It recommended, for instance, the adoption of safeguards goals that focused on the detection of significant quantities of weapons-usable material and meaningful warning times for its diversion to a weapons program.[123] SAGSI also played an important role as a conduit for government views on safeguards' "rigor" (whether to tighten or loosen them), as well as helping to "condition" governments for changes being mooted. In addition, SAGSI was instrumental in developing the reporting format for the SIR.[124] The Safeguards Department was sometimes wary of SAGSI's involvement, especially the operations divisions, which were "pretty conservative."[125]

A continuing challenge was the difficulty of objectively assessing the effectiveness of safeguards. Member states pressured the Safeguards Department to become more self-critical and innovative. In 1975, the Director General established an evaluation unit, which subsequently became a division, in order to ensure "continuing, independent, internal evaluation of the effectiveness of safeguards."[126] In 1976, on U.S. initiative, the Board asked the Secretariat to submit an annual SIR on how the system was working and what steps were needed to improve it.[127] According to Shea, this resulted in "a clear definition of the minimum requirements necessary for the safeguards applied at each facility, in standardizing the safeguards for similar facilities between regional sections, and for providing a clear definition of outstanding problem areas."[128] The SIR was never as revealing as originally intended, and was distributed only to the Secretariat and member states, with summaries publicly released in later years. Nakićenović accused the Agency of packing the SIR with trivialities to distract from its own failures in safeguards implementation.[129]

Unfortunately, after the first SIR was released in 1977, the U.S. Nuclear Regulatory Commission (NRC) was so alarmed by what it perceived to be the weaknesses of IAEA safeguards that it announced it would refuse to continue to certify their adequacy when considering whether to approve U.S. nuclear export licenses.[130] While this appears to have resulted from a confusion between physical protection and safeguards, it nonetheless triggered a new round of U.S. skepticism about the system and indicated the risks of transparency.

Despite these shortcomings, Shea told the U.S. Senate Foreign Relations Committee in 1981 that "fundamental improvements have been made in the technical effectiveness of IAEA safeguards in the period since 1977."[131] He claimed that "the requirements are more clearly specified, the equipment more efficient and more reliable, staff levels have increased some fourfold, and the staff is far more professionally trained than in the past." The IAEA inspectorate was "held accountable for their activities to a much greater extent than before."[132] He said he was not aware of any undetected diversions of nuclear material from 1977 to mid-1981.

Meanwhile, the International Fuel Cycle Evaluation (INFCE), an exhaustive international evaluation of the technical assumptions concerning the development of the nuclear fuel cycle which had begun at U.S. urging in 1977, ended in March 1980 with a voluminous report that largely endorsed the validity of safeguards. Its working groups on different parts of the fuel cycle "did not identify significant problems with the capability of methods and techniques [of safeguards] as applied to existing operating plants."[133] They did, however, conclude that "further development and improvement of existing methods and techniques was necessary to meet safeguards objectives at reasonable costs," regarding technologies for uranium enrichment, industrial-scale reprocessing of irradiated fuel, and mixed oxide fuel fabrication for light-water reactors and breeder reactors. The report assumed, however, that the main problem lay with declared facilities, rather than the undeclared facilities that Iraq was seeking to build as the report was being compiled.

By 1980, David Fischer, the widely respected head of external relations at the Agency, concluded that the application of NPT safeguards was:

> tending to become more and more of a routine quasi-industrial operation with well-defined and well-understood objectives based increasingly on precise numerical operations which permit [the] IAEA to reach "quantified" conclusions, i.e., conclusions based entirely upon mathematical calculations rather than the inspectors' own judgement, however objective that may be.[134]

Although he undoubtedly intended to be complimentary, Fischer was inadvertently pointing to problems with safeguards that would be revealed by the Iraq case: routinization and increasingly blind pursuit of criteria-based technical approaches designed to eliminate subjective judgments. Fischer himself acknowledged, however, that "in an increasingly complex operation of this kind, it will never be possible to eliminate entirely the element of judgement."[135]

In the late 1980s, this trend became entrenched with the introduction of a "check list" or criteria-based approach.[136] The leak of a safeguards report to the magazine *Der Spiegel* in early 1988, which described discrepancies in the German nuclear program, embarrassed the Agency not only because it was a lapse in safeguards confidentiality, but also because the safeguards report was of such poor quality. Director General Blix asked David Kay to produce a template as a "model for future reporting."[137] Criteria were established with prescribed technical objectives for safeguards activities. These specified the scope, normal frequency, and extent of inspection activities for each type and category of material at each location. They were first used to plan safeguards operations. After they proved effective in that regard, they were then also employed to evaluate whether inspection goals had been attained.

During his tenure as DDG for Safeguards from 1978 to 1983, Grümm says he sought to further strengthen "the quality of verification" through the creation of a training section, the improvement of instruments, the computerization of inspection reporting, and the introduction of stringent performance criteria.[138] Yet Canadian Jon Jennekens, DDG for Safeguards from 1987 to 1993, recalls that in 1988, just three years before the Iraq case broke, he was still pursuing his predecessors' "effort to develop and to promulgate unified planning, implementation, and evaluation criteria" across the department.[139] What he did achieve was an improvement in inspection reporting forms "in the interests of consistency, completeness, and reduction of the narrative component."[140]

Jennekens also sought to develop "a more cooperative, more efficient, and more effective set of arrangements with State Systems of Accounting and Control [SSACs]," which were not working well. In most instances, he records, the initiatives were welcomed and supported by states, but in other cases, they were met with "disinterest [sic], obfuscation, opposition, resentment, and even hostility." Clearly Iraq did not have an effective state system from the IAEA's perspective, as it was essentially failing to report all its undeclared fuel cycle elements. Jennekens records that it was not until June 1991, some three years later—and, notably, after the Iraq case had broken—that "we managed to achieve the broad agreement on a more meaningful set of partnership arrangements with SSACs."[141]

Technology

In the 1970s and 1980s, the Agency struggled to adopt the most advanced and robust safeguards technology. This was partly due to budgetary constraints on both acquiring technology and deploying it in the field and partly due to a lack of technical capability to use and maintain it. The Secretariat was traditionally provided with only a minuscule budget for conducting research and development of safeguards technologies. Inspector General Allan McKnight complained as early as 1971 that "the level of scientific support for safeguards has been disappointing."[142] From the outset, the Agency relied heavily on member states, above all the United States, for scientific and technical support. State Safeguards Support Programs, as they became known, proved invaluable in exposing the Secretariat to new technological possibilities, although sometimes these were beyond the ability of the Agency to absorb and use.[143] New technical means of verification, such as cameras, video, and advanced seals, were gradually introduced as funds became available, but these deployments were slow, uneven, and never cutting-edge.

By 1982, nuclear expert Manning Muntzing noted that in the past decade, there had been "substantial research and development world-wide on new instruments and on techniques for verifying the identification, location, and quantity of [nuclear] materials."[144] The need was for extensive applications engineering, in concert with the operation of facilities, to ensure practical relevance. This identified another challenge for the IAEA: not only governments but also facility operators needed to accept new safeguards techniques and technologies. Muntzing listed several seemingly mundane but important needs at the time: equipment maintenance and repair infrastructure; a continuing miniaturization of equipment; standard reference materials for calibration of field and laboratory equipment and support of quality control and assurance programs; the pursuit of redundancy through cameras and other surveillance devices; and development of a new generation of rugged and inexpensive security seals. For Myron Kratzer, the central technical controversy of the 1980s was not about which technologies to deploy, but between the advocates of better materials accountability and those who believed that containment and surveillance could increasingly replace supposedly mundane inspection tasks.[145] The debate was notably not about detecting undeclared facilities.

Some technologies that could have helped detect Iraq's illicit activities, such as satellite imagery or wide-area environmental sampling, were, however, simply not available to the IAEA in the 1980s. Not only were they expensive, but member states would also have objected to their use. Politics again intruded. Environmental sampling, even at declared sites, would have been particularly powerful in the Iraq case, but member states would not have allowed it.[146] Both technologies were subsequently adopted.

One area where the Agency could have advanced more quickly prior to the Iraq case was in the use of off-the-shelf IT in both general management and administration and safeguards work. John Carlson recalls that, contrary to the Agency's reputation for technical excellence, it came late to computerization. He recalls that the IAEA's offices in the 1980s reminded him of the Australian Public Service of the 1960s—"no computers and every manager [invariably male] with a female secretary who typed his letters."[147] Inspector reports were typed by a typing pool. There was no standardized reporting system or computerization of reports or other records, making previous reports difficult to retrieve.[148] Planning for future inspections was mostly by word of mouth from inspector to inspector. William Lichliter confirms that when he arrived in 1981, he had to use a typewriter.[149] The Safeguards Department had a mainframe, but no personal computers. Operations C had two card readers. Among Lichliter's first "battles" were to get a word processor for each floor, then for each section, and only then for each section employee.

Efforts that had begun as early as 1968 to process safeguards data systematically were repeatedly delayed. Nakićenović attributes this partly to politics. A 1975 task force that was mandated to plan computerization was, against the advice of safeguards staff, led by a Soviet national whose field was not data processing. It ended in failure. In 1976, another Soviet national without data experience was appointed head of a new Safeguards Information Division, again against the advice of experts. The system became operational in August 1981, but was "still not much use for inspectors" because "there were hardly any inspector reports in the system's memory."[150] By the end of 1983, however, the IAEA Safeguards Information System (ISIS) database contained approximately three million records.[151]

Attempts by Grümm and Tempus to induce inspectors to file inspection reports electronically in the early 1980s were resisted by some. Tempus, DDG of Safeguards from 1983 to 1987, recalls that although "it was clear to me that computers would play a central role in safeguards in the future," some inspectors "were afraid . . . that they would be squeezed into a fixed scheme, not allowing them to do what they felt they should do at the facilities"—which, again, sounds like a cultural issue.[152] Some inspectors were apparently more freewheeling than others and were not constrained by the prevailing cautiousness. As Shea and Thompson note, "A wide range of personalities is represented in the inspectorate, which is both normal and healthy."[153]

Lichliter has a different recollection: that the computer "revolution" in the Safeguards Department was ultimately driven not by "computer people," who wanted to protect the centrality of their mainframe, but by inspectors who "went out and bought their own computers."[154] This also speaks to the independent spirit of at least some of the inspectors.

By 1990, an inspection reporting form, commonly called a "log sheet," was being used to record all information required for computerized inspection reports. The processing and handling of inspection data was "for the most part computerized." The *IAEA Bulletin* reported "significant progress" in "development of criteria for the evaluation of inspection goal attainment."[155] Shirley Johnson says that the new approach helped transform safeguards, as inspectors now had a format to follow, resulting in much higher-quality reporting.[156]

The safeguards section of the Annual Report of the Agency for 1990 gives an impression of continuing technical improvement and innovation in safeguards, much of it geared to large-scale facilities such as enrichment and reprocessing plants and geological waste repositories. Some innovations were applicable to the type of relatively small-scale declared facilities operated by Iraq, including the installation of closed-circuit television and the replacement of old Minolta cameras with new models. Specific state recipients were not identified for these programs, so it is impossible to say whether they were deployed in Iraq.[157] Future instrumentation developments that might have affected Iraq included the preparation of standard measurement procedures; testing of a plutonium isotopic measurement technique based on state-of-the-art hardware and software (introduced routinely in 1991); and the replacement of photographic optical surveillance units with the compact surveillance and monitoring system (COSMOS), also beginning in 1991.[158]

An evaluation of inspection goal attainment for the safeguards system in 1989 was conducted according to the now familiar safeguards criteria and the results reported to the Board in the SIR, but of course this remained confidential.[159] A total of 2,286 inspection reports and 2,485 inspection statements were reviewed, and computerized quality control checks were applied. The average time between an inspection and the dispatch of the results to the state concerned was an unimpressive 52.5 days (64 in 1988 and 50 in 1989).[160] In preparation for the introduction of safeguards criteria in 1991–1995, work had begun by 1989 on rewriting computer programs and "documenting" the computerized inspection report system.[161] As for research and development activities, the 1990 Annual Report noted that due to increased collaboration between member state support programs, there had been "an improvement in programme effectiveness," implying that there had previously been coordination challenges.[162] Notably, in view of what was about to befall the Agency, none of the technical measures discussed in the report were said to be applicable to the detection of undeclared materials, facilities, or activities.

Although the 1990 Annual Report provides a snapshot of the state of safeguards techniques and technology for that year, it is impossible to assess the overall effectiveness of the safeguards system due to both the lack of detail provided and the opacity of the official assessments resulting from safeguards confidentiality. Moreover, after the flurry of independent external assessments of the technical effectiveness of safeguards in the early 1980s following the bombing of the Tamuz-1 reactor, there was a dearth of them in the rest of the decade, ensuring that there was little information and analysis to challenge that which was provided by the IAEA itself.[163] As noted above, Blix bemoans the fact that "despite its questioning of the safeguards system's reliability, neither the U.S. nor any other government took the initiative to strengthen the system during the 1980s."[164] There would at that time, he says, have been "insuperable resistance to more intrusive inspections."[165]

The Inspectorate

The notion that international inspectors could undertake independent activities on the territory of sovereign states was truly revolutionary, and it took years for the IAEA to organize itself to carry out this role. Originally, there had been skepticism about the possibility of an effective and efficient international safeguards inspectorate. A U.S. committee that included no less a luminary than Robert Oppenheimer, who had led the scientific effort to produce the atomic bomb under the Manhattan Project, and which was chaired by David Lilienthal, Chair of the U.S. Atomic Energy Commission, questioned whether it would be possible to recruit the "very large and very highly quali-fied organization of experts and administrators needed."[166] They assumed that "the work itself, which would be largely policing and auditing and attempting to discover evidences [sic] of bad faith, would not be attractive to the type of personnel essential for the job. The activity would offer the inspectors a motive pathetically inadequate to their immense and dreary task."[167]

Ignoring the prognostications of these early U.S. thinkers, the IAEA Secretariat decided at an early stage that safeguards inspectors should have degrees in nuclear science or engineering. These were about the only clear qualifications for the job. It is not apparent who made this decision, but it was presumably approved by Director General Eklund, himself a nuclear physicist, who was determined that the Agency have a technical and scientific orientation. This decision would raise expectations about the technical qual-ity of safeguards and the professionalism of the inspectorate and profoundly shape safeguards culture. Responding to a claim by "a distinguished scientist" that "any science graduate recruited off the streets could perform inspections," Inspector General McKnight retorted that international inspections called for special knowledge and skills.[168] A 1975 SIPRI study speculated that in addi-tion to "an adequate professional and academic background,"

> inspectors should possess a great deal of tact, be adaptable to difficult situations, be willing to travel for long periods to remote and not always attractive locations (nuclear plants tend to be constructed outside urban areas); they should preferably speak several languages and they should be willing to do a job which may keep them away for a long time from their purely technical or research interests, although those who have the knack can contribute much by working out procedures and helping to develop instruments.[169]

This ideal picture did not always comport with reality, and certainly did not describe the "selection criteria" for employing inspectors because, at the time, there were none. Typically, those recruited in the early decades of safe-guards were, as advertised, engineers, physical scientists, or mathematicians, many from the nuclear industry. Most were from Western countries and all were men; there were no female inspectors until 1982. There were surpris-ingly few accountants, despite safeguards being based on nuclear materials accountancy.[170]

Prior to the reforms of the 1980s, IAEA inspections remained surpris-ingly freewheeling, depending more on the reliability and reputation of the individual inspector than on strict procedures. Reporting on inspection outcomes was "done in a relatively simple format that summarized inspection activities and their results," including the "depth" of the inspection.[171] Shirley Johnson recalls that when she joined the Agency in 1982, as the Iraqis' clandestine program was getting underway, inspectors were still giving verbal reports on their findings and submitting notes on scraps of paper, which "may or may not be filed."[172] Safeguards documents were transported in commissary bags like those used by diplomatic couriers. Her first job at the Agency was to put in order seven years of safeguards reports.[173] As for inspectors' use of field equipment, Shea recalls that:

> Most often, or more often, it wouldn't work when they got there and if it did it would either show results that they didn't understand—they're looking at lights flashing and trying to give the impression that they knew what was going on and if they got a result that differed from an operator declaration they would often kick the equipment to be sure—or wonder why it was different.[174]

In their 1982–1983 ACDA report, "Human Factors Affecting Inspector Performance," which was written less than a decade before the Iraq case broke, Shea and Thompson remarked that "the current level of inspector performance is below what is possible and what is intended."[175] Moreover, "there is no requirement regarding the quality of the actions implemented, and no means to determine that quality, other than on the basis of what the inspectors choose to report."[176] Inspectors' reports were reviewed, usually on a sampling basis, by the Division of Safeguards Evaluation, but this was "mostly limited to checking completeness" and did not "delve into the quality or correctness of the information included in the inspection report."[177] Shirley Johnson openly says that when she first joined the Agency, there were a lot of "ineffective inspectors, unmotivated, time wasters," the inevitable result of low recruitment requirements and little oversight and training.[178] Shea and Thompson concluded that the system suffered from "deficient inspector performance due in part to limitations in the ability of inspectors to perform as intended but mainly due to motivational problems."[179] Without using the word, they had identified a serious problem with safeguards culture. Johnson says that "much of the problem was that unless they came out of a nuclear weapon state program they had no concept of a security culture. And no feel for the severity of the spread of nuclear weapons. It was just a good paying job that got them out of the country."[180]

Inspector recruitment was one source of the problem. Inspectors were originally recruited simply through a written application, without interview, after nomination by their government.[181] Astonishingly, there was essentially no IAEA involvement in the recruitment process. Member states determined who would be nominated and who could be reached by job vacancy advertisements. These were written "in the most obtuse, international bureaucratese, with no clear indication of the job described, the benefits provided, or career opportunities."[182] Once on the job, there were no performance reviews.

Invariably, this system produced some under-performing or non-performing staff. The situation was exacerbated by the perverse effects of the Agency's UN-style recruitment policy, which required valiant efforts to achieve "geographical balance," not just suitable qualifications, in the appointment of professional staff.[183] The Agency's Statute itself provides that its "paramount obligation" in recruiting and employing staff is to "secure employees of the highest standards of efficiency, technical competence and integrity," but that "due regard shall be paid to the contributions of members to the Agency and to the importance of recruiting the staff on as wide a geographical basis as possible."[184] This resulted in lobbying by member states for appointment of their nationals to maintain their "quota" and quiet manipulation of the system by the Secretariat to preempt criticism from member states. Director of Development Adolf von Baeckmann, perhaps only partly tongue in cheek, advised Lichliter, who worked in both safeguards and management between 1981 and 2007, that "if an ambassador says this is a good man, he is a good man."[185]

A further difficulty was that the Statute obliges the Agency to keep its long-term staff to a minimum, resulting in a reliance on short-term contracts and a slow process of advancement to long-term positions. Lichliter noted that the position of inspector was "pioneering" in that there were initially no job descriptions, no job classifications, and no promotions, so a career pathway was impossible.[186] He never succeeded in getting real long-term contracts for inspectors, but did get *de facto* ones. Eventually, he says, inspectors would get nine-year terms "unless you spit in the face of the DG."[187] Unfortunately, the first contract renewal had to come one year before the first contract expired, just as inspectors were settling into their jobs. In addition, politics once again affected the safeguards regime. Member states remained opposed to the growth of a permanent inspectorate. Lichliter tellingly notes that "the Board always considered staff a liability rather than an asset."[188] Therefore, the development of an experienced career inspectorate was impossible in the early decades, before the Secretariat found ways to work around these strictures, and even today it remains a problem.

Reforms in the recruitment process were gradually made. By the mid-1980s, inspectors were being recruited by the Agency through direct advertisement, although member states, especially those with sophisticated nuclear enterprises and training programs that produced good candidates, still sought to place their personnel in key positions. Blix learned, however, that he could "go for competence without revenge from great powers" by using his power of appointment.[189]

Training was another challenge. Dimitri Perricos believed training to be "the most important item in the making of an inspector."[190] They need to know "the reality" of what needs to be done, the proper technical skills, and know the safeguards agreement "as good as a lawyer." The first training course for IAEA inspectors did not occur until 1968–1969, seven years after the first inspection was carried out.[191] By 1972, there was a three-month initial training course, but by the end of it, Perricos observes, "everyone was eager to get out into the field."[192] By 1975, SIPRI was reporting that all new inspectors first underwent several months of on-the-job training at headquarters, familiarizing themselves with the main documents, safeguards procedures, codified practices, and the operation of such devices as surveillance cameras, measurement instruments for non-destructive assay of nuclear material, and seals.[193] Refresher courses were designed for existing inspectors.

Further improvements were instituted following the 1981 Coopers and Lybrand report.[194] A section was established in the Safeguards Department that was devoted exclusively to inspector training; training manuals were developed; and a more intensive training program was inaugurated. The new program included field exercises at nuclear plants for all new inspectors, advanced courses in inspection procedures, and refresher courses at regular intervals for seasoned inspectors.[195] The scope of training was extended beyond technical aspects to include recognition of non-compliance "indicators" and other types of observation skills.[196] Performance reviews were instituted in the mid-1980s. By 1982–1983, Shea and Thompson were reporting that "the specialized training provided by the IAEA for new inspectors has improved very significantly in recent years," but with little career definition for an inspector, there was "little in the sense of organized training" beyond that.[197] Skeptics about the quality of the inspectors remained. Tariq Rauf has remarked that "safeguards inspectors have the misplaced perception of being 007s or sleuths—in fact, safeguards inspectors are bean counters: measuring nuclear material as called for under safeguards agreements, [but] when it comes to big picture assessments safeguards inspectors' skills are open to question."[198]

How the state of safeguards in the 1980s affected inspections in Iraq is impossible to specify due, once again, to safeguards confidentiality. In the view of Fischer and Szasz, "The existing constraints on the IAEA's choice of inspectors and their freedom of action can nevertheless, in some cases, seriously affect the efficacy of its safeguards operation."[199] Shea and Thompson noted that:

> an errant inspector (or group of inspectors) may miss a critical signal, causing the safeguards system to fail to detect a diversion, thereby permitting the acquisition of nuclear weapons by a nation to go unchecked. A similar effect would arise if an inspector chose to ignore or set aside such a signal, for example, avoiding the added burden required to resolve such signals, or yielding to pressures exerted by supervisors not wishing to create politically awkward situations, or responding to threats or bribes by the state in question.[200]

The fact that only Soviet and Hungarian inspectors were permitted to inspect Iraq does not indicate any lack of training, professionalism, or dedication to the safeguards task. On the contrary, Eastern European recruits tended to be highly capable technically. They presumably undertook the same Agency training as other IAEA inspectors. It is not known, of course, what their inspection reports contained or what their performance evaluations revealed. It is also not known whether familiarity with the Iraqis and their program had lulled them into complacency.[201]

What is known is that Eastern Bloc inspectors operated in response to different personal and professional incentives. Their governments took part of their salaries, and they were not permitted to stay beyond five years; therefore, they could not become eligible for IAEA pensions.[202] On the other hand, they were living in Vienna, a comfortable Western city, and were able to travel internationally, a privilege that was not enjoyed by their fellow citizens. All of this changed when the Soviet Union collapsed in 1989–1990, resulting in Russian and other Eastern European inspectors eventually becoming more like their Western counterparts. But this occurred too late to have an impact on the Iraq case.

Safeguards confidentiality also prevents researchers from learning whether one or two inspectors were sent to Iraq on each visit, whether some inspections were curtailed due to lack of available "inspector hours," or whether some visits to Iraq were canceled or postponed due to a shortage of inspectors or budgetary constraints. Alternatively, it is not known whether the Secretariat pursued its rights in Iraq to the maximum amount allowed and conducted additional inspections there. While it turned out that this would not have mattered in terms of Iraq's declared facilities and materials, it may have made a difference in the inspectors' observation of possible indicators of Iraq's illicit activities.

How Much Do Politics and Organization Explain the Iraq Case?

The traditional theories about what went wrong in the case of Iraq, which focus on politics and organizational capacities, go a long way toward explaining what happened. On the political side, the negotiations on the comprehensive safeguards system produced numerous compromises driven by politics, economics, and practicalities. This resulted in a regime that was far from optimal. While purporting to forestall any nuclear material being used for nuclear weapons by NNWS, the system was in truth designed only to detect the diversion of nuclear material from declared peaceful nuclear activities. Iraq would presumably have been caught if it had attempted a "breakout" of this type. That the system did not detect Iraq's alternative, illicit nuclear enterprise is a direct consequence of the system's design, which in turn can be blamed largely on politics.

As for the organizational explanation, the IAEA's safeguards system had improved considerably with the advent of comprehensive safeguards. Overall, by 1990, DDG Jennekens had concluded that the prospects for IAEA safeguards were "quite bright, albeit with a not unexpected degree of uncertainty."[203] Although the management and operation of safeguards was imperfect in its early decades, it had measurably improved by 1990 following periodic bouts of reform. The Agency's technical prowess had advanced considerably by 1990, but was still far from state-of-the-art. Financial constraints prevented it from carrying out the expected number of inspections and deployment of sufficient quantities of even standard safeguards equipment such as cameras, video, and seals. Modern technology, such as satellite imagery and wide-area environmental sampling, would have helped detect Iraq's activities, but were not available to the Agency at the time for political, technical, and financial reasons. The Secretariat, while it may have had suspicions about Iraqi behavior and intentions, had no firm technology-derived data on which to base a challenge to the Iraqi authorities or a report to the Board. Various reforms in the way inspectors were recruited, trained, and deployed had also been implemented, but these were obviously insufficient.

Presumably, although it is impossible to document, improvements in safeguards over the years were applied to IAEA activity in Iraq, despite the country's reported foot-dragging in facilitating inspections. While it is difficult to measure precisely the impact of management and operational reforms, they should, in theory, have resulted in a greater likelihood that some aspect of Iraqi non-compliance would be discovered, which may have triggered further questions, additional inspections, and, ultimately, an unraveling of Iraq's deception. There appears to be no reason to believe that the inspectors assigned to Iraq were not as well managed, trained, and competent in their tasks as any other. It appears that they did the job expected of them. Despite the dearth of detail, the IAEA's safeguards reports make clear that inspections in Iraq were carried out in line with the accepted criteria-driven approach and that no significant anomalies were detected.

Whether the inspectors' verification tasks were the appropriate ones is another question. Because Iraq had not yet attempted to divert material from its stockpile or from its reactors but was only in the planning stages of its "crash" program, the safeguards system had nothing to detect. The crash program is highly likely to have been detected, once it began, through normal safeguards procedures and perhaps through export controls on nuclear-related technology. On the other hand, undeclared facilities and materials were, by their nature, off limits to IAEA inspectors, so they were not included in inspectors' reports even if they had been observed inadvertently.

Standard safeguards procedures were followed, not deviated from. In fact, the very reforms designed to improve the effectiveness of safeguards in the 1980s may have helped blind the Agency to the unique nature of Iraq's nuclear activities. Former DDG for Safeguards Herman Nackaerts has retrospectively described the safeguards approach at the time as "narrow, prescriptive and criteria-driven."[204] Iraq was not, then, a case of the ineffective implementation of a system as devised, but precisely the opposite: the system was followed to the letter. Unlike NASA's *Challenger* disaster, this was not a case of the "normalization of deviance," in which relatively minor departures from procedures were, over time, normalized.[205] Rather, it was a case of standard operating procedures being followed but not capturing the whole picture because they were not designed to do so in the first place.[206]

For the Iraq case, the question is not just whether safeguards were technically capable of detecting Iraqi non-compliance, but whether the Iraqis *believed* that they were, so that the safeguards regime would have been sufficient to convince them not to make the attempt. We now know that there was debate within the Iraqi nuclear establishment about the likelihood that they would be detected by the IAEA if they tried to divert material from their declared facilities and stockpiles. Many seemed convinced that they would be caught. As for the separate secret parallel program, which did not rely on diversion of declared material, it appears that the Iraqis felt that they could "get away with it."

Unfortunately, the IAEA signaled this likelihood by stating in publications such as its 1980 "Introduction to Safeguards" that "the IAEA's system can do no more than detect diversion of safeguarded material or misuse of safeguarded plants, and trigger international action."[207] From their "infiltration" of the Safeguards Department, the Iraqis would have confirmed, to their own satisfaction, the Agency's public position that it did not go on "fishing expeditions" for undeclared material, activities, or facilities. Iraq would also have been aware of other weaknesses in the system.

For years the Secretariat had, to its credit, been transparently telling the international community that it had "serious problems."[208] As Shea noted, "I do not think even one member of the technical staff of the IAEA would consider the existing arrangements satisfactory."[209] He described the IAEA's abiding dilemma in revealing its limitations to potential non-compliers thus:

> It has always struck me that the Agency was very conscientious in maintaining a balance. On the one hand, it must satisfy its member States that the conclusions and assurances it provides are credible, especially when it reports that no diversions have occurred. On the other hand, the system is known to be imperfect: the agency must keep the support of its sponsors, without unjustified claims, hinting strongly enough at its deficiencies to provide the driving force to improve those flaws.[210]

As for the ultimate question of whether IAEA inspectors could have realistically been expected to uncover Iraq's secret program, given the constraints under which they operated, Fischer concluded that:

> The inspectors that the IAEA had chosen for Iraq and that Iraq had accepted under the procedure for designating inspectors may only have worked by the book, but even if they had been much more curious, it is doubtful whether they would have been much the wiser about what was going on at the Tuwaitha centre. The site was very large; the reactors and associated facilities under safeguards, as well as the entrance to the reactor site, were at one end of the centre and much of the rest of the centre, where important parts of the Iraqi nuclear weapons program were carried out, was hidden by a large

berm—a high earthen dike—which was an internal extension of the berm that surrounded the entire centre. Only by flying over the site at a low altitude would it have been possible to obtain a picture of the extensive operations under way.[211]

A textbook on safeguards by eminent safeguards specialists published by Brookhaven National Laboratory in 2019 appears to contradict Fischer. The authors record that Iraq's nuclear weapons program included enrichment activities "both in buildings adjacent to facilities where the IAEA had routinely conducted inspections under INFCIRC/153 and at undeclared locations."[212] Not only had Iraq pursued a clandestine nuclear weapon program, "but some of its clandestine activities were carried out 'under the nose of inspectors' at the Tuwaitha Research Center where inspectors were present on a routine basis."[213]

Doubts remain about what would have happened if the inspectors had been more probing and inquisitive and had taken steps to find out more. As Blix tantalizingly puts it: "It is not likely that any of the measures the agency could have taken would have led to discoveries, but they might conceivably have led to controversies—alarms."[214] If inspectors' suspicions were then raised, would they have inquired of their Iraqi interlocutors about evidence, however slight, that Iraq had not been entirely forthcoming in its safeguards declarations? Shouldn't the sheer scale of the facilities at Tuwaitha that inspectors were not permitted to see have rung alarm bells among both the inspectors in the field and the safeguards managers at IAEA headquarters? Did the parts of buildings that inspectors were not permitted to access arouse curiosity? If the Soviet and Hungarian inspectors were the only ones permitted on Iraq inspections, did their increasing familiarity with the Iraqi program not cause them to wonder about all the activity they undoubtedly glimpsed? Or did it lull them, along with the rest of the Safeguards Department, into complacency? Was the entire Agency, including its member states, acculturated to the notion that safeguards were not really meant to detect *any* noncompliance with the NPT but only the diversion of declared materials and facilities? To address these questions, we turn next to the role of safeguards culture.

Three

The Explanatory Power of Culture

ONE OF THE criticisms leveled at the IAEA after Iraq's nuclear weapons program became apparent was that the IAEA's organizational culture, specifically its safeguards culture, was at fault. It was not that international politics had rendered IAEA safeguards inadequate to the task, nor that the Safeguards Department and its inspectors were bumbling Inspector Clouseaus who were poorly managed, recruited, and ill-trained. Rather, the critique was that the Secretariat was in a type of cultural amnesia about the real purpose of safeguards and the possibility of significant non-compliance. As former Australian ambassador Michael Wilson noted at the time, critics accused the IAEA of being "complacent and unobservant."[1]

Since the revelations about Iraq, there have been hints from many sources, including the IAEA itself, that organizational culture may have played a role in the case, although no specifics have been mentioned. Some Agency officials and outside observers identified the need for the IAEA to develop a completely new safeguards culture.[2] Under the heading "IAEA organizational culture and 'mindset,'" the U.S. Office of Technical Assessment (OTA) called for change in April 1995, claiming that "many feel that the IAEA is more conservative and cautious than it should be or needs to be, and that it cannot adapt to a new, more ambitious agenda."[3] Even before the Iraq case, external reports on IAEA management had underlined the importance of culture, although usually without using the word. A 1974 Group of Experts Report emphasized that "changes in organizational structure will, as such, not result in improvements in operational efficiency unless there is cooperation, trust, confidence and respect between the international civil servants functioning in an international organization."[4]

This chapter considers what the organizational culture approach can bring to consideration of the Iraq case and investigates the elements of IAEA safeguards culture that may have contributed to the Secretariat's failure to detect Iraq's non-compliance.

Organizational Culture Theory

Organizational culture theory emerged in the 1980s as a means of understanding the way organizations truly behave ("the way we do things around here"), as opposed to the way they officially say they do. It had its intellectual roots in the anthropological study of native cultures beginning in the nineteenth century.[5] Management schools in the 1980s adopted the concept as a tool to study the effectiveness and efficiency first of commercial corporations and later of other types of institutions, including international organizations.

Such studies have become established practice.[6] Case after case indicates that "culture matters," not just to tribes, societies, and nations, but to every type of organization. The language of culture has steadily made its way into corporate managerial practice and popular discourse. Employees promote slogans proclaiming that "Safety is our culture," and professional football teams debate changing their cultural attitudes toward violence, women, and racial minorities. Engendering a better culture is seen as the solution to significant dysfunction in the U.S. nuclear "missileer" force, the U.S. Secret Service, and, recently in the corporate world, the ride-sharing company Uber.[7]

Organizational theorists describe culture as a "lens" through which human behavior, motivation, and practices can be examined. Edgar H. Schein was a pioneer in organizational culture studies. He explains that culture matters because "it is a powerful, latent, and often unconscious set of forces that determine both our individual and collective behavior, ways of perceiving, thought patterns, and values."[8] His current "dynamic" definition of culture is as follows:

> The culture of a group is the accumulated learning of that group as it solves its problems of external adaption and internal integration, which has worked well enough to be considered valid and therefore to be taught to new members as the correct way to perceive, think, feel and behave in relation to those problems. The accumulated learning is a pattern or system of beliefs, values, and behavioral norms that come to be taken for granted as basic assumptions and generally drop out of consciousness.[9]

Schein sets out three levels of culture from the "very visible to the very tacit and invisible," which he designates as Artefacts, Espoused Values, and Underlying Assumptions.[10]

Artefacts are an organization's visible structure, processes, and symbols. Espoused values are those that an organization publicly proclaims—strategies, goals, and philosophies.[11] The underlying assumptions are those unlikely to be articulated publicly but taken for granted within the organization, including unconscious thoughts, presumed beliefs, perceptions, thoughts, feelings, and habits.

In investigating the culture of an organization, John Carroll suggests seeking the answers to the following key questions:

- How strong and pervasive is the culture? What do people think of as right, proper, moral, and fair?
- What do the mission, strategy, goals, and rewards mean to people?
- What subcultures exist across groups or between managers and workers?
- What are the consistencies and contradictions in work practices, norms, rituals, role models, symbols, stories, training programs, rules, incentive plans, and so forth?[12]

Investigating culture can be problematic, especially at the level of underlying assumptions, because those within the culture may be reluctant to reveal uncomfortable or discordant beliefs, attitudes, or prejudices, especially when they conflict with the organization's espoused values. Probing such issues can be the cultural equivalent of psychoanalysis for individuals. Another difficulty in exploring culture is determining the boundary between objective facts and subjective perceptions of those facts. If culture is understood as the way in which organizations adapt to operational challenges, often without realizing it, then such a blurring of boundaries is to be expected. There is also likely to be dissent over what is a "good" culture and what is a "bad" one. All organizations have cultures, whether they are aware of it or not. The key question is whether its culture contributes to achieving the organization's goals and, not incidentally, to the well-being of its personnel. Notwithstanding these complexities, it is possible to identify cultural elements that affect organizational behavior and that can be changed if necessary.

Defining Safeguards Culture

There is no agreed definition of a desirable safeguards culture to date, either within the IAEA or elsewhere. In seeking such a definition, one naturally turns for inspiration to internationally agreed definitions of nuclear safety culture and nuclear security culture. *Safety culture* has been defined by the IAEA as "that assembly of characteristics and attitudes in organizations and individuals which establishes that, as an overriding priority, protection and safety issues receive the attention warranted by their significance."[13] *Security culture*, in a similar vein, is considered "that assembly of characteristics, attitudes and behavior of individuals, organizations and institutions which serves as a means to support and enhance nuclear security."[14] It should be noted that these are definitions of ideal, optimal cultures that are to be striven for, not descriptions of existing safety and security cultures, which may, of course, be problematic.

Mimicking these models, safeguards culture may be defined as "that assembly of characteristics, attitudes and behavior of individuals, organizations and institutions which supports and enhances nuclear safeguards."[15] This definition has the advantage of comparability across the three areas of nuclear governance (safety, security, and safeguards—sometimes known as the 3Ss). One difficulty, though, is that safeguards is a technical term that does not have a self-evident meaning to the uninitiated (and, in fact, is widely confused with safety). Therefore, it is probably preferable to include in the definition a reference to the purpose of safeguards—which is to prevent, or more accurately deter, the proliferation of nuclear weapons. It would also be helpful to put safeguards into the broader context of international peace and security from which it has sometimes been disembodied.[16] For the purposes of this book, I define an optimal IAEA safeguards culture as "that assembly of characteristics, attitudes, and behavior, exhibited by the Agency's personnel, which supports and enhances effective and efficient nuclear safeguards as a vital contribution to the non-proliferation of nuclear weapons and to international peace and security."

This definition goes beyond what some in the IAEA Secretariat have called safeguards "working culture," which confines safeguards culture to attitudes, beliefs, and assumptions about the technical aspects of safeguards— inspections, nuclear accountancy, technical means of verification, and data analysis. Such a narrow focus probably reflects an unwillingness to accept the broader concept of culture and a belief that the technical elements are the easiest to understand and potentially change. The drawback of the "working culture" concept is that it ignores the governance and management aspects of IAEA safeguards culture that are so influential, as well as the broader organizational culture in which safeguards are embedded. As we shall see, it is impossible to separate the inspector-level, "working" elements of safeguards culture (for example, inspectors' willingness to pursue suspicions of non-compliant activity) from the organizational culture of the Safeguards Department and the IAEA as a whole. As John Carlson has noted:

"Safeguards culture is a product of the institutional context—the terms and interpretation of the relevant treaties and agreements; the available verification methodologies; the shared attitudes and values of the Agency; recruitment, training and retention of inspection staff; management practices; and so on."[17]

The Institutional Context: The UN and the IAEA

Although organizational culture continually evolves, its basic elements coalesce during an organization's formative years. In the IAEA, this occurred in the late 1950s and early 1960s.[18] By the time the Iraq case emerged in 1991 to challenge some of its fundamentals, the IAEA's organizational culture had gestated for almost thirty-five years into a self-reinforcing set of beliefs, attitudes, and behaviors. Operating within the broader IAEA culture, safeguards culture naturally came to reflect the many factors—political, technical, operational, and financial—that had shaped the IAEA's culture itself. The IAEA is in turn nestled, like a Russian *matryoshka* doll, within a broader UN organizational culture (see Figure 3.1 below). Teasing out what is unique to safeguards culture, rather than general UN and IAEA organizational cultures, can be tricky. When interviewed for this project, some Agency personnel sought to deflect criticism of safeguards culture by pointing this out. The implication for changing safeguards culture in isolation is clear: at least some changes may have to be part of Agency-wide or even UN-wide cultural change.

Safeguards
Culture

IAEA
Organizational
Culture

UN
Organizational
Culture

Figure 3.1
The Institutional Context of Safeguards Culture.

Although the IAEA regards itself as "independent" (or autonomous) within the UN system, it is deeply affected by the same organizational culture that affects UN headquarters, the specialized UN agencies, and all other UN-type bodies. Along with the UN's diplomatic, managerial, administrative, and financial practices (including common pension and salary scales) comes the organizational culture it embodies. The diplomatic rituals involved in negotiating treaties, convening meetings and conferences, and dealing bilaterally with member states all draw on the standard UN organizational culture. The UN inherited large parts of this culture from its predecessor, the League of Nations, which had been established in 1920.[19] The bureaucratic operations of the League were in turn strongly influenced by civil service traditions of the United Kingdom.[20] This history means that some aspects of IAEA organizational culture originated almost one hundred years ago. They are deeply ingrained and difficult to change, as many reform efforts at UN headquarters have revealed.

Some of the IAEA's artefacts signify its UN organizational culture. Its logo is UN blue and duplicates the laurel wreath of the UN's symbol. Since 1979, the Agency has been located, along with other organizations in the "UN family," at the Vienna International Center (VIC). Its staff, like those of all other UN bodies, has from the outset been multinational (currently from more than one hundred countries), with all the advantages and disadvantages this brings in personnel management and creating a unified organizational culture.

Alisa Clarke describes the UN managerial culture as "often characterized by a short-term focus, influenced by factors such as the limited tenure and national interests of governments, specified financial timeframes for donors, and the usually biennial programming and budgeting cycle."[21] Mac Darrow and Louise Arbour, the former UN High Commissioner for Human Rights, note the "troubling reality that UN internal incentives, accountability systems and support structures do not generally tend to encourage courageous decisions. To the contrary, the more discernible tendency has been toward risk aversion."[22] The need for approval at higher levels and from various offices "serves as a disincentive for creativity and means that responsiveness to dynamic situations is somewhat circumscribed."[23] Such a culture would appear problematic for calling out states in non-compliance with their safeguards obligations. Risk aversion is not a desirable quality in an IAEA inspector, but it may be the inevitable consequence of the IAEA's operating within a broader UN culture.

One striking example of the impact of UN organizational culture on the IAEA that directly affects safeguards culture is staff recruitment practices.[24]

These result in a widespread assumption that among UN staff there will be a certain amount of what is referred to as "dead wood"—people who are employed because of their nationality rather than their demonstrated competence.[25] Such an assumption, whether true or not, is common in the IAEA's culture, including in the Safeguards Department.

While the IAEA is deeply influenced by the broader UN organizational culture, it has also evolved elements of a distinct organizational culture due to its location and its unique mandate, responsibilities, and status. From the outset, the IAEA was largely a Western project, inspired by the United States, that came to reflect Western cultural values. The Agency's original member states, mostly Western, and the first employees of the Secretariat, again mostly Western, had a disproportionate and lasting impact.

As the first sizable international organization located in Vienna, rather than in the traditional UN cities of Geneva and New York, the IAEA's organizational culture developed in relative isolation in its formative years. John Hall says that in those days, the IAEA had a strong sense of its own special identity, fostered in part by its home for its first twenty-two years—the Grand Hotel, located on the famous Kärntner Ring in downtown Vienna.[26] After extensive lobbying by the Austrian government and the city of Vienna to host the IAEA, the organization was treated as a prize catch by the government, the foreign ministry, and the Vienna city authorities.[27] Dieter Goethel, IAEA Director of Personnel from 1989 to 2002, says that the Austrians regarded the IAEA as "their Agency."[28] Sixty percent of general employees were Austrian; the Austrians had made it known that they expected the IAEA to employ significant numbers of Austrian support staff. Its original location made it easily accessible to the Viennese. IAEA car number plates signified "status in society."[29] There was an annual IAEA spring ball at the Hofburg Palace (which continues to this day), and the Director General had a box at the opera that any staff member could apply to use when he could not attend.[30] The Agency was thus well integrated into Austrian society, despite occasional racist incidents experienced by staff from developing countries.[31] Things changed somewhat in 1979, when the Agency was relocated to the VIC, the new UN center on the Danube in the outskirts of Vienna, a move that was "vainly resisted by some senior members of the Secretariat."[32]

Even after this move, the Agency retained a European sensibility that made its organizational culture more deliberative and slower-paced, more akin to that of the European Union, than the breakneck-pace and crisis-driven culture at UN headquarters in New York. The annual IAEA ball at the Hofburg Palace is a cultural artefact in more than one sense of the word.

The IAEA's Self-Portrayal

One of the key cultural artefacts of the IAEA is its official self-description as "an independent, intergovernmental, science and technology-based organization in the United Nations system that serves as the global focal point for nuclear cooperation." This self-description, or elements of it, arises constantly in the Agency's official and unofficial self-portrayals. The term "autonomous" is also often used by IAEA officials, although the difference between independent and autonomous is not immediately apparent in the international organizational context.[33] The IAEA's declaration of "independence" implies not so much independence from its member states (although there is an implicit element of that), but from other parts of the wider UN system.[34] Although the Agency is required by its Statute to report annually to the UN General Assembly, this is a mere formality. It is answerable neither to that body nor to the UN Secretary-General or the UN Security Council. Certainly—and this is a source of pride—it does not answer to the hapless UN Economic and Social Council (ECOSOC), like the UN specialized agencies do.[35]

The Agency is, however, truly independent of neither the UN nor its member states. The Security Council, the most powerful body in the UN system, has, in its role as guardian of international peace and security, on occasion ordered the IAEA to do things it might not necessarily wish to do. In the case of inspections in Iraq, the Security Council forced the Agency to collaborate with and in some respects rely on what the IAEA regarded as an upstart neophyte verification body, the UN Special Commission (UNSCOM), and its later incarnation, the UN Monitoring, Verification and Inspection Commission (UNMOVIC). In the case of North Korea, the IAEA was "requested" to conduct verification activities in regard to the 1994 Agreed Framework between the United States and the Democratic People's Republic of Korea (DPRK), without having been consulted during the negotiations on its likely involvement.[36] It could hardly have refused. In adopting Resolution 1540 in 2004, the Security Council appeared to ignore the mandate of the IAEA by requiring member states to report to the Council rather than the Agency on steps that they were taking to combat nuclear terrorism (without consulting the Agency, and much to its annoyance).[37]

A notable element of the Agency's self-description—"the global focal point for nuclear cooperation"—implies that the IAEA is the true keeper of the "nuclear file" for the international community and the key global governance institution in the nuclear field. It was on these grounds that the IAEA asserted its right to manage and conduct the nuclear inspections in Iraq after 1991, in competition with UNSCOM, a bespoke verification body created by the Security Council.[38] The Agency sought to assert its primacy by insisting that joint inspections be designated "IAEA/UNSCOM," rather than the other way around. For internal purposes, it dropped the UNSCOM designation altogether.[39] The IAEA also adopted a querulous and initially uncooperative attitude toward the Security Council's 1540 Committee, seeing it as duplicating the Agency's own work in the nuclear field.[40] In the Libya case in 2003, the Agency sought to reassert its prerogative to verify Libya's declared nuclear capabilities over the attempt by the United Kingdom and the United States to exclude the Agency until the case was essentially closed.[41]

Safeguards culture has inevitably drawn upon all elements of the broader IAEA organizational culture. Those aspects with the greatest impact have been the Agency's self-portrayal as a guardian of international peace and security; its scientific and technical orientation; and the balance it has struck between deference to its member states and its claim to organizational autonomy. Emphasizing the Agency's special contribution to international peace and security has motivated the inspectorate and created a kind of mystique about the Agency's safeguards work. Stressing the Agency's scientific and technical competence has boosted the technical reputation of safeguards and its claim to impartiality. Paying due deference to member states has helped to ensure continuing political and material support for safeguards, while the Agency's assertion of autonomy in managing its internal affairs has helped ward off political interference and threats to impartiality. These key elements of the IAEA's organizational culture will be considered in detail below in the context of their contributions to safeguards culture. First, however, it is useful to explore the key stakeholders in IAEA safeguards culture—the so-called safeguards community.

Purveyors of Safeguards Culture: The Safeguards Community

Culture "belongs" to a definable group of people who, wittingly and unwittingly, devise and grow it, guard it, and purvey it to others. In addition to the general sources of safeguards culture already identified, it is the sprawling safeguards community, illustrated in Figure 3.2 below, which has naturally had the greatest impact in shaping the specifics. Cultural influences flow in all directions, although not equally in each. The IAEA itself is the nucleus around which all the other contributors orbit.

The Influence of Member States

As in the case of IAEA organizational culture generally, it is the Agency's founding states, mostly Western, that have had a disproportionate and lasting

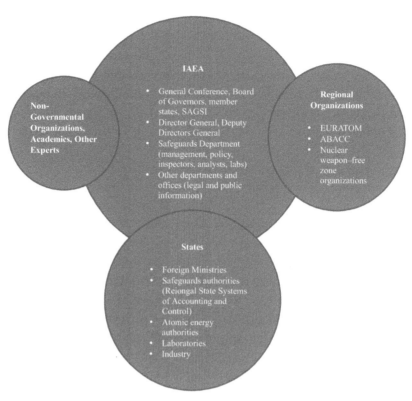

Figure 3.2
The IAEA Safeguards Community.

impact on safeguards culture. The greatest external influences on IAEA safeguards culture over the years have come from certain member states. This has occurred either through the example set by their national activities or through their direct involvement with the Agency—at the General Conference and in the Board of Governors, or in bilateral interactions with the Secretariat. The United States, which devised the earliest nuclear safeguards to prevent recipients of its bilateral nuclear assistance from misusing it to acquire nuclear weapons, has been the most important player. Other Western states, including Australia, Canada, Germany, Japan, the United Kingdom, and the Nordic countries, as well as a small number of non-Western states, notably the former Soviet Union/Russia and Brazil, Egypt, India, and South Africa, were influential at the outset.[42]

States that had the first opportunity to set the tone and parameters of safeguards culture were those that crafted the original draft of the Agency's Statute (the United States and its allies) and negotiated the final text (notably the Soviet Union and India, which joined the original drafters).[43] Next came the fifty-six original IAEA member states that participated in the 1957 Preparatory Commission that established the Agency.[44] At various times, some of these states sought to constrain safeguards rather than boost them, which itself has affected safeguards culture. The Soviets, supported by India and other developing countries, were active in the early years in rejecting attempts to strengthen safeguards, and even tried to derail preparations for launching the Agency's safeguards program. Even states such as Australia, Canada, Germany, and Japan, which have since become staunch supporters of safeguards, originally sought to narrow the parameters of the system for their own national reasons.

After the negotiation of the Nuclear Non-Proliferation Treaty (NPT) in 1968, a handful of states was highly influential in shaping the new comprehensive safeguards system based on IAEA Information Circular INFCIRC/153, especially Japan, the Soviet Union, the United Kingdom, the United States, and West Germany.[45] The compromises reached between them resulted in "elimination of an earlier IAEA authority to approve or review the design of facilities to be safeguarded, [a] stress on automation of safeguarding, and the maximum reliance by [the] IAEA on national accounting and control systems."[46] The system that emerged emphasized the use of instrumentation and other mechanical devices whenever and wherever possible so as to "minimize direct physical intrusion by inspectors."[47] This had profound implications for the evolving safeguards culture.

States that have played a significant role in supporting, funding, and helping run the IAEA since then have had a disproportionate influence on the subsequent evolution of safeguards culture, notably through State Support Programs that provide training and cost-free experts. Nuclear research laboratories, especially those of the United States and the United Kingdom, but also in states such as Germany and Finland, have contributed to safeguards culture by generating safeguards concepts and technologies that may (or may not) be absorbed by the IAEA.

From the outset, the safeguards system has been a largely Western project that came to reflect Western cultural assumptions, especially about the value of transparency, accountability, science and technology, verification, and compliance, as well as the sanctity of sovereignty. The domination of the Safeguards Department by Western staff is an enduring cultural artefact, despite valiant Agency efforts in recent years to diversify it. The post of Deputy Direc-tor General (DDG) for Safeguards is widely regarded as being "assigned" to the West (or at least to Western neutrals). The details of safeguards were designed in the IAEA Secretariat by mostly Western international civil ser-vants and experts from supportive member states, whether nuclear engineers, regulators, or legal and technical specialists.[48] The wide latitude given to the Secretariat to design the system came about because of the deliberate avoid-ance of detail by the INFCIRC/153 negotiators. This in turn was due to their desire to make safeguards as acceptable to as many states as possible. The dan-gers of nuclear weapons proliferation were seen as too pressing to delay the outcome with endless detail: "We wanted the widest possible implementation of the [NPT] soonest."[49]

Western domination and prioritization of the safeguards enterprise has reverberated throughout the IAEA's history, notably in increasing demands by the developing countries for managerial control of and equal budgetary treatment for their own priority programs—promotion of the peaceful uses of nuclear energy and technical assistance through the Technical Cooperation (TC) program.

Regional, Non-Governmental, and Academic Stakeholders

In addition to the IAEA, there are two regional bodies that apply nuclear safeguards—the European Atomic Energy Community (EURATOM, also known as the EAEC), and the Argentine-Brazilian Agency for Accounting and Control of Nuclear Materials (ABACC). They have evolved different safe-guards cultures from that of the IAEA, one of their assumptions being that neighboring states know each other better than outsiders and thus can do a better verification job. The IAEA, in contrast, assumes that an independent multinational inspectorate, removed from bilateral or regional associations, will perform verification better.

The IAEA and EURATOM have had a periodically fraught relationship over the years, partly due to their different cultures, which probably led, at least in the early decades, to reinforcement of their distinct cultures rather than cross-fertilization. EURATOM has a much closer relationship to the industry that it monitors and possesses more powers to directly bring mem-bers into compliance, whereas the IAEA acts overwhelmingly through its member states—a strong cultural norm at the Agency. Several key safeguards personnel have come to the Agency from EURATOM over the years. Accord-ing to Tariq Rauf, the Agency has "benefited immensely" from EURATOM's technical experience and safeguards technology, but it is not clear that this has affected the Agency's safeguards culture.[50] Rauf says that competition between the two verification bodies is healthy.

As for ABACC, a bilateral Argentine-Brazilian safeguards body whose verification work the IAEA monitors, presumably it regards the IAEA's involvement as unnecessary, paternalistic, and signaling a lack of trust. As the senior, more experienced, and better-resourced multilateral organization, the IAEA reportedly, but unofficially, considers ABACC a flawed mutual inspection regime that does not necessarily keep the two parties "honest." Any cultural influence is therefore likely to be one-way: from the IAEA to ABACC.[51]

Finally, there are non-governmental experts and academics who contribute to the IAEA's safeguards culture. Almost all of them are Western. The most influential have come from the European Safeguards Research and Development Association (ESARDA), the U.S.-based Institute of Nuclear Materials Management (INMM), and Germany's Forschungszentrum Jülich GmbH.[52] In addition, there are academic research units such as the Project on Managing the Atom at Harvard Kennedy School, the James Martin Center for Nonproliferation Studies in Monterey, California, and the Stockholm International Peace Research Institute (SIPRI), as well as the Verification Research, Training and Information Center (VERTIC), a non-governmental organization which is based in London. Rauf says that some of the inputs from the non-governmental community have been "problematic and introduced biases and simplistic world views, while other inputs have brought in useful new tools and concepts."[53]

The Role of the IAEA's Governing Bodies

Notwithstanding the contributions made by individual member states and other external players, the IAEA itself is rightly seen as the main shaper and custodian of safeguards culture generally, and of the Agency's own safeguards culture more particularly. The Agency's governing bodies play a key role in setting the broad political and legal parameters within which IAEA safeguards culture develops. They also affect the tone of safeguards culture through their commentary on the Secretariat's performance and decisions (and non-decisions). The General Conference, through its annual meetings and resolutions, reflects the membership's view of safeguards and sends political signals to the Director General and the Secretariat that have an impact on safeguards culture. One example is the universal support among member states for "non-discrimination" in the application of safeguards. Another is the demand that increases in budgetary support for safeguards should be matched by increased support for TC. When the General Conference could not agree on its annual safeguards resolution in 2012, it signaled to the safeguards community that political support for the enterprise should not be taken for granted, reinforcing a long-standing cultural assumption of the Secretariat.

Although the General Conference sets broad parameters in which safeguards operate, it is the 35-nation Board of Governors that has the most power in IAEA governance and hence has the greatest impact on safeguards and its accompanying culture, not least in approving IAEA budgets. The Board negotiates and approves model safeguards agreements, makes major decisions to strengthen or not to strengthen safeguards (if and when the Secretariat seeks Board endorsement), approves the appointment of inspectors, and, perhaps most important, makes decisions about states' compliance with their safeguards obligations.

The IAEA Director General

Although culture theory suggests that cultural formation is both a top-down and a bottom-up phenomenon, an organization's leadership plays a critical role in culture formation. Cultural signals from leaders reflect not only organizational imperatives but also their own personal, professional, and national proclivities. The Director General of the IAEA thus plays a critical role in framing safeguards culture, whether he (so far, all have been men) is aware of it or not. Over the years, this role has become increasingly important as the Agency has been buffeted by highly visible international crises involving safeguards non-compliance controversies and nuclear accidents. The 24-hour news cycle has elevated the Director General to the erroneous status of "chief

Table 3.1. IAEA Directors General, 1957–2021.

Director General	Nation of Origin	Term of Office
W. Sterling Cole	United States	1957–1961
Sigvard Eklund	Sweden	1961–1981
Hans Blix	Sweden	1981–1997
Mohamed ElBaradei	Egypt	1997–2009
Yukiya Amano	Japan	2009–2019
Rafael Grossi	Argentina	2019–present

Note: Rafael Grossi was still IAEA Director General as of this writing in 2021.
Source: www.iaea.org.

weapons inspector" or the exaggerated one of chief multilateral nuclear negotiator.[54] This personalization of the Agency began with the Iraq issue (under
Hans Blix and Mohamed ElBaradei) and has continued with the Iran case
(under ElBaradei, Yukiya Amano, and currently Rafael Grossi).[55]

To date, four of the six Directors General have been lawyers: Sterling Cole,
Blix, ElBaradei, and Amano. This may help to account for the legalistic tone of
the Agency's culture, including its safeguards culture. One example is the idea
that states are "innocent until proven guilty" (although there are also political
and institutional reasons for this).[56] Surprisingly, only one Director General
so far, Sigvard Eklund, has been a scientist. A nuclear physicist, he held the
position for twenty years and is credited with putting the scientific and technological stamp on the Agency and on safeguards that persists to this day.

Given that two Swedes were Directors General for thirty years—most of
the Agency's history—it is a matter of conjecture as to how Nordic the culture
of the Agency is. The strong Swedish values of effectiveness and efficiency
were brought to the Agency by both Eklund and Blix, and were constantly
iterated as making the IAEA different from other members of the UN organizational "family."[57] Most of the Agency's modes of operation, however, were
inherited from the UN and its predecessor, which were in turn strongly influenced by British civil service traditions.[58]

As mentioned earlier, Berhanykun Andemicael and John Mathiason,
both former UN officials and management specialists, contend that at least
the first three IAEA Directors General were "quiet, technologically competent and politically adept in their relations with Member States"; they seem to
attribute this partly to the fact that they had been career officials.[59] More con-
troversially, they claim that "the IAEA, consciously through its management
training, as well as through its culture, has sought to inculcate a participative leadership style and has largely achieved it."[60] This is belied by the widely
acknowledged bureaucratic silos at the Agency that prompted ElBaradei to
launch his "One House" policy, designed to create a more integrated, unified organization.[61] Moreover, the management styles of both ElBaradei and
Amano were rather more directive than participatory. Given the well-known
tendency of IAEA departments to act as autonomous "fiefdoms," it is an open
question as to how much influence successive Directors General have had on
the culture of individual departments.

The Secretariat

If culture is a creature of habit, then the most powerful source, institutional
home, and day-to-day repository and guardian of safeguards culture is the

Secretariat. The number of IAEA employees who work within safeguards culture, "embody" it, are affected by it, and help shape it is surprisingly small. In 1990, just before the Iraq non-compliance revelations, there were only 500 staff in the Safeguards Department, including approximately 250 designated inspectors.[62] At the end of 2020, the number of regular staff in the Safeguards Department was about 770, of whom 285 were inspectors.[63] In addition, a handful of individuals in senior management and other parts of the Agency devoted at least some of their time to safeguards issues. Most IAEA staff—which numbered 2,175 in 1990 and approximately 2,500 in 2021—deal not with safeguards, but with other aspects of the Agency's operations.[64]

Naturally, the Safeguards Department is a key progenitor of and stakeholder in safeguards culture. It is by far the most publicly well-known of the IAEA's departments, in function if not in name, due to the high-profile violations of safeguards by Iraq, North Korea, and Iran. Live coverage of inspectors searching for "weapons of mass destruction," as the media would have it, is compelling, even if misleading. None of the other programmatic departments of the IAEA comes close to safeguards in its importance to international peace and security. As the largest, best-funded, and most newsworthy department, it thus has an outsized influence on the Agency's image and overall organizational culture.

The department is headed by a DDG who is also the most prominent of his fellow DDGs. Media coverage of the head of safeguards accompanying the Director General to crucial meetings in Baghdad, Tehran, and Washington, D.C., illustrates the point. Some heads of safeguards have regarded themselves—not the Director General, as the press often insists—as the "world's top nuclear inspector."[65] This has created some discord between particular Directors General and their DDGs for Safeguards. One interlocutor from the executive branch has spoken of "some fantasy of the 19th floor" (where the Safeguards Department is located) about the department's preeminence in making safeguards decisions.

In theory, the background, attitudes, and management style of the DDGs should have determinative sway over the culture of their department. Yet all DDGs have faced difficulties in establishing an appropriate safeguards culture, partly due to resistance from Directors General, section heads, and independently-minded inspectors. DDGs are neither all-powerful nor necessarily popular, and must operate within the traditional constraints of international secretariats.

If the generation of culture is both a top-down and a bottom-up phenomenon, as culture theory suggests, then it is the Safeguards Department's staff, whose entire professional time is devoted to safeguards, who embody it, propagate it, and evolve it at the "working level." Among Secretariat staff with some responsibility for safeguards, there will always be varying views about what safeguards culture is and should be. No culture is monolithic. Moreover, there is great variation in the ability and willingness of individuals to affect the form and content of the culture.

The pioneering Western safeguards personnel were from the outset reportedly a tightly knit group, with a generally guarded attitude toward their Soviet and Eastern European colleagues.[66] Personnel from developing countries were largely absent while safeguards culture was being formed. Also, it can be assumed that the maleness of the department reinforced its scientific and technical bent and had a direct bearing on safeguards culture, as female interlocutors have attested.[67] Prior to the Iraq case, only 12.2 percent of professional personnel at the Agency were women, there was no female in the senior ranks, and there was only one female safeguards inspector.[68]

Only gradually did the Safeguards Department, including the inspectorate, become more representative of the IAEA's membership, partly due to pressure from developing countries for greater representation for their nationals. By 1990, although the Agency was still predominantly Western, its personnel were drawn from seventy-five countries.[69] Absorbing non-Western personnel into the safeguards enterprise was not always smooth. Dieter Goethel, who became IAEA Director of Personnel in 1989, reports that there was little guidance at the time about how to handle multicultural differences, so one just learned by trial and error.[70]

Safeguards procedures and processes were designed by mostly Western, male IAEA employees and cost-free experts from supportive Western states; these types have continued to dominate the Safeguards Department through the present day. Again, it is the United States, with its abundant financial, personnel, and technical capacities, that has played the key role.

Outside the Safeguards Department, three other types of officials reporting directly to the Director General—dealing with external relations, legal affairs, and public information—have had lasting effects on safeguards culture.[71] The officials that handle the IAEA Secretariat's official relationships with member states and the wider international community—now embedded in the Director General's office, but formerly a separate Office of External Relations and Policy Coordination (ExPo)—have a notable role in advising the Director General on how to deal with significant safeguards issues that are brought to the attention of senior officials, most importantly allegations of safeguards violations. The way these cases are handled establishes policy and precedents for future cases that ultimately inform safeguards culture. One example is the decision to call (or not call) for a special inspection when serious allegations of safeguards non-compliance arise.

The drafting of legal documents, notably safeguards agreements, by the Office of Legal Affairs constrains or expands what is legally and bureaucratically possible, and thereby also sends cultural messages. The content of the "Inspectors' Document," the 1961 edict that originally set out inspectors' rights and responsibilities, has, for example, shaped safeguards culture ever since.[72] Interpretations by the IAEA's lawyers also help define the authorities that the Safeguards Department perceives it has, as well as influencing the way it parses the language of safeguards. This is notably the case in the Agency's non-compliance reports, such as when to use or not to use the word "non-compliance."[73]

Although the Agency's public affairs officials, in what is currently called the Office of Public Affairs and Communication, should not have an independent view of safeguards, the way they portray the system, for better or for worse, affects not only the way it is perceived by the outside world, but also the way it is viewed within the organization—especially the relative importance accorded to safeguards compared to the Agency's other missions and programs. The messages emerging from public affairs about safeguards do not always conform with the preferences of the Safeguards Department.[74] One example is the erroneous representation of the IAEA as the "custodian" of the NPT. While the Agency is responsible for verifying compliance by the non–nuclear weapon states (NNWS) with the treaty and fostering the peaceful uses of nuclear energy pursuant to it, the Agency is not responsible for the disarmament provisions of the NPT's Article VI. A second example is the use of the term "watchdog," which is abjured in the Safeguards Department but occasionally deployed by public affairs.[75]

Traditional Safeguards Culture Prior to the Iraq Case

Traditional, pre-Iraq safeguards culture, in addition to being nested in UN and IAEA organizational cultures, was shaped by the political trade-offs reached by the states that negotiated the IAEA safeguards regime. It mirrored the struggle among IAEA member states about how extensive and intrusive nuclear safeguards should be. Compromises that prevented the establishment of a more optimal verification regime were especially influential in shaping the taboos of the culture, hence the prominence of cultural motifs concerning respect for state sovereignty; the principle of non-discrimination toward member states; and the emphasis on technical, impartial, apolitical solutions. The officials who devised and managed safeguards over the years were cognizant of this political tug-of-war from the outset. Carlos Büchler, the first designated safeguards inspector and an active participant in developing the original system, noted in the late 1990s that "the entire history of safeguards is characterized by the struggle between its advocates and its detractors which, in some ways, continues up to this date."[76]

Employing Schein's schematic of organizational culture, Table 3.2 sets out the most important espoused values of traditional safeguards culture as reflected in its key cultural artefacts. The most significant artefacts are the IAEA Statute and the NPT, as well as the various types of safeguards agreements, the "Inspectors' Document," and other safeguards documentation.

Table 3.2. Traditional Safeguards: Artefacts and Espoused Values.

Safeguards contribute to international peace and security.	**Statute, Art. 1:** "The Agency shall seek to accelerate and enlarge the contribution of atomic energy to peace." **Statute, Art. II.B:** "The Agency shall: Conduct its activities in accordance with the purposes and principles of the United Nations to promote peace and international co-operation, and in conformity with policies of the United Nations furthering the establishment of safeguarded worldwide disarmament and in conformity with any international agreements entered into pursuant to such policies." **NPT Preamble**: "The States concluding this Treaty . . . Believing that the proliferation of nuclear weapons would seriously enhance the danger of nuclear war . . . Undertake to cooperate in facilitating the application of IAEA safeguards on peaceful nuclear activities."
The purpose of safeguards is to prevent the diversion of nuclear energy from peaceful uses to nuclear weapons and other explosive devices.	**Statute, Art. III.1:** "To establish and administer safeguards designed to ensure that special fissionable and other materials, services, equipment, facilities, and information made available by the Agency or at its request or under its supervision or control are not used in such a way as to further any military purpose." **NPT, Art. III:** "Each non-nuclear-weapon State Party to the Treaty undertakes to accept safeguards . . . with a view to preventing diversion of nuclear energy from peaceful uses to nuclear weapons or other nuclear explosive devices."

(continued)

Table 3.2. *(continued)*

Safeguards should respect state sovereignty and equality.	**Statute, Art. III:** "The activities of the Agency shall be carried out with due observance of the sovereign rights of States."
Safeguards should be based on the principle of non-discrimination.	**Statute, Art. IV:** "The Agency is based on the principle of the sovereign equality of all its members."
Safeguards should avoid hampering the economic and technological development of states.	**NPT, Art. IV:** "Nothing in this Treaty shall be interpreted as affecting the inalienable right of all the Parties to the Treaty to develop research, production, and use of nuclear energy for peaceful purposes without discrimination."
Safeguards should be technically-based and cost-effective.	**INFCIRC/153 Safeguards Agreements, Art. 6:** "The Agency shall, in implementing safeguards pursuant to this Agreement, take full account of technological developments in the field of safeguards and shall make every effort to ensure optimum cost-effectiveness and the application of the principle of safeguarding effectively the flow of nuclear material subject to safeguards under this Agreement by use of instruments and other techniques at certain strategic points to the extent that present or future technology permits."
Safeguards information should be confidential.	**Statute, Art. VII:** "The Director General and the staff shall not disclose any industrial secret or other confidential information."
Recruitment of safeguards staff should be competence-based, while also considering geographical distribution.	**Statute, Art. VII:** "The paramount consideration in the recruitment and employment of the staff . . . shall be to secure employees of the highest standards of efficiency, technical competence, and integrity. Subject to this consideration, due regard shall be paid to the contributions of the members to the Agency and to the importance of recruiting the staff on as wide a geographical basis as possible."

Sources: Statute of the International Atomic Energy Agency, 1957; and Treaty on the Non-Proliferation of Nuclear Weapons, also known as the Nuclear Non-Proliferation Treaty (NPT).

Culture theory tells us that an organization's artefacts and espoused values are deliberately designed to portray the organization in the best possible light and to act as rallying cries for its personnel and supportive stakeholders. It is only in the culture's underlying assumptions that "the way we do things around here" is truly revealed. The following section dissects prominent espoused values of IAEA safeguards culture in light of the underlying assumptions held by IAEA personnel prior to 1990 (to the extent that these can be determined in the absence of a scientific survey) and considers how they might have contributed to the Iraq denouement.

The IAEA's Contribution to International Peace and Security

One striking aspect of the Agency's organizational culture is the pride it takes in its perceived special role in international peace and security. Article II of the IAEA Statute announces that: "The Agency shall seek to accelerate and enlarge the contribution of atomic energy to peace, health and prosperity throughout the world."[77] Its original motto, "Atoms for Peace," was derived from U.S. President Dwight Eisenhower's landmark speech in 1953, which envisioned a world in which atomic energy would no longer be viewed as a destructive force but as a tool for enhancing international cooperation and peace and security.[78] The IAEA's unprecedented inspection activities on member states' sovereign territories and direct access to the UN Security Council in case of safeguards violations make it unique among international organizations.[79] As Blix put it, "the IAEA occupies a special place amongst organizations in the United Nations family in that part of its work is directly related to peace and security in the world through verification of commitments by States to use nuclear material and installations exclusively for peaceful purposes."[80]

Because the Safeguards Department directly carries out this part of the Agency's mandate, one of the most widely-shared espoused values among safeguards staff and IAEA personnel generally is the belief that safeguards contribute to international peace and security. Interviews conducted for this project, and those by the Pacific Northwest Nuclear Laboratory and the University of Vienna, as well as memoirs of IAEA personnel, confirm the ubiquity of such a view. It has traditionally been a strong motivation for joining the Agency, specifically to work in the Safeguards Department, whether as a manager or an inspector. It has been used explicitly by the Secretariat in recruiting and training staff.

From the outset, however, one of the underlying assumptions of traditional safeguards culture was that there was a disconnection between the Agency's high-flying rhetoric about the contribution of safeguards to international peace and security and the day-to-day work of the department. This was epit-omized by a cartoon on an inspector's office wall that said: "Safeguards is like wetting your pants while wearing a dark suit—you get that warm feeling but no one seems to notice."[81] This speaks to several phenomena of the culture: mixed messages from member states and the Agency's leadership about the importance of safeguards; awareness of technical flaws in the system; and the overarching political context of what was often promoted as a purely technical activity in a technical organization. This skepticism even extends to the suspi-cion among some safeguards personnel that if non-compliance were detected, political pressure would be brought to bear by member states or the Agency's leadership to ignore or at least delay revealing it for as long as possible.

Part of the skepticism derives from the murky origins of safeguards. Even Eisenhower's speech was not entirely clear about how spreading the peaceful uses of nuclear technology around the globe would avoid nuclear weapons proliferation. The "safeguards" he mentioned remained to be invented. In the early days of the Agency, no one, including the newly recruited safeguards staff, had any idea what safeguards were and how they would function in an international organization. A hurriedly convened safeguards conference held in August 1955 in Geneva proved "something of a disaster" when the U.S. del-egation struggled to explain how nuclear safeguards would effectively prevent weapons proliferation.[82] As Allan McKnight once pointed out, the negotiators of the IAEA Statute produced "no clear enunciation of the concept of the non-proliferation of nuclear weapons."[83]

Even the states that were most enthusiastic about non-proliferation, notably the United States, had little idea how to proceed beyond replicating its bilateral safeguards arrangements. Only with the Agency's gradual develop-ment of safeguards concepts and techniques did the link between safeguards activity and global peace become clearer. Even then, the contribution of safe-guards seemed more symbolic than real. Initially they were applied purely on a voluntary basis to a handful of willing states such as Norway and Japan. Described as "amateurish," early techniques were rudimentary, technology was basic, and inspectors were not carefully recruited or trained.[84] The con-tribution of safeguards to peace in this era was more theoretical than actual.

Only with the advent of the NPT and comprehensive safeguards agree-ments (CSAs), along with professionalization of the inspectors and devel-opment of better techniques and technology, did the potential and actual contribution of safeguards to peace and security crystalize. This was espe-cially true as increasing numbers of states became NPT parties and adopted comprehensive safeguards. Even so, safeguards personnel have always known that their mission does nothing to disarm the official nuclear weapon states, as provided for in Article VI of the NPT, or those states that have remained outside the NPT and acquired nuclear weapons. Safeguards staff have also been aware of the inadequacies of safeguards in states in which they were applied, especially in detecting undeclared activities and in safeguarding bulk-handling facilities such as enrichment and reprocessing plants. Furthermore, they have been conscious of the contradictions and trade-offs involved in respecting all espoused safeguards values simultaneously: states' concerns with protecting their sovereignty; their "inalienable right" to the peaceful uses of nuclear energy (as sanctified in Article IV of the NPT); the principle of non-discrimination versus pursuing the most effective, efficient, and technically advanced safeguards; and detecting non-compliance without fear or favor.[85]

More existentially still, safeguards personnel have been aware that NPT states parties are perfectly entitled to acquire the entire nuclear fuel cycle if they wish, providing that they do so under safeguards. A mere three months' notice is required to withdraw from the treaty, albeit only on the grounds of a threat to their "supreme interests," whatever these are claimed to be. In a domestic regulatory regime, government inspectors—in industrial safety, for example—cannot be turned away because the corporation has withdrawn from the inspection system due to alleged threats to its interests. In the international sphere, that is entirely possible.

It is the great dilemma of international affairs: there is no overall governing authority with enforcement powers, but instead a relative anarchy that is tempered only by cooperative instincts that may or may not ultimately prevail. Unlike domestic systems, there is no international police force or threat of fines or jail sentences for regulatory infractions. This puts a somewhat different spin on the notion that safeguards help protect international peace and security: they do so only with the indulgence of states; deal with only a small part of the proliferation challenge; and, no matter how effective and efficient, they may not catch the most likely suspects.

From the Agency's inception, inspectors also sensed that the system did not give them sufficient authority or tools to ensure effective safeguards. They were required to "make accommodations" to meet the Agency's "lofty goals" by making the best of whatever was provided and "somehow reach the conclusions sought."[86] McKnight noted an assumption from the beginning that safeguards needed to be "sensibly administered," presumably meaning that the arrangements should not be too intrusive and that states should not be pressed too hard on infringements. The "Inspectors' Document" declared that inspections should cause "the minimum possible inconvenience to the State and disturbance to the facilities inspected."[87] CSAs state that implementation should be "consistent with prudent management practices required for the economic and safe conduct of nuclear activities."[88] In 1975, SIPRI observed that "a 'guiding principle' for the Agency ever since it started its safeguarding task had been that safeguards should not become "an undue burden."[89]

In addition, as noted above, there was a "pattern of parsimony in relation to IAEA activities and particularly the safeguards function."[90] There was never enough funding to carry out all the inspections envisaged, especially as increasing numbers of states acquired CSAs. McKnight concludes that "this constant line produced over the years a timidity and lack of positivism in the IAEA Secretariat."[91] In the early 1980s, informal discussions by Tom Shea and David Thompson with inspectors, past and present, showed that they tended to "feel that what they and their colleagues are able to achieve is significantly less than the importance of their work should merit, and that variations in the inspector corps is a significant cause for that condition."[92] Inconsistency in the safeguards system, they somewhat dramatically concluded, "often poisons an inspector's morale, leading to reduced expectations and poor performance."[93]

Although it is unclear how representative his views were, Roger Richter's testimony to the U.S. Senate Foreign Relations Committee in June 1981 is the most detailed description of the frustrations that inspectors may have felt in that era about the real import of their job.[94] His comments on reporting back to Vienna after conducting an inspection were alarming:

> You will now return to Vienna and report that your inspection disclosed no discrepancies between the operator's records and that of the Agency. You will report that you verified the un-irradiated fuel with your stabilized assay meter and that you verified the irradiated fuel in the core by counting the elements and by virtue of seeing the reactor in operation. The difficult part of the job is that you must prepare yourself mentally to ignore the many signs that may indicate the presence of clandestine activities going on in the facilities adjacent to the reactor, facilities which you were not permitted to inspect unless the host country has informed the IAEA that fuel elements from the reactor were transferred there. You will now complete a standard report. Filling in the blanks, you will try to forget that you have just been party to a very misleading process.[95]

It is not known if Richter felt unable to share his concerns with his departmental line managers, if there was a specific case he had in mind, or if he did speak up and was rebuffed. Nor is it known whether he made representations internally to anyone about Iraq specifically (he was not an Iraq inspector but was in the section that dealt with Iraq). Even so, he apparently felt compelled to resign and tell the U.S. mission to the IAEA about his concerns. The rarity of an inspector speaking out publicly in this way indicates that the culture strongly discourages it. The response of his Agency colleagues would probably have been that he did not understand the true nature of safeguards, including the limitations, and was naïve to have expected otherwise, given the way the system had been set up.

More positive from a safeguards culture perspective was Richter's demonstration that at least some inspectors, notwithstanding the prevailing stereotype, did have a critical and questioning attitude toward their tasks. He had not absorbed the supposed culture of being "unobservant and inattentive" and was frustrated by the culture of silence. At least some IAEA personnel privately saw Iraq as a country of proliferation concern before the Israeli attack, and there had obviously been internal discussions about this in the Safeguards Department, despite the espoused value that demanded, in the absence of firm evidence, non-discrimination in the treatment of member states.[96]

In addition to cynicism about the contribution of the Agency to peace and security, doubts about the true value of the safeguards enterprise helped produce a culture of defensiveness and opacity. The Secretariat became disinclined to answer critics publicly or engage in public debate about safeguards' shortcomings, presumably out of fear that it might be revealed as a house of cards. Even the response to Richter came via the U.S. State Department talking to IAEA officials, not as an official rebuttal from Vienna.

At the level of individual safeguards staff, especially inspectors, the disconnection from the espoused value that lauded the Agency's role in international peace and security encouraged a culture that focused narrowly on bureaucratic and technical tasks and valued process rather than outcomes. It is revealing that in defending their performance after the Tamuz incident, the Secretariat described its inspectors as technicians, not diplomats, implying that they were not expected to negotiate access with the Iraqis but simply carry out a technical task. Such an underlying assumption, which did not comport with the espoused values of the "Inspectors' Document," contributed to the weaknesses in safeguards that were revealed by the Iraq case.

Safeguards as Technical, Not Political

A related espoused value of the IAEA's organizational culture with deep implications for safeguards culture is the Agency's self-identification as a technical organization, or, in grander terms, a "science and technology-based" organization.[97] From the outset, the IAEA promoted this image, presumably to contrast itself with what it saw as "political" or "politicized" international bodies. Reinhard Loosch, Governor for the Federal Republic of Germany from 1972 to 1992, recalled that "in the perception prevailing during the early years of the Agency's life, until well into the 1970s, the Agency was primarily a technical organization . . . basically different in character from the United Nations and other members of the United Nations family."[98] Discussions in the Board of Governors were:

> dominated by personalities who had been chosen by their governments on the basis of their achievements in nuclear science or their professional involvement in the development of peaceful uses of atomic energy, and who did not consider themselves as primarily diplomatic or political representatives of their countries but rather as people responsible for the advancement of the statutory objectives of the Agency.[99]

Despite being an intergovernmental agency established by governments, the IAEA's founding fathers were mostly nuclear scientists, engineers, and heads of national nuclear agencies; they collectively set the tone of the new organization.[100] As Director General for twenty years, Eklund, a distinguished Swedish physicist, reportedly "changed the course" of the Agency by stressing the scientific and technical aspects of its work after it appeared to drift too far into politics.[101] Three of the first DDGs of Safeguards were scientists.[102] In 1958, the convening of a standing Scientific Advisory Committee (SAC) reportedly exerted a "powerful influence" on all of the Agency's programs, leading to the establishment of the Agency's scientific laboratories at Seibersdorf, outside Vienna.[103] The Agency hosted the scientifically-oriented third Geneva Conference on the Peaceful Uses of Atomic Energy in 1964. In the same year, it established a Center for Theoretical Physics in Trieste, set up a joint division with the UN Food and Agriculture Organization (FAO), extended its collaboration with the Oceanographic Institute in Monaco in studying the effects of marine radioactivity, and expanded its own Seibersdorf facilities.[104]

This scientific and technical orientation was reinforced by the cooperation and compromise of the "Spirit of Vienna," which IAEA historian David Fischer once called a "benevolent genie that presided invisibly over the Board, the delegations to the IAEA and the Secretariat."[105] This view presumed that IAEA delegations, especially those sitting on the Board, were independently minded and inclined to see their role as technical more than political. Decisions were to be taken collegially, on rational, scientific grounds.[106] The fact that Governors represented governments was downplayed. The Spirit of Vienna was a cultural totem if ever there was one—part espoused belief, part underlying assumption. The extent to which this "cultural myth" ever existed is disputable. Blix says it reigned "by and large."[107] At times, the Agency certainly operated on the assumption that it did—which, in cultural terms, is what matters.[108]

The scientific and technical orientation and the mutually reinforcing Spirit of Vienna had their political uses. They bolstered the IAEA's claim to be independent and impartial in its relations with member states because its judgments, by implication, would be empirically based, well researched, rational, and apolitical.[109] Modern international organizations like the IAEA, Barnett and Coleman theorize, make "a mixture of authority claims—rational-legal, delegated, moral, and expert—that give [them] the space to act in relatively autonomous ways."[110] Such claims often turn on their belief that they are "impersonal and neutral, that is, that they are *not* exercising power but instead are using depoliticized, objective criteria to fashion their policies."[111] International secretariats, in this fashion, ultimately seek to "protect their autonomy, and minimize organizational insecurity."[112]

One of Eklund's motivations in emphasizing the technical nature of the IAEA was apparently to placate Soviet opposition to his appointment and "create a more conciliatory atmosphere in the Board of Governors, conducive to operational efficiency."[113] Emil Keblúšek, Czechoslovak ambassador to the IAEA and Chair of the Board from 1982 to 1983, states directly that "the possibility of restricting the Agency's scope to technical matters has brought effectiveness to the activity of the staff and to the utilization of financial resources. This principle has been one of the secrets of the stable nature of all the programmes of the IAEA."[114]

Yet even though the culture presents itself as science and technology oriented, the IAEA is not primarily a scientific organization but an international bureaucracy that helps govern a field which happens to have a large scientific and technical component. Through their permanent diplomatic missions in Vienna, most of the Agency's business is with governments, not with scientific stakeholders or even with the nuclear power industry; hence, the culture is a hybrid technical and bureaucratic one. The scientific reputation of the Agency mostly derives from its non-safeguards missions dealing with the peaceful uses of nuclear energy in medicine, agriculture, and the generation of electricity, not from safeguards. While safeguards are technical in nature, they are not scientific in the sense of pioneering research. The Agency's budget for safeguards research is minuscule, and it relies mostly on its member states for such efforts.

The Secretariat's self-declared technical orientation is undoubtedly reinforced by what has been described, critically, as a "legalistic" approach to carrying out its monitoring and verification mandate and the non-compliance issues arising from that role. Two of the Agency's three Directors General before 1991—Cole and Blix—were lawyers. Blix reportedly has denied that the IAEA has a legalistic culture, but the whole edifice of safeguards, at least, is built on a cascading series of legal documents from the IAEA Statute down. Moreover, international organizations, especially those in the UN system, do not need much encouragement to adopt a legal approach to any issue. It is the principal means by which international agreements are codified and implemented, and it is hard to imagine an alternative.

The criticism underlying the charge of legalism is that over the years, the Secretariat became too rigid and narrow in interpreting its verification mandate, retreating to the letter rather than the spirit of its Statute, the NPT, and safeguards agreements. This was true at the level of judging compliance with individual safeguards agreements, as in the Iraq case, where the Safeguards Department went "strictly by the book" and no further in its nuclear materials accountancy activities. The non-use of special inspections is another case in point: only after the Iraq case did the Secretariat appear to rediscover its clear legal rights to conduct such inspections. Similarly, the Secretariat has never asserted its statutory right to conduct virtually "anytime, anywhere" inspections or to include safety considerations in its inspections' purview.[115] On the contrary, it quickly retreated from those rights when it encountered member state opposition soon after its foundation. Like other bureaucracies, the IAEA Secretariat uses legal interpretations to pursue the goals it has already chosen.

The Deference-Autonomy Trade-Off

Another key aspect of IAEA organizational culture particularly pertinent to safeguards culture is the underlying assumption of the Secretariat that it can only do what member states permit it to do. This is not an espoused value of the Agency, but an underlying assumption. It derives from interpreting the zeitgeist of several espoused values, including the sovereignty of states; the taboo against discrimination between them; and the Statute's expectation that the Board of Governors makes policy, elects the Director General, and expects the international civil servants of the Secretariat to do its bidding. This deferential posture appears in various commentaries by safeguards officials and is a constant refrain in private discussions with them. Sometimes it is part of an attempt to shift the blame for safeguards' shortcomings to member states.

The espoused value of non-discrimination between member states looms large in this culture of deference. Despite wide variation in the willingness of states to be cooperative, transparent, and fully compliant with their safeguards obligations, safeguards staff feel obliged to at least purport to treat all states with equal deference. This is not just because this accords with the espoused values, but also because the Secretariat relies on states for practical cooperation in applying safeguards. State cooperation is required in approving visas for inspectors, admitting them through customs, accommodating and accompanying them, and ensuring that they can conduct their inspections successfully. Increasingly, the Agency has relied on states to set up effective and efficient State Systems of Accounting and Control (SSAC) to ensure that the state itself provides accurate and timely data as a basis for the inspectors' verification activities. This has helped to further encourage a culture of caution in dealing with its chief "customers."

The underlying assumptions of any organization may of course be contradictory, forcing it to devise compromises and "work-arounds." Such trade-offs are rarely explicit because they are not openly discussable, but become part of the culture. Hence, despite protestations that it can only do what member states allow it to do, the IAEA Secretariat also asserts its autonomy. While careful not to get too far ahead of states' preferences, it has from the outset used its creativity and technical capabilities, and on occasion artfulness, to advance its objectives without explicitly seeking member states' approval. The *quid pro quo* is that it remain apolitical, technical in nature, operate as effectively and efficiently as possible, and show due deference to its political masters when this is unavoidable.

Principal-agent theory is illuminating in this regard. It holds that principals—in this case the member states—contract an agent—the IAEA—to carry out functions that they are unable to perform by themselves.[116] This requires the agent to pay due deference to its principals and not over-reach itself, but in return it is granted a certain degree of autonomy in carrying out its functions. A key insight of the theory is that over time, the agent becomes so expert and knowledgeable that the principals do not understand the agent's functions in all their complexity. They increasingly cede, whether knowingly or not, autonomy to the agent. Most states' delegations accredited to international organizations, including the IAEA, are unable to follow, for example, the complexities of the budgetary process, much less the technical details of the various nuclear-related programs. Many are simply not interested or are consumed by other priorities. A key insight of principal-agent theory here is that international secretariats spend their days focusing on the needs of their organizations, whereas their putative masters, the member states, have short attention spans, change their diplomatic personnel regularly, and struggle to cope with the mountains of documentation emerging from the range of international organizations. The agent in this case, the Secretariat, is usually inclined to keep it that way.

Far from automatically doing what member states allow it to do, there is in fact a constant low-level power struggle between principal and agent.[117] From its inception, the IAEA Secretariat, adhering to this pattern, both had autonomy thrust upon it and sought to carve out additional autonomy as its mandate expanded. As its membership grew and the issues it confronted became more complex, its opportunities for autonomy increased.

The ultimately unresolvable tension between deference and autonomy is deeply embedded in the culture of all international organizations. It is, however, thrown into particularly sharp relief in an organization like the IAEA, which is mandated to monitor and verify compliance by its member states with their legally binding international obligations. It is made even starker in the IAEA's case because the Agency deals with the matter of highly sensitive and jealously guarded national security. NNWS have renounced nuclear weapons and allowed an international organization to verify that undertaking.

The dilemma in navigating between deference and autonomy is that it is circular: in theory, member states want effective and efficient safeguards that can readily detect non-compliance and hold transgressors to account. But the Agency can only hope to achieve this if it is careful not to offend its member states—including those suspected of non-compliance and their allies—because they literally help provide the means to carry out its mandate. Challenging the sovereignty of member states could ultimately deprive the Agency of its livelihood. As ElBaradei himself has put it:

> Our position is somewhat schizophrenic. On the one hand, Member States pay the Agency's salaries and set its direction and mandate; on the other, we are charged with judging their compliance with international commitments under a treaty they have pledged to uphold. Even though Member States might understand this IAEA role intellectually, in the abstract, and when it applies to others, we inevitably encounter some resistance when we report a government's failure to comply with their obligations.[118]

The deference-autonomy dilemma compounds the cultural disposition that bureaucracies typically exhibit toward opacity, legalism, and formalism. Safeguards is no exception to these phenomena. Hence, in safeguards culture, along with the deference shown to member states, there is an unspoken assumption that the Director General and the Secretariat have significant autonomy in managing safeguards implementation. Over time, an element of the safeguards culture has thus emerged that implies: "Trust us, we know what we are doing. You, the member states, have entrusted us with running the verification system. Let us get on with it." Director General Amano was remarkably blunt about this in 2019:

> I periodically remind IAEA Member States of the importance of respecting established safeguards practices. The bottom line is that Member States should not intervene in our work of safeguards implementation. We, for our part, do not attempt to intervene in policy decisions of the Board and General Conference.[119]

Such an attitude is supported by the IAEA Statute's invocation of the traditional assertion of the autonomy and independence of international civil servants from their states of origin: "in the performance of their duties, the Director General and the staff shall not seek or receive instructions from any source external to the Agency."[120]

Long before the Iraq case, the Safeguards Department had itself secured significant autonomy in establishing, managing, and strengthening the safeguards system. Although all of the Agency's programs exhibit a certain degree of autonomy, it is more striking in the case of safeguards because the stakes are so high. The department was initially able to carve out significant autonomy because its fledgling staff, assisted by a small number of member states, had to practically invent international safeguards from scratch.[121] The United States, which was operating bilateral safeguards at the time, was of great assistance to the Secretariat, but the majority of member states had only limited experience and were not much help.[122] The first general safeguards document, INF-CIRC/26, was produced by the Secretariat in 1958 even "in the face of what was doubtless extreme discouragement."[123] The Secretariat also "did most of the groundwork for the 1965 and 1970 safeguards systems."[124] Mohamed Shaker, former Egyptian ambassador to the IAEA and author of a seminal three-volume study on the NPT, bemoaned that "we did not understand safeguards proposals put during the Article III NPT negotiations. We started to learn more when the proposal went to the IAEA to sort out the details of verification. For us it was like learning a foreign language."[125] Archival documentation reveals that, rather than member states directing its efforts, the Secretariat fielded correspondence from a range of puzzled member states as to exactly what comprehensive safeguards would entail. Its officials more or less devised answers on the spot.[126]

David Fischer and Paul Szasz noted in 1985 that:

> The tendency of the Secretariat to take initiatives in matters of policy that were once the province of the Board has already been evident for some time in the negotiation of safeguards agreements and, to an even greater extent, in the development and refinement of safeguards philosophy, approaches and practices. This is a natural consequence of the evolution since 1970, which gradually made the IAEA Secretariat the world's most experienced and authoritative body in the art of applying safeguards, but it has been hastened by the growing immobility of the Board.[127]

They concluded that "under the very broad (and not always clear-cut) guidance of the Board and of the Standing Advisory Group on Safeguards Implementation (SAGSI), the Secretariat alone bears the burden of transforming safeguards documents into a complex and generally effective operation for safeguarding the nuclear industries of most NNWS."[128] Over the years, the Safeguards Department has also taken various initiatives to enhance safeguards, not necessarily at the prompting of or with the permission of member states. One example was the amendment of safeguards agreements to rule out exemptions for so-called peaceful nuclear explosions after India's 1974 test.[129]

Departmental autonomy is reflected in the fact that the Safeguards Department alone negotiates the details of its bilateral safeguards agreements with each member state. The Board simply takes note. The subsidiary agreements and facility attachments are negotiated and concluded by the Secretariat and the details known only to restricted Secretariat personnel and the member state concerned. One reason why confidentiality is so prized at the IAEA is that it not only protects the member states, but it also strengthens the autonomy of the Secretariat in running the system. As principal-agent theory predicts, no single member state knows all of the system's details—only the Agency does. Preserving the mystique of safeguards by means of confidentiality and opacity is a cultural imperative that supports the autonomy of the Secretariat in carrying out its mission.

To preserve its autonomy, in its first few decades the IAEA Secretariat sought to keep politics away from its safeguards judgments through a criteria-based approach and a data-oriented reporting system. This, in turn, led to a fixation on the technical aspects of implementing safeguards to the exclusion of the broader non-proliferation implications: process versus outcome. There was no formal "risk assessment" involved in applying safeguards, but only private speculation by individual inspectors and managers. Safeguards were formulaic, based on the amount of nuclear material and the type of facilities a state operated. Inspector activity was focused only on declared material and associated facilities. No official thought was given to broader political indicators that a state might violate its safeguards agreement, and no "whole of state" assessment was done of nuclear and associated activities.

The language of the IAEA's safeguards reporting has always had an apolitical, technical quality that sanitizes and obfuscates. The 1990 Annual Report intoned that "it is considered reasonable to conclude that the nuclear material under Agency safeguards in 1990 remained in peaceful nuclear activities or was otherwise adequately accounted for."[130] The annual Safeguards Implementation Reports are parsimonious with detail, do not "name names," and are in any case confidential.[131] State Evaluation Reports on individual safeguarded states are also secret, even though they rarely contain anything truly confidential or embarrassing. Although there has been support in the Safeguards Department for the annual Safeguards Implementation Report to be made public, presumably to bolster the case that safeguards are effective, the Secretariat has been wary of certain member states' reactions, especially in the Board.[132] Veteran safeguards expert Rich Hooper has explained that "the conventional wisdom [i.e., culture] in this house, forever, but certainly through the Blix years, is that you avoided at all costs being reprimanded by the Board for overreaching. And secondly, that you never accept the renegotiation of an already accepted measure."[133] Hooper described the quality of safeguards reporting as generally rather low, "and this is how the Board of Governors wants it."[134]

Reporting a state to be in non-compliance with its safeguards agreement brings the deference-autonomy dilemma to the forefront. Olli Heinonen, in asking "when to ring the bell?" on non-compliance, says the timing, manner, and nature of any release of incriminating safeguards information is always tricky.[135] During his time at the Agency, he noted a reluctance in the Secretariat to report emerging safeguards implementation problems to member states, a tendency to let non-compliance problems become "more wicked" before alerting the Board, and an aversion to publicly responding to criticism or rebuttals by member states.[136]

Tariq Rauf contends that for Director General ElBaradei,

> the issue was always about the authenticity and veracity of the
> third-party information, and the limited capacity of the Agency to
> verify/authenticate it. Hence a reluctance to report prematurely to
> the Board in the absence of such independent confirmation espe-
> cially when the third-party provider stipulated a restriction from
> sharing the provided information with the accused State and with
> the Board. Hence a delay in reporting to the Board to avoid politi-
> cization, protect the Agency from disinformation attempts (as with
> Iraqi uranium import, centrifuge tubes, etc.).[137]

Rauf explains that most implementation problems were of a technical or operational nature, inconvenient for the Agency but not rising to the level of non-
compliance. For example, he says in some "good" countries, operators denied
requests or offered alternative dates for complementary access or short-notice
inspections by citing inconvenience for their production schedule. As for the
Agency's aversion to responding to public criticisms, sometimes its hands
were tied by restrictions placed on the release of information that had been
provided by third parties.

The implications for the Iraq case are that the Secretariat was trusted by
its principals to make the correct judgments about compliance, but that trust
regrettably proved misplaced. Given that it was solely the Secretariat that
had access to confidential safeguards information, combined with the lack of
transparency and detail in its reporting and the narrow technical focus of the
information provided, there was no opportunity for the Agency's "principals"
to second-guess the Secretariat's conclusions. To be fair to the Secretariat,
none of the principals provided any information from their National Technical Means, if indeed they had any, to challenge its conclusions about Iraq.[138]
As Blix has noted, the international community, in largely siding with Iraq
in its 1980–1988 war with Iran, appeared to turn a blind eye to Iraq's ques-
tionable nuclear activities and would not have relished the IAEA Secretariat
making a fuss over the issue.[139] Sometimes having a less than effective agent is
in the principal's interests.

Inspector and Other Safeguards Sub-Cultures

While individual personnel may share basic cultural values on safeguards,
they will often have different perspectives deriving from their varying professional backgrounds, roles, work practices, and experiences. As Schein

notes, "It is especially important to understand some subcultures, especially engineering and top management, because the reference group—the group to which members compare themselves—lies outside the organization in the occupational community."[140] In other contexts these have been called "epistemic" communities, which are those that share a particular expertise and knowledge and ultimately an identifiable culture.[141] Complicating the picture further, as Errol Meidinger notes:

> The correlation between organizational position and regulatory orientation might be taken to suggest that agency socialization is more important than background in determining an actor's role in a regulatory culture. Unfortunately, an alternative explanation cannot be ruled out: perhaps individuals with orientations consistent with particular offices choose naturally to gravitate toward, or are chosen by such offices.[142]

Inspector Culture

The inspectorate has traditionally exhibited the strongest organizational subculture at the IAEA. A high degree of group solidarity is apparent, undoubtedly due to the bonding forged by experiences in the field, where there is a need for mutual support and collaboration.[143] Even during routine inspection operations, inspectors work in remote locations in trying circumstances. They may need to deal with uncooperative, suspicious, or poorly organized interlocutors and endure long absences from headquarters. Shea and Thompson noted in the early 1980s that:

> In states with a strong positive outlook towards non-proliferation, the national view might be that the inspectors must be given every cooperation to perform their required functions, but even within such nations, the attitudes of plant operators vary widely and the inspectors must cope with conditions which often impede the inspector's ability to carry out the activities authorized, under stressful conditions.[144]

There is nothing like outside attacks or criticism to encourage a cultural "circling of the wagons" and cultural bonding. Shirley Johnson explains that inspectors had to be physically fit and "mentally secure."[145] Away from their families and friends, they created "adopted families," "their own little society."[146] Johnson recalls her regular informal Friday Afternoon Discussion

Group (otherwise known as the FADGE), in which the entire project staff, including both professional and support staff, would gather in her office to swap experiences, learn from each other, and "let off steam."[147]

Despite early skepticism about the challenges of creating an international inspectorate, IAEA inspectors came to be regarded as the star performers of the Safeguards Department and even of the entire Secretariat. Johnson says that in the early 1980s, they were treated like "gods."[148] For Heinonen, they were the most important element of safeguards: "the eyes, ears, legs and brains."[149] As one participant in a Harvard Nuclear Safeguards Culture Workshop in 2014 put it, "If the Agency did not have a devoted cadre of inspectors it could not have coped with Iraq, North Korea or South Africa."[150]

The inspectors were also perceived by other IAEA personnel, legitimately or not, as privileged. Inspectors were only in the "office" in Vienna approximately one-third of the time, and spent the rest of their time "in the field," either traveling to and from an inspection, working on-site, or at a meeting or training. After sixty days in the field, they could claim additional compensation ("a jealousy point" among other staff, says Johnson).[151] They enjoyed air travel "one class above Economy," according to the Agency's rules, which in those days usually allowed First Class. When airlines started introducing Business Class, that fit the rule. As for accommodation in the field, inspectors were given a per diem and could choose their own. As a result, they would try to find the "nicest, cheapest, most convenient" accommodation.[152] But some of the government guest houses they were obliged to stay in, as in India and North Korea, were "scary."[153] Overall, Johnson calls it "a romantic thing: travel to exotic places, inside nuclear facilities that cannot be accessed by others." She confessed, though, that "the reality was not quite as romantic or exotic. When recruiting [new inspectors] do I overstate it?"[154]

Also, as suggested above, inspector culture was formed partially in response to external attacks and criticism. As Shea and Thompson put it in the early 1980s, "Instead of expressions of support and encouragement, inspectors are confronted with events which attack the [safeguards] program and belittle their efforts."[155] In the early 1980s, five U.S. congressional hearings conducted into nuclear non-proliferation were primarily motivated by concerns about the adequacy of IAEA inspections. In compiling a reader on safeguards in 1983, the U.S. Congressional Research Service identified forty-six contributors who were generally supportive of safeguards, and thirty-six who expressed doubts.[156] Seven took both views. Media coverage of IAEA inspections during the early decades was also critical or skeptical.[157]

A strong inspector sub-culture was facilitated by a shared technical language and background in the physical sciences, predominantly engineering, which, as noted, comes with its own culture—especially an emphasis on calculation and measurement.[158] This was reinforced, in the view of some, by inspectors being overwhelmingly male and Western.[159] Perhaps paradoxically, while demonstrating strong group solidarity, the inspector sub-culture has also traditionally been highly individualistic, due perhaps to its engineering bent. A 2002 external consultants' report (the MANNET Report) opined that the Agency's "professional/technical culture" was "geared to individual achievement," "the organizational culture of the Agency is very individualist," and co-operative mechanisms are "counter-cultural."[160] One would expect such rebellious, independent-minded inspectors to be more likely to "blow the whistle" on an errant country in defiance of their managers' predilections. The MANNET Report, however, saw inspectors' individualism as a negative quality that needed rectifying with "dedicated resources and support from senior managers."[161] This indicates the challenges involved in pinning down exactly what an ideal safeguards inspector culture should be.

The quirkiness of the inspectors' culture has long produced a cultural gulf between the inspectorate and IAEA management. Slobodan Nakićenović explained that "the delicate and complicated duties and hard life of the inspectors, who are the Agency's real safeguards capital, have never been properly understood and respected."[162] Shea and Thompson pointed to management problems as the reason for some inspectors resigning.[163] Any attitude survey of inspectors, they said, would "indicate significant problems in this area, both in management philosophy [i.e., culture] and the manner in which management functions are carried out."[164] In contrast to management, which must take into account a wide variety of considerations, including political and financial objectives and the greater good of the organization, the culture of the inspectors inclined them to "let the facts speak for themselves" and "let the chips fall where they may," regardless of the political or institutional implications, a trait exhibited publicly by Roger Richter and later Scott Ritter, who purported to blow the whistle on UNSCOM inspections in Iraq.[165]

The dichotomy between inspectors and management recalls C.P. Snow's now somewhat old-fashioned characterization of the "two cultures": those schooled in the humanities and those with a scientific education.[166] For Snow, scientists "are inclined to be impatient to see if something can be done: and inclined to think that it can be done, unless it's proved otherwise."[167] Such an insight, if true, would make inspectors more inquisitive, more willing to experiment with safeguards techniques and approaches, and more willing to "call it like it is" regardless of the political implications. Snow contends that "non-scientists have a rooted impression that the scientists are shallowly optimistic, unaware of man's condition."[168] If we take this to imply a certain political naïveté, then Snow's two cultures, stereotypes though they are, may have affected the way the IAEA leadership handled evidence of non-compliance presented to them by inspectors, including decisions not to ultimately charge a state with non-compliance.

Beyond these stereotypes, there is an entire theoretical literature devoted to the relationship between management and inspectors and to the role of inspectors beyond "ticking boxes." While usually based on the experience of national regulators and inspectors working for domestic agencies, such as those dealing with customs, law and order, and health and safety, the insights seem uncannily applicable to the IAEA. It appears that there is a natural structural tension between inspectors' desire for autonomy, trust, and respect for their technical expertise and local knowledge, and the need of management to maintain control of the inspectors to ensure that they abide by rules and protocols, comply with efforts at standardization and quality control, represent the best interests of the agency, and help implement organizational (and cultural) change when required. An Organization for Economic Cooperation and Development (OECD) guide to best practice principles in regulatory policy proposes that:

> It is essential to ensure that inspectorates are able to set and follow their own work priorities based on professional decisions and expertise, with political decision as much as possible intervening at the level of overall strategy and resource allocation, but not in day-to-day operations and operational work. It is also essential that staff and management feel free to pursue the agency's objectives without fear of political interference or politically driven management changes.[169]

Interestingly, the literature on regulators and inspectors rarely appears to deploy the term "culture" to explain its findings.[170]

Along with apparently generic traits of inspectors, individualism among IAEA inspectors is shaped by the unique way they are appointed, and the powers and responsibilities assigned to them. An IAEA inspector is officially defined as "an IAEA officer appointed by the Director General and approved by the Board of Governors to perform safeguards inspections."[171] After designation by the Board, the name of the individual is proposed to the state or states to which the Secretariat proposes to assign him or her. If the state agrees—and, as we have seen, states are entitled to repeatedly reject proposed inspectors within certain limits—the IAEA "effects" the inspector's designation.[172] The individual, as provided for in the IAEA Statute, is granted privileges and immunities comparable to those offered to diplomats. These are necessary for performing their functions, notably in their dealings with member states' border and customs officials, but they also give inspectors a certain status. IAEA safety helmets and white uniforms that are occasionally donned add to the mystique.[173]

The desired values of the inspectorate in conducting inspections were outlined by Blix during and after the Iraq case:

- Driving and dynamic—but not angry and aggressive
- Firm—but correct
- Ingenious—but not deceptive
- Somewhat flexible—but not to be pushed around
- Calm—but somewhat impatient
- Keeping some distance—but not arrogant or pompous
- Friendly—but not cozy
- Respectful of those you deal with—and also demanding of respect yourself.[174]

Blix reminded participants in an inspector training course that "a light tone or a joke may sometimes break a nervous atmosphere."[175]

These values do not really explain why IAEA inspectors' culture did not lead to their uncovering Iraq's rampant non-compliant activity. Being courteous and respectful is not the same as being unwilling to be inquisitive, investigative, and skeptical, none of which were qualities listed by Blix. Even when there were grounds for suspicion about Iraq before the 1991 invasion, inquisitiveness was not encouraged.[176] There were reportedly incidents where some "over-zealous" inspectors sought to raise concerns, but were told by senior management not to "rock the boat." Clearly some individual inspectors had tried over the years to be more inquisitive, but the overarching IAEA culture discouraged this. Over time, it was only natural for the inspectors to absorb the cultural zeitgeist of the Agency as espoused by the Secretariats' leadership and focus on the narrow tasks assigned to them.[177]

With the early *ad hoc* approaches gradually abandoned, inspector tasks were organized in a mechanistic and routine way and focused specifically on the verification of nuclear material accountancy and facility design.[178] To verify an inventory of nuclear material, inspectors would count the items present and take measurements, using radiation detectors and taking samples for more detailed analysis at headquarters. Inspectors also verified international or domestic transfers of declared nuclear material from one state to another. Facility design information provided by the state was examined and verified by observation. Containment and surveillance devices (tamper-proof seals and cameras and detectors installed at facilities) were checked and replaced if necessary. Crudely put, the culture became to "tick the boxes," change the batteries on the remote monitoring equipment, and return to Vienna.

Most inspections were presumed to be routine, and even those that were not, like short-notice or unannounced inspections, assumed a routine character. Meetings between inspectors and state officials and facility operators were intended to be an "important" forum for communications between the state and the Agency on operational issues, but were not expected to be interrogations. Inspections were not meant to be adversarial; inspectors were not supposed to start with a presumption of non-compliance or evince suspicion or doubt. Compliance was expected and presumed. After the Tamuz incident, the Agency conceded the "difficult nature of the job of being an IAEA inspector," noting that inspectors were "generally technical people and not diplomats" who were "in the forefront of a unique international experience in going into nuclear facilities in a country to verify that the country is telling the truth about its nuclear activities."[179] As we have seen, however, prior to the Iraq case, inspectors were not recruited, trained, or equipped to determine the whole truth.

Carlson says that traditional inspection culture had positive attributes: it was "thorough, systematic, unbiased, non-discriminatory, auditable and inspector-independent—to a large extent an inspection produced the same result regardless of the personal characteristics of the inspector performing the inspection."[180] It would have taken the equivalent of a whistleblower, someone willing to go beyond the call of duty, to look over the berm, as it were. As Johnson says, "Unfortunately we missed a lot." After Iraq, "we now had a much better handle on what we were doing. . . . Iraq opened our eyes."[181]

Other Safeguards Sub-Cultures

During the research for this project, several interlocutors drew attention to other identifiable sub-cultures in the Safeguards Department besides that of the inspectors. New DDGs often saw their first task as attempting to unify the quasi-independent divisions and sections under their charge. Some Agency personnel claim that even the operational divisions have their own unique cultures, resulting from the varied countries for which they have responsibility. U.S. safeguards officials noted that, at least under the pre-Iraq culture, the operations divisions tended to be run as little fiefdoms, planning and carrying out inspections but not well integrated into the whole Safeguards Department.[182] Johnson says that "with three different safeguards divisions there were almost three different inspectorates."[183] Sections that handled large, sensitive facilities (involving plutonium, enrichment, reprocessing, and advanced research) needed more experienced and knowledgeable inspectors, who usually came from developed countries. They were bound to develop different attitudes toward safeguards compared with sections dealing with, for example, Africa, which still has only one nuclear power station (in South Africa) and where safeguards are relatively parsimonious.

A diversity of divisional managers, with no departmental standardization or oversight, also naturally tended to engender different work practices and cultures. Johnson notes that the Safeguards Department had a rotation policy to ensure "cross-fertilization" across departmental divisions.[184] Yet the Agency's external auditor confirmed as recently as 2013 that "the processes adopted for implementing safeguards by these divisions were not uniform and were sometimes at variance with the laid down Departmental procedure."[185]

Unique sub-cultures were also bound to develop in the divisions dealing with Concepts and Planning, Information Management, and Technical and Scientific Services, in part because of the different professional backgrounds of their staff, but also because of their different roles. Pier Roberto Danesi, Director at the Agency's Seibersdorf laboratories from 1986 to 2002, told Elisabeth Roehrlich that his operations were split by two cultures: "experimental" in the labs and "programmatic" at headquarters.[186] These sub-cultural differences do not matter as long as they are appropriate to their roles in each case, and as long as they comport with the overall safeguards culture, which, in the case of nuclear non-proliferation, should be geared toward the most effective safeguards implementation.

Some interlocutors have suggested that having different cultures within the safeguards enterprise is healthy, and that it would be a mistake to try to ensure uniformity and risk "group think." Diversity should help to ensure that a healthy skepticism is brought to bear before safeguards judgments are made and sent higher up the organizational chain where political dangers lurk. Diversity should also help to ensure that a non-compliance case like Iraq will be caught. Some policy and management differences may, of course, be due to different personalities and have nothing to do with the underlying culture.

Cultural Drift: Correctness without Completeness, Diversion Deified

The cultural key to unlocking the mystery of how the IAEA managed to miss Iraqi non-compliance may not be found in the Agency's UN heritage, its over-all artefacts or espoused values, or the existence of competing sub-cultures, but in the singular implicit notion, pervasive throughout the Agency at the time, of "correctness without completeness." A basic underlying assumption of traditional safeguards culture, fully revealed by the Iraq case but always in plain sight, was that safeguards should focus only on nuclear materials, facilities, and activities declared by a state, rather than also considering what might not have been declared. This assumption, not openly expressed or written down in IAEA documents—which is typical of culture—was that inspectors and other safeguards personnel should be concerned only with the *correctness* of a state's declaration of its nuclear holdings, rather than its *completeness*. Diversion of declared materials from declared facilities to weapons purposes was to be considered the primary non-compliance threat. The notion that a state might set up a fuel cycle and nuclear weapons program completely sep-arate from its declared program was not perceived as a concern. Roger How-sley recalls an incident that illustrates this point. After a week-long SAGSI meeting, his boss, the Director of British Nuclear Fuels Limited, asked him what he thought of safeguards. "And I remember saying to him I really don't understand why it is limited to the verification of declared material because who would divert declared material? And with a sort of slight air of superi-ority he sort of tapped me on the head and said 'well, eventually you'll get to understand what this is all about.'"[187]

This cultural drift was not what the original concept of safeguards had envisaged, and certainly not what the espoused values of the Agency pro-claimed. The NPT says that safeguards procedures in NNWS "shall be fol-lowed with respect to source or special fissionable material whether it is being produced, processed or used in any principal facility or is outside any such facility."[188] The whole ethos of safeguards was to prevent the acquisition of nuclear weapons by states through early detection of nefarious activities involving any nuclear material.

Long-time participants in devising and implementing safeguards are mystified as to how safeguards moved away from its original intent, as par-ticipants in the evolution of a culture often are. Hooper, one of the "fathers" of INFCIRC/153, explains that in 1993, as consideration was being given to strengthening safeguards,

I went to considerable effort and I was certainly in a position to find out why did things develop the way they did. Why the focus on declared vis a vis [sic] undeclared, where the basic undertaking of states clearly provided the possibility of both. Arrangements were made for me to spend a couple of weeks with David Fischer, who from the Agency side certainly was a very experienced and knowledgeable observer and participant in all of this. But in the end, the conclusion I came to was [that] all these questions are largely rhetorical—that there is no answer, it just is that way.[189]

Myron Kratzer described it in somewhat over-wrought terms as "the mystery of the ages":

How anyone could pick up this document that says in Paragraph 2 that the Agency's obligation—right and obligation—is to apply its safeguards to all material and to end up saying, well all we really have to do is apply our safeguards to declared material, is to me the mystery of safeguards. Which has not been solved, which I think has led to great difficulties; certainly, it was a critical factor in the Iraqi experience of 1991.[190]

IAEA inspector Jacques Baute also inadvertently explained the failure in Iraq in cultural terms: "I don't think it was a failure of the Agency as such. It was a global misunderstanding translated into these limitations of . . . technical implementation."[191]

Many factors conspired to produce this "global misunderstanding." In fact, the obsession with correctness and the neglect of completeness, along with the focus on diversion to the exclusion of other non-compliance scenarios, was over-determined—which is why it became such a "sticky" part of safeguards culture. It is a fascinating case of an original espoused value being undermined over the years by a growing, underlying, culturally-embedded assumption.

U.S. Bilateral Agreements Set the Precedent

Concern over diversion dominated the safeguards discourse from the outset of the nuclear era. This was because the first nuclear power, the United States, was determined to keep its nuclear monopoly, and it presumed that the only way other states could ever develop the bomb would be by "diverting" nuclear knowledge, expertise, technology, and materials that it provided to them. U.S.

worries over diversion permeated the earliest discussions of international control of nuclear energy. The word "diversion" appears to have been introduced into the nuclear lexicon in the seminal March 1946 Acheson-Lilienthal Report on the International Control of Nuclear Energy.[192] A 1962 report on U.S. policy toward the IAEA that had been requested by President John F. Kennedy also highlighted the problem of "diversion."[193]

Bilateral safeguards agreements signed by the United States with recipients of its technical assistance, materials, and technology were concerned only with preventing its diversion to weapons purposes, not with the recipients' entire nuclear fuel cycle, because they usually did not have one. Similarly, the bilateral inspections that the United States put into place were concerned only with U.S-supplied material and technology. This was logical in that most recipients' fuel cycles at that time consisted only of what the United States had supplied.

Kratzer further explains that a relatively early event in the life of safeguards in the late 1960s was the widely publicized inability to account for a large quantity of highly-enriched uranium at a conversion and fabrication plant in Apollo, Pennsylvania.[194] He says: "This accountability incident was a truly watershed event . . . in the evolution of safeguards, since it brought to public attention in a way that could not be ignored a fact long known to safeguards experts: that accounting for nuclear material is, in most cases, more akin to measuring bean stalks than to counting the beans."[195] This prompted even greater emphasis in the United States, and ultimately at the IAEA, on nuclear materials accountancy and, by extension, ensuring that the accounts were correct.

It was only natural, then, that when the IAEA took over implementation of U.S. bilateral safeguards agreements and began modeling its new agreements on U.S. precedents, diversion also became the singular preoccupation.[196] Such influence was compounded by the key role that U.S. safeguards personnel played in devising the international system. In drawing on U.S. bilateral agreements as precedents and models, the IAEA adopted both their focus and their language.

Negotiators Further Embed Diversion

Negotiations on the NPT did not help. The treaty does not mention the problem of undeclared, clandestine materials, facilities, or activities and foresees no role for the IAEA in this regard.[197] The NPT's Article III speaks only of preventing "*diversion* [emphasis added] of nuclear energy from peaceful uses to nuclear weapons or other nuclear explosive devices."[198] In the treaty negotiations, much less attention was focused on the challenge of undeclared activities than on how safeguards would deal with nuclear material used for non-explosive military purposes, principally submarine propulsion.[199] Nonetheless, the NPT does require NNWS parties to accept IAEA safeguards "on all source or special nuclear materials in all peaceful nuclear activities."[200] A formerly confidential U.S. report on the comprehensive safeguards negotiations calls this "the unmistakeable and explicit obligation of the NPT."[201]

As the body charged with handling safeguards for the NPT, it was left to the IAEA to flesh out the details. The negotiations over safeguards at the IAEA compounded the diversionary trend. Rudolf Rometsch, the first head of the Safeguards Department (1969–1978), recalled that negotiations on the comprehensive safeguards system in the Committee of the Whole in 1970 "led . . . to a sort of dogma for field work—if not a taboo. . . . It was the question of whether inspections should be designed also to detect undeclared facilities. The conclusion was clear at the time: looking for clandestine activities was out of the question and the inspection system was designed accordingly."[202]

Yet former IAEA legal advisor Laura Rockwood cites the negotiating record to point out that the drafters explicitly rejected a South African proposal that "safeguarding and inspection . . . shall be concerned solely with the material reported upon by the state concerned."[203] The proposal was formally opposed by Hungary (presumably at the behest of the Soviet Union), received no support from any other participating state, and was omitted from paragraphs 1 and 2 of INFCIRC/153.[204] Such an amendment, it was widely recognized, would have rendered verification woefully incomplete. Instead, INFCIRC/153 gives the IAEA "the right and the obligation to ensure that safeguards will be applied . . . on all source or special fissionable material"— whether declared or not.[205]

In addition, the negotiating committee understood the phrase "nuclear material subject to safeguards" to mean not just material that was being safeguarded, but that which was *required* to be safeguarded to fulfill the fundamental purposes of the NPT.[206] Moreover, on occasion, according to a 1984 report by the U.S. Arms Control and Disarmament Agency (ACDA), "the more explicit and emphatic term 'nuclear material required to be safeguarded' was used" in INFCIRFC/153. ACDA concluded that through the use of such terms, the "key specific safeguards obligations are made applicable not merely to material being safeguarded pursuant to a state's 'declaration,' but to that material which *ought* [emphasis in the original] to be safeguarded pursuant to the fundamental undertakings of Paragraphs 1 and 2 [of the NPT]."[207]

ACDA concluded that "most persuasive in establishing the right of the Agency to apply safeguards to all nuclear material, regardless of whether it is declared or not, is the authorization of the Agency to take action when it finds it can no longer verify that there has been no diversion of nuclear material to be safeguarded."[208] Although "as a practical matter," ACDA confirmed that the Agency was not given the capability to search out activities or facilities, special inspections were provided as a means for the Agency to request access to such facilities.[209] Despite these understandings, ACDA also lamented that "considerable dissension has existed for some time, both within the Agency's Secretariat, and between the Agency and member states, as to the Agency's rights and responsibilities to apply safeguards to 'undeclared material,' a term which, not unimportantly, is found nowhere in INFCIRC/153."[210]

At the outset, IAEA officials worried about how to handle undeclared activities. Fischer and Szasz recalled that:

in several cases [the Agency] has had to begin its operation by what amounts to an act of faith in the government's (or EURATOM's) initial report. . . . Even if one assumes that the initial reports of the 1970s were comprehensive, this would not by itself prevent the countries concerned from subsequently building secret plants or secretly producing nuclear material.[211]

Hans Grümm records that in 1979, safeguards officials attempted to discuss verification of the completeness of the initial inventory submitted by states, but they encountered "substantial resistance."[212] He notes that "only by bypassing them and other institutions did we succeed in incorporating suspicious terms like 'undeclared facility' and 'undeclared material' in the safeguards glossary in 1990."[213]

Other safeguards personnel may well have always assumed that they were mandated to be concerned with completeness as well as correctness, but simply did not have the resources to pursue what was undeclared. But this was not emphasized by the Agency in its official documentation, its practice, or, perhaps most important, its culture. Rockwood reports that when she joined the Agency in 1985, she was surprised by the prevalence of the assumption, even in the Office of Legal Affairs, that safeguards only concerned declared materials and facilities, and that special inspections could not be used to clarify suspected cases of undeclared activities.[214] She fought a rear-guard battle to turn the culture around.

Over time, member states began interpreting the obligations in INF-CIRC/153 as being for the "exclusive purpose of verifying that nuclear material is not *diverted* [emphasis added] to nuclear weapons or other explosive devices."[215] Even the strongest state supporters of safeguards sometimes wavered. After Israel bombed the Tamuz-1 reactor in 1981, U.S. Ambassador to the UN Jeane Kirkpatrick famously declared that safeguards inspectors were "not policemen; they can only inspect what has been declared."[216] Expert opinion also drifted. No less an expert than Kratzer asserted baldly in 1967 that the "sole" purpose of safeguards was "to determine whether any material has been diverted to unauthorized use."[217] Safeguards veteran Joseph Pilat argued in 1982 that "IAEA inspections can verify inventories of nuclear material, and in principle detect diversion of safeguarded material or misuse of a safeguarded facility on the basis of these inventories. Inspection cannot prevent diversion of safeguarded material and can do nothing about unsafeguarded material or facilities."[218]

Linguistic Ambiguities

Compounding the political and practical drivers of the diversion focus was a linguistic issue that is seldom noticed. In English, the word "divert" means to "turn aside (a thing) from its path or original purpose."[219] In the case of safeguards, this implies moving nuclear material physically from a declared location to another so it can be used for illicit purposes. Yet material could be misused *in situ* without being "diverted." Illegally imported material is, logically, not being diverted from one purpose to another but brought into the country and used for illicit purposes. Indigenously mined, processed, and enriched nuclear material is not being diverted so much as brought into being. Undeclared nuclear facilities, such as those built by Iraq, are not being diverted but constructed.

Such linguistic ambiguities may further explain why, over time, the assumption seemed to arise that safeguards were only concerned with diversion from what already existed and had been declared by a state, and not with any misuse of any nuclear material that the state had acquired by other means. Rather than using the term "diversion" of nuclear material to explosive purposes, a better term, in theory, would have been "any use" of nuclear material for such purposes. Unfortunately, this was not as memorable or catchy. It is not clear whether it occurred to the negotiators or not, but there is no indication in accounts of the negotiations that the use of the word "diversion" was debated.[220]

A further ambiguity in safeguards language that reinforces the pervasiveness of the diversion obsession is that rather than obliging a state to declare and place under safeguards *all* nuclear material in its possession, the safeguards requirement is to declare *all nuclear material in peaceful uses*. Unless read in conjunction with the NPT's prohibition on acquiring nuclear weapons, this could be read as permitting a state to retain undeclared material for non-peaceful purposes while only declaring material intended for peaceful purposes. As ACDA notes, the reference to peaceful uses was required, as the NPT permits non-explosive military use of nuclear material, as in naval propulsion. The application of Agency safeguards to military uses, even non-explosive ones, it says, "would be not only inappropriate for the Agency but unacceptable to most states."[221]

Once diversion was adopted as the term of art for safeguards, it became pervasive. While it is beyond the scope of this book to provide a complete literature survey and textual analysis of its use, some evidence may be drawn from the one-thousand-page *Safeguards Reader* compiled by the U.S. Congressional Research Service in 1983.[222] The collected papers, from a wide range of sources both supportive and critical of IAEA safeguards, reveal scores of references to the problem of diversion but only a handful to the challenge of undeclared, clandestine, secret facilities or activities. Most of the latter refer to the IAEA's not having the authority or capability to seek out such phenomena.

The language of diversion was also institutionalized in Agency reporting on its safeguards findings. The Director General would report that "there has been no diversion" or that "nuclear material was adequately accounted for." As is the way with bureaucratic culture, once this compliance language had been adopted, it persisted—and in some contexts remains unaltered to the present day. The term "completeness" did not appear in IAEA parlance, including its documentation, until the South Africa case in 1991, when, responding to the lessons learned from Iraq, the General Conference requested that the Board of Governors instruct the Director General to ensure the "correctness and completeness" of South Africa's initial report on its nuclear holdings.[223] As a comprehensive account of the South African nuclear program notes:

> The failure of the IAEA in Iraq and the long-standing suspicions of South Africa's nuclear capability suddenly gave rise to new rules: the IAEA was not only expected to start verifying the initial inventory immediately, as they did for all countries, but in addition had to verify the correctness and completeness of South Africa's declaration of its nuclear material and nuclear facilities.[224]

The fact that the South Africans got the impression from the IAEA inspection team that "nobody was quite sure how this was to be done" is further evidence that the Agency had until then never been truly concerned with "correctness and completeness."[225]

Safeguards Staff Go With the Flow

Once the obsession with diversion was embedded in the NPT and in safeguards documentation, it was inevitable that the Agency's safeguards staff, especially the inspectors, would develop an "appropriate" accountancy mentality. From the point of view of safeguards staff, verifying declared materials and facilities was an identifiable task, relatively easy to accomplish, and already enough to cope with without complicating matters and upsetting member states by worrying about the perceived low probability of "unknown unknowns." Criteria and inspector performance measurement demand objective measures, while subjective observations were not measurable or readily reportable. An espoused value that upheld that safeguards were merely technical and without subjective judgments had a reinforcing effect. Robert Kelley describes what was asked of inspectors: "Did you do this and this and this, not what did you see."[226]

Over time, satisfying the safeguards criteria at declared facilities became the purpose of the work, rather than the goal of determining whether there was any indication of nuclear weapons proliferation. Heinonen says that once safeguards criteria were introduced, the process became too mechanistic: "it was numbers count, but it went too far."[227] The espoused values of non-discrimination and respect for state sovereignty reinforced such tendencies. This is a classic case of culture becoming dysfunctional, of means becoming ends, of outcomes dominating process. As Carlson describes it:

> A comprehensive and enduring culture developed around the use of quantitative and relatively mechanistic procedures. This culture was self-reinforcing through a particular interpretation of policy imperatives—specifically, the requirement to avoid discrimination—which resulted in the approach of uniformity in safeguards applications, and a distrust of exercising judgment. Judgment was thought to be subjective, prone to bias and error.[228]

Not only did member states discourage a concern with undeclared activities, signaled by both their rhetoric and their lack of support for institutionalizing and funding such work, but also neither the Agency's leadership nor safeguards managers encouraged inspectors to pursue such activities or rewarded them if they tried. Inspectors were not personally incentivized to find non-compliance. As Heinonen says, "If you found an anomaly it was your problem!"[229] Inspectors' reports were reviewed by the Division of Safeguards Evaluation, but according to Shea and Thompson, "for the most part this was limited to checking completeness" (in the sense of having ticked all the boxes, not in terms of ensuring that undeclared material might be indicated), and did not "delve into the quality or correctness of the information included in the inspection report."[230] In their telling, even correctness was not that much of a concern at that time; the requirements for adequate safeguards, sent to section heads by management, established the minimum number of activities to be completed during the year.[231] If those requirements were met and there were no unresolved anomalies, the Secretariat would conclude that its inspection goals were met at that facility.

Such an evolution in the culture was reinforced by a natural human tendency to focus on what is visible and measurable. Carlson notes: "After all, nuclear weapons cannot be produced without nuclear material, and nuclear material has the virtue of being easy to characterize and measure."[232] Charles Duelfer noted this phenomenon at work in UNSCOM's neglect of Iraq's biological weapons program in the first three years of its mission: "We behaved a bit like a children's soccer team, on which the mass of players simply chases the ball. We had been pursuing missiles and chemical weapons because we could see them."[233] This recalls the famous experiment in which volunteers were asked to watch a video of a basketball game and count the number of passes made by players wearing white, while ignoring those wearing black.[234] In the middle of the game, someone dressed as a gorilla runs across the court. Roughly half of the participants failed to notice the gorilla. Even Blix compared the Agency's traditional safeguards to "a man looking for a lost key near a lighted street-lamp who, when asked whether he was sure he had lost the key there, said 'No, but it's easier to look here.'" The Agency, he said, should "look for any keys lying away from the street-lamp in darkness."[235]

Confirmation bias also occurred. Experience seemed to reinforce confidence in the system: no serious non-compliance cases arose until Iraq.[236] As ElBaradei concedes, "Before Iraq it was smooth sailing at the Agency. The good life."[237] Other psychological phenomena also likely played their part in reinforcing a culture of confidence that diversion was the main game.[238] Avoidance of cognitive dissonance could, for instance, have led the Secretariat to discount media reports of Iraqi proliferation activities.

Some safeguards officials absorbed the pervasive diversion culture so well that they resisted the occasional efforts by the Director General to make exceptions to it. Nakićenović reports that in 1970, his inspectors were instructed, presumably by the Director General, to collect information about unsafeguarded facilities. He claims that despite his objections "and those of my inspectors," this "irregular" and "illegal" instruction was not withdrawn "until I informed the United States and Soviet representatives to the Agency that I would resign."[239] He claims that there was another attempt to collect information about an "unpublished" nuclear program and individual unsafeguarded facilities in 1973. He says he was "bypassed" by this instruction. Director General Eklund apparently refused to halt this activity or consult the Board about it until Nakićenović once more informed the U.S. and Soviet representatives.

Special Inspections and the Small Quantities Protocol

Related to the issues of correctness and completeness, and demonstrating a similar cultural drift, was the unwillingness of successive Directors General to use special inspections. Such inspections were intended to clarify doubts about a state's compliance in various situations, not just when suspicions were aroused about a major safeguards violation. Yet notwithstanding the low-key, unthreatening manner in which the possibilities were described in safeguards documents, a special inspection was never instigated prior to 1990.[240] Between 1985 and 1990, the Office of Legal Affairs crafted two separate legal opinions on special inspections for Blix that would, it was believed, "fly politically" with the Board.[241] Of course, a special inspection may not have been warranted, as Rockwood notes: "It's a question of information and opportunity and I am not certain those existed between 1985 and 1990."[242]

Over time, having never been instigated, special inspections became elevated to the status of "challenge inspections" in other arms control regimes, such as the 1993 Chemical Weapons Convention (CWC) and the 1996 Comprehensive Nuclear Test Ban Treaty (CTBT), where they may be utilized in cases of suspected non-compliance. These have a more confrontational form than special inspections because they are only used when a state has been accused of a possible violation. In these two regimes, it is another member state that is entitled to ask for a challenge inspection to be carried out in a fellow member state. In the IAEA's case, it is the Secretariat that is authorized to make such a request. Clearly, the Secretariat felt unable to risk asking for one.

As Carlson notes, "Reasons for the lack of formal use of the special inspection provisions are not clear, but the situation appears to have become self-reinforcing, i.e., the longer the provisions were not used, the more they came to be regarded as being available only in very exceptional circumstances."[243] Blix's view was that "once a special inspection is requested it is a problem."[244]

In any event, an understanding gradually arose within the Secretariat that a request for a special inspection was too politically charged and should not be sought, implicitly removing an important tool from the inspectors' tool-box. Blix has noted that "the moment you demand a special inspection, you are likely to have passed the stage of mild diplomacy and cooperative open-ness."[245] The taboo against special inspections thus entered safeguards cul-ture. When the Agency finally requested one, in 1993, the country concerned, North Korea, refused.[246] The Agency has never requested one since.

A similar phenomenon seems to have occurred with the Small Quanti-ties Protocol (SQP), which holds safeguards in abeyance for states with little nuclear material. Hooper, despite being one of the leading lights of safeguards, has no idea how it developed the way it did, who drafted it, or who approved it: "It was never discussed and when it surfaced, about 1972/3, it was simply presented to the Board as it was."[247] He was "simply horrified" when he saw it, presumably because it envisaged up to two-thirds of states under safeguards as having few safeguards obligations. This too became part of "the way we do things around here."

The State of Safeguards Culture by 1990 and its Role in the Iraq Case

What can one conclude, then, about the state of safeguards culture by 1990 and its role in the Iraq case? Clearly it had been shaped by political and orga-nizational realities, many dating from the earliest years of the Agency's exis-tence. It also embodied many of the espoused values that comported with the IAEA Statute, the NPT, and the various safeguards documents pursuant to them. A primary value, with strong resonance among all IAEA employees but especially safeguards staff, was that safeguards were carried out in the service of a higher purpose, namely international peace and security. Other key values included a science and technology orientation; a regard for the sovereign rights of states; adherence to the principle of non-discrimination between states in applying safeguards; the need for safeguards to be effective and efficient; and the importance of safeguards confidentiality.

As we have seen, however, there were also unspoken elements in the culture—"the way we do things around here"—that did not comport with the espoused values, and which could be detrimental to effective safeguards. Most important from a non-proliferation perspective was the widespread cultural assumption that safeguards only had to do with detecting diversion of declared nuclear material and that the Agency did not concern itself with undeclared materials or facilities. The characteristics of the UN's bureaucratic system, such as formalism, legalism, and the niceties of international diplomacy—to the extent that the Agency had adopted them—all helped to shape a safeguards culture that did not prize risk-taking but instead rewarded process over the prevention of nuclear non-proliferation.

Despite lauding the assumed contribution of safeguards to international peace and security, the culture also embodied worries about what the real contribution was, especially because of perceived treaty loopholes and weaknesses in safeguards implementation. "Over-expectation" about what safeguards could achieve was a concern, not least because of the widespread assumption that the Agency's UN-style recruitment and promotion system invariably produced a certain amount of "dead wood" among IAEA staff, including inspectors. In addition, the culture was affected by cynicism about member states' much vaunted support for safeguards when they were so miserly in providing the tools and resources for safeguards personnel to do their jobs properly.

While a strong element of the culture was the need to show due deference to member states' preferences, there was also, paradoxically, a strong sense of autonomy in managing the safeguards system day-to-day. This reinforced a culture of opacity which had arisen due to the perfect storm of "safeguards confidentiality," the defensiveness of the Secretariat in devising and running a clearly imperfect system, and the normal tendencies of bureaucracies to over-classify information and generally regard information as power.

Within the Secretariat, there were identifiable sub-cultures with contrarian elements that did not necessarily align with the organization's espoused values. Among safeguards inspectors, there was a strong element of group solidarity due to the nature of their work and a streak of individualism deriving from their largely scientific and technical backgrounds. These augured well for a robust, effective inspectorate. But there was also skepticism verging on scorn about the ability of "management" or the "bureaucracy" to run the safeguards system to the inspectorate's liking. The inspectors' culture, in addition to being skeptical that findings about non-compliance would be taken seriously and acted upon at the highest levels, signaled resentment that they were not being given the necessary resources to do their jobs or the recognition they believed they deserved. The culture of management, meanwhile, was something of a mirror image: there was an element of disdain for inspectors (mere "bean counters"), or at least an assumption that they needed to be closely managed and that their raw findings about anomalies must be considered in the context of international politics and the organizational needs of the Agency.

Assessing the precise role of culture in the Iraq case, despite calls afterward for the culture to be "improved," is nonetheless problematic. Because much of culture is about atmosphere and perceptions, the notion that it can be blamed for a particular event is difficult to substantiate. For instance, it is not even clear whether the Hungarian and Soviet inspectors who visited Iraq had absorbed the general inspector culture. Although cultures tend to be pervasive, they do not automatically predict whether or to what extent individuals will embody it. The Agency's Soviet and Eastern European employees may indeed have operated outside of the main culture, which was decidedly Western. In the absence of evidence that they treated Iraq differently than other inspectors might have, the effect of culture cannot be assessed definitively. In any event, Iraq had only planned to divert material but had not actually done so, so there was nothing for the inspectors to detect.

Yet the broader effect of safeguards culture on the Iraq case can be identified with respect to its undeclared parallel nuclear weapons program. The safeguards culture clearly and systematically turned a blind eye to the possibility that a state would violate safeguards in such a fashion. While such a possibility was raised periodically by individuals within the Secretariat and by outsiders, especially in the U.S. Congress and the non-proliferation community, the culture simply disallowed the idea. Safeguards was not perceived to be in the business of detecting the proliferation of nuclear weapons, but rather in detecting the diversion of declared nuclear material to weapons or purposes unknown. It is in this respect that one can see, in retrospect, an effect on Iraq's violation of safeguards. As a member of the IAEA, a participant in Board meetings, and with their nationals employed in the Secretariat, the Iraqis would have accurately sensed the culture of "correctness without completeness" and the obsession with diversion to the exclusion of other possibilities. While there was no guarantee that their illicit activities would go undetected, they correctly judged that exposure of their activities by the Agency was only likely to happen inadvertently rather than by deliberate Agency effort. The risk was apparently judged worth taking.

As for the Agency, the cultural taboos against concerning itself with undeclared activities, using non-safeguards-derived information in assessing a state's proclivity to proliferate, or utilizing special inspections to resolve compliance questions applied to Iraq as much as to any other member state. This was despite the international attention paid to concerns about the intentions of Iraq's nuclear program both before and after the Israeli bombing of the Tamuz-1 reactor in 1981. The culture did not permit the Agency to try to assess a state's intentions or to discriminate in seeking additional safeguards measures, such as more inspections, for a particularly suspicious member.

When asked prior to the Iraq case about the issue of undeclared facilities or materials, the Agency tended to declare that it "did not go on fishing expeditions," as if that were the only way that undeclared activities might be detected. As Blix put it, the IAEA could not scour the territories of its numerous NNWS members "in a blind search" for undeclared nuclear plants or material.[248] This is somewhat misleading. Most states, it could safely be assumed, would as a matter of course comply with their safeguards agreements. Agency personnel, as in the Iraq case, would presumably always have had private suspicions about the small number of states that may not have been compliant, based on media reports and other information in the public domain, if only derived from rumors. Individual inspectors could certainly have been curious about suspicious activity at or next to declared facilities.[249] Only a vanishingly small minority of states would have to be "scoured."

Blix also said that the NPT safeguards system prior to the Iraq case had "few teeth," which begs the question of what teeth he thought there were and why they had not been used—or why new teeth had not been sought.[250] ElBaradei, in explaining the Iraq case, claimed that "the Agency was only *expected* [emphasis added] to verify what a country declared. We had little authority, and few mechanisms, to search for undeclared nuclear materials or facilities. If this sounds frighteningly naïve, it was. For regimes that chose to conceal their illicit activities, the IAEA was a beat cop with a blindfold."[251]

Blix, pondering whether the Secretariat could have done more to detect Iraq's significant and systematic non-compliance over many years, suggests that it could have conducted more inspections at Iraq's declared installations, scanned the media for information, and found a few suspicious Iraqi imports, "which might conceivably have led to useful controversies—alarms."[252] Blix told Gudrun Harrer that Iraq had been asked questions about, for instance, the krytrons discovered at Heathrow Airport, but in response was shown invoices from the University of Baghdad, suggesting that they were for academic research.[253] But, revealingly, he makes no mention of special inspections or other ways of detecting undeclared activities. His description of safeguards as essentially confidence-building rather than verification is culturally rather than legally determined.[254] Veteran inspector Baute explains that even without additional rights, presumably meaning special inspections and greater access,

> If the Agency had looked at open sources in the late 80's there were clear indications that Iraq was interested in centrifuge enrichment . . .

so that was something that could have been done, if only someone had the idea to push the Agency to do it. Among the footprints, of course, there were . . . ones . . . left by the procurement attempts by Iraq to have an accelerated centrifuge program through buying everywhere whatever they could buy. That was a big difference in their approach from the EMIS (the electromagnetic isotope separation) where actually security and confidentiality were their priority and pretty much it seemed that nobody knew except the core group in Iraq that were working on it, what was going on. We could have even without additional rights put the finger on the fact that something was happening, but it would have only been partial. Only a change in the rights would have made the difference.[255]

Baute's reference to the Agency needing to be "pushed" to do something reflects the passive approach of safeguards culture at the time; some "rights" already existed, but were not exploited.

Clearly the failure of the Agency to detect Iraqi non-compliance resulted from a combination of politics, organizational deficiencies, and culture. As the foregoing analysis has demonstrated, the three are intertwined, often to the point of inseparability. The utility of culture as a lens to examine the organization's behavior should nonetheless be obvious from the novel, often surprising perspectives it reveals. Such a conclusion is reinforced by the evidence of culture shock that the IAEA experienced when confronted by the legacy of its previous safeguards myopia in Iraq.

Four

Culture Shock: The Impact of Iraq on Safeguards Culture

ORGANIZATIONAL CULTURE THEORY posits that cultural change is often only possible when a crisis threatens an organization's well-being or existence. For the IAEA, that crisis was Iraq. The episode delivered several blows to the IAEA's self-image, its safeguards system, and its prevailing culture, constituting "culture shock." Cultural discombobulation was registered at several levels: the broad organizational level; in operational planning and management; and among inspectors in the field. These developments ultimately led to a self-styled "revolution" in the way safeguards were conceptualized and implemented, with inevitable implications for safeguards culture.

This chapter first considers the cultural shock experienced by the IAEA as the Iraq case unfolded. Second, it explores how this led to the strengthening of the safeguards system, especially through the so-called 93+2 process and the adoption of the Model Additional Protocol in 1997.[1] Third, the chapter considers the IAEA Secretariat's attitude toward safeguards culture in the wake of the Iraq case and examines why a deliberate, explicit strategy of cultural change was not pursued. Finally, the chapter examines how the strengthened safeguards system, even in the absence of a deliberate cultural change strategy, inevitably led to cultural change through the introduction of new or amended artefacts and espoused values and the emergence of new basic underlying assumptions, which will be discussed in more detail in Chapter 5 of this volume.[2]

Organizational Culture Shock

The first cultural shock experienced by the IAEA after the Iraq case came when it belatedly learned that the UN Security Council (UNSC) had considered bypassing it for the job of conducting nuclear verification in Iraq and assigning the task instead to a new verification body, the UN Special Commission (UNSCOM). This body, it was envisaged, would also handle the chemical, biological, and missile missions.[3] Some UNSC members doubted not only the ability but also the will of the IAEA to act aggressively enough.[4] Gudrun Harrer, author of a comprehensive account of IAEA inspections from 1991 to

1998, claims that the implied criticism was that the Secretariat under Hans Blix had been too passive and lenient with Iraq, presumably because of the Agency's culture of deference to member states.[5] British UNSCOM advisor Tim Trevan records that "certain parts of the US administration were furious that the IAEA inspectors had, in the course of their inspections in Iraq before the Gulf War, failed to notice the huge weapons program going on under their noses, details of which were now available from high-level Iraqi defectors."[6] Furthermore, U.S. critics claimed that the IAEA "had not done anything to win over critics by seemingly refusing to admit that anything was wrong with its record in Iraq. The worry was it had not learned its lesson."[7]

Particularly vocal was U.S. Ambassador Robert Gallucci, who later became UNSCOM's Deputy Executive Chairman. He claimed it was inconceivable that the Agency could handle a situation where intelligence information suggested that "if you can surround this building and cut them off and search them fast, you might find the design of a nuclear weapon."[8] The United States apparently also feared that the IAEA Board of Governors, whose membership periodically included Iraq or Iraq's allies, might sabotage the Agency's technical work by "politicizing" it.[9] The Secretariat was judged, by implication, powerless to resist such pressure. Given these views, the foremost global nuclear governance body was perilously close to being sidelined. Even the threat of such an outcome was a blow to its self-perceived status, hard-won international reputation, and organizational culture.[10] The U.S. proposal to exclude the IAEA reportedly "caused uproar in Vienna" among diplomatic missions accredited to the Agency. It was only after pressure from France, the United Kingdom (belatedly), and former U.S. IAEA Governor Richard Kennedy that the United States relented. The main argument in the Agency's favor was that it was the only body technically capable of carrying the nuclear file at short notice (it was ready within a week to conduct its first inspection).

As a compromise, to exclude the "politicized" Board and the allegedly feckless Safeguards Department, UNSC Resolution 687 of April 3, 1991, gave the nuclear verification task in Iraq to the IAEA Director General himself. Notwithstanding the novelty of this approach, Blix quickly formed an Iraq Action Team at IAEA headquarters, comprising some of the Agency's most senior and experienced safeguards personnel, as well as seconded experts from member states. The team was headed by Professor Maurizio Zifferero, a former Deputy Director General (DDG), and included Dimitri Perricos, one of the "ablest and most experienced members of the Safeguards Department."[11] The deputy head of the team, beginning with David Kay, was always from the United States.[12] The group was established outside the Safeguards Department to preserve the fiction that it was the Director General himself who was carrying out the UNSC's mandate. The team reported directly to both Blix and Rolf Ekéus, the head of UNSCOM. The on-site inspection teams that would go to Iraq included IAEA safeguards personnel, outside experts, and UNSCOM representatives. Despite having long kept nuclear weapons experts at arms length, the Agency now began employing them from certain nuclear weapon states to handle any weapons-related information that was discovered. Robert Kelley, a deputy leader of the Iraq Action Team, notes that Iraq was not inspected by "ordinary IAEA inspectors," but mostly by Western experts brought in to provide capability that was not available within the Secretariat.[13] Regular IAEA inspectors were assigned to nuclear materials verification.

The second cultural shock for the Agency came from being compelled to collaborate closely with, and in some respects act under the direction of, UNSCOM.[14] Although the IAEA would conduct its own inspections, UNSCOM would handle field logistics, call in U-2 overflights to survey potential nuclear locations, and have the right to designate inspection sites, often based on U.S. intelligence information. These novelties were an affront to the long-standing IAEA organizational culture, which covets both its independence from other organizations and its autonomy in managing its verification operations. Perricos described the initial discussions with Gallucci about the IAEA-UNSCOM relationship as "some smooth, some not," but "we came up with the guarantee that the Agency would do the nuclear part, and there would not be too many interferences from the other part of the organization."[15] Yet, as Blix puts it, "we at the IAEA felt UNSCOM sought to treat the agency as a dog on a leash."[16]

The different natures and radically different cultures of the two verification bodies made such a clash almost inevitable. The IAEA was a decades-old, traditional UN-style multilateral organization governed by its member states, employing time-worn bureaucratic and technical procedures. UNSCOM was a new, upstart verification organization with a cultural blank slate. On its first day of operation, it had two employees and no offices. It was answerable to the UNSC but led by an Executive Chairman, Rolf Ekéus (coincidentally, like Blix, a Swede), who was determined to minimize UN bureaucratic constraints and the meddling of UN member states.[17] Based in New York, UNSCOM was closer to the political action at the UN General Assembly and the UNSC and to Washington, D.C., while the IAEA was several time zones away in much slower-paced Vienna. UNSCOM was dominated—some would say infiltrated—by U.S. and other Western intelligence officers, while the IAEA team was determinedly multinational.

The legal cultures of the two organizations were also contrary; while the IAEA took it for granted, as it had traditionally done, that a country was innocent until proven guilty, UNSCOM's motto was avowedly "guilty until proven innocent."[18] An example of this was Gallucci's decision to immediately tell the UNSC, without consulting Blix, about the discovery of details about Iraq's plutonium separation activities during the fourth IAEA inspection (designated IAEA 4).[19] Traditional safeguards culture would have required due consideration and analysis of the findings before revealing them in a carefully constructed report. The U.S. administration apparently considered using this incident to seek to remove the IAEA from the Iraqi nuclear disarmament file and entrust it all to UNSCOM.[20]

The cultural difference between the two organizations regarding the utility of intelligence information was particularly stark. According to a former inspector, intelligence briefings in New York to UNSCOM staff were like "teachers feeding the pupils" compared with the "sometimes unruly atmosphere in Vienna," where presumably skepticism was rampant.[21]

As political scientists Jean Krasno and James Sutterlin note, bringing the cultures of the IAEA and UNSCOM "into some kind of working conformity was difficult."[22] Former UNSCOM inspector Scott Ritter claims that the cultural distance seemed to grow over time, as UNSCOM became more capable and confident and started "compartmentalizing away from the IAEA."[23] Paradoxically, in view of the Agency's long experience with safeguards confidentiality, this was reportedly in part because of the IAEA's perceived "cavalier attitude towards operational security and protecting confidential information," presumably meaning intelligence supplied by member states.[24] UNSCOM was apparently worried that the multinational IAEA would be prone to leaking inspection plans to the Iraqis. Ignoring the fact that UNSCOM was itself a UN body, Ritter reveals further cultural misunderstanding by asserting that:

The IAEA was an honest-to-God United Nations organization that believed in a thing called transparency: "We're the UN; we can let anybody in. We got an Egyptian sitting here. We trust him as an Egyptian. His country is part of the IAEA; they sit on the Board of Governors. Therefore, when he sits around with us, we can't hide anything from him because we have no right to hide from our Egyptian brother." Although your Egyptian brother is going to his Egyptian ambassador and reporting everything that you say and he feeds it to the Iraqis.[25]

UNSCOM insisted on acquiring its own nuclear inspectors, rather than relying on the IAEA's experienced staff.[26] For its part, the IAEA feared being second-guessed by UNSCOM in its inspection findings. Sensing the different mandates, responsibilities, and perhaps even cultures of the two organizations, Iraq attempted to play them off against each other, agreeing to cooperate with IAEA inspectors, whom they were long used to dealing with, but seeking to undermine UNSCOM's.[27]

The third cultural shock, which was probably the most unsettling to the Agency's culture, came in 1991, when the Iraq Action Team discovered evidence at Abu Ghraib, Fallujah, and elsewhere that Iraq was seeking to hide evidence of an extensive nuclear weapons program.[28] Iraq's decades-long deception of the Agency began unraveling. The Agency, culturally indisposed to believing that its safeguards system could be thwarted despite its well-known deficiencies, had difficulty accepting what it was seeing. Ekéus recalls that when "we detected components for the production of nuclear weapons . . . there was a lot of heartache before the IAEA would accept that there indeed was a nuclear weapons program in Iraq."[29]

Traditional safeguards culture assumed that the safeguards system resulted from a "gentleman's agreement"—a commonplace phenomenon in solidifying culture—whereby states would solemnly abjure nuclear weapons in exchange for a verification system that was not too demanding, intrusive, or expensive and which did not greatly affect the economic viability and competitiveness of their peaceful nuclear industries. It was further assumed that no state would go to all the trouble of negotiating a safeguards agreement only to violate it. Therefore, a case such as Iraq would be highly unlikely. A 1975 Stockholm International Peace Research Institute (SIPRI) report on safeguards questioned "the likelihood of a state which has voluntarily accepted the application of safeguards consciously breaching its obligations, rather than having recourse to the termination clause contained in each safeguards agreement."[30] It even expressed skepticism about the need for unannounced inspections at declared facilities:

> It would be a great coincidence if an inspector should catch an operator in the process of diverting a small part of his material. The "cops and robbers" approach of making frequent spot checks to detect diversion or at least deter it through the risk of being caught red-handed, if it was ever applied by the Agency, has certainly now been abandoned.[31]

Western cultural biases were at work in sustaining such assumptions. As Blix noted, in retrospect, "the original system was designed primarily with open, advanced industrial countries in mind, and was aimed primarily at creating confidence.... Over time, this system proved too weak to ensure the discovery of clandestine installations in a closed society."[32] Those who had devised the safeguards system had seemed to assume that a proliferating state would model its program on the U.S. Manhattan Project, requiring massive facilities and huge numbers of personnel that would be readily observable with or without safeguards. Even worse, there appeared to be an assumption that developing countries were too technically backward ever to master nuclear weapons technology. This has repeatedly been proven wrong, beginning with early U.S. assumptions about China. Iraq was likely assumed—except by Israel—to have been incapable of developing nuclear weapons, surreptitiously or otherwise. Such considerations contributed to a tacit assumption by Agency officials that states' declarations would be complete.

Furthermore, it was assumed that if illicit plants were built, they would be detected by means other than IAEA safeguards, notably states' own "National Technical Means," especially satellite reconnaissance or "third party" intelligence.[33] The hidden assumption was that the United States, with its superior intelligence-gathering capabilities and military prowess, would come to the rescue. This would, it was supposed, lead to a special inspection request, which would, it was naïvely assumed, be accepted and reveal all. If not, the state would be reported to the UNSC, which, led by the United States and supported by the other depositary powers of the Nuclear Non-Proliferation Treaty (NPT), the United Kingdom and the Soviet Union, would take enforcement action.[34]

Confidence in the power of safeguards created high expectations that were communicated to member states, the media, and the public, creating disappointment and even shock when the system failed to perform as advertised. Veteran safeguards expert Lawrence Scheinman observed that some contemporary critics of safeguards "appear to be guided in their assessment of safeguards effectiveness by the stringent spirit of early efforts to define international safeguards, and their approach tends to guide public perceptions of the purpose of safeguards."[35] Former IAEA officials David Fischer and Paul Szasz agreed that:

> Some of the media criticism of the IAEA reflected inflated—and therefore disappointed—belief in the power and authority of the IAEA. It came as a shock to many to learn that IAEA inspectors may,

as a rule, visit only nuclear plants notified to the IAEA, that the IAEA usually has to give advance notice of the arrival of inspectors and that it must get the government's agreement to the list of inspectors that it sends to any country.[36]

Non-proliferation expert Warren Donnelly even asserted in 1976 that on the whole, the greatest difficulty faced by the IAEA was "over-expectation."[37] The first Director of Safeguards Operations, Slobodan Nakićenović, claims that he warned Director General Sigvard Eklund about the limits of verification and personally regaled members of the Board and other member states, but to no avail.[38] He cautioned that the Director General's reports were more unconditional than the conclusions of its inspectors, who "indicated a degree of reliability based on their quantitative measurements."[39]

The fourth cultural shock for the IAEA emerged as the revolutionary nature of its new mission in Iraq became apparent. For the first time, the Agency was being required to inspect, verify, and help destroy the nuclear infrastructure of a militarily defeated country as part of enforced disarmament imposed by the UNSC.[40] The UNSC had demanded Iraq's total cooperation and transparency. Its goal was a complete accounting for and destruction of Iraq's nuclear capabilities, whether military or peaceful, either by the Iraqis themselves with IAEA supervision or by the Agency itself. The Agency was authorized to conduct intrusive inspections, at least in theory, "anytime, anywhere." More unprecedented activities emerged as the work began, including the need to search for evidence of weaponization. The Agency also had to deal with new concepts, one being UNSC Resolution 687's reference to "nuclear weapons–usable material." This contrasted with the IAEA's usual language, based on that of the NPT: "special source material," which only implied the potential use of material for nuclear weapons, as well as "direct-use" or "directly usable" material.[41] Also unprecedented was the instruction to concern itself with Iraq's research and development work, including aspects of nuclear weapons beyond nuclear material, such as non-nuclear components, and to track nuclear procurement efforts. This would inevitably lead to concerns about the weaponization of Iraq's nuclear capabilities.

Other novel aspects of the situation soon emerged. Never before had the IAEA been offered intelligence information from member states to plan inspections. Perhaps most significantly, never before had the IAEA been required to deal with a state that blatantly resisted its inspection process through prevarication, procrastination, deception, concealment, and intimidation. These circumstances were radically different from the Agency's previous experience with multilateral, collaborative verification resulting from voluntary safeguards agreements. As the IAEA reported to the UNSC:

> Iraq's persistent practice of only limited acknowledgement of activities exposed through inspections, its concealment of evidence in such critical areas as uranium enrichment and nuclear weapons development, its denial of unrestricted access to certain sites, its detaining of the Agency's team on one occasion and its confiscating of documents from inspectors have made it rather difficult for the Agency to discharge its duties.[42]

Founded to deal exclusively with the peaceful uses of nuclear energy, the IAEA's sudden confrontation with a weapons program—the phenomenon it was designed to prevent—was inevitably culturally discordant. Some people both inside and outside the Agency asked whether the IAEA's mandate permitted such activity. Some member states regarded the Agency's role in Iraq as disarmament, not safeguards, with no relevance for the existing safeguards system. There were also concerns in some quarters that the Agency's contacts with nuclear weapons materials and information risked proliferation to unauthorized individuals, groups, or even member states.

Traditional Safeguards Culture Proves "Sticky"

Indicating how "sticky" the existing IAEA safeguards culture was, the first reaction of the Secretariat to the new situation was to adhere to its tried and true approaches. Initially, it gave the impression that its standard procedures would suffice in determining the extent of Iraq's nuclear weapons program. Although Blix told the Board that the new Iraq inspections were designed to deal with a unique situation, he said that they were "inspired" by the Agency's typical safeguards procedures and techniques.[43] The Secretariat's instinct was also to employ the standard legal framework and terminology embodied in comprehensive safeguards agreements (CSAs).[44]

The first IAEA inspection, IAEA 1, was conducted from May 15 to May 21, 1991 at Iraq's declared nuclear complex at Tuwaitha, which the Agency had inspected uneventfully for decades.[45] It was, tellingly, described as a "normal" IAEA inspection, even if the "background was quite different."[46] Here we can see traditional safeguards culture and resistance to change at their clearest. The report to the UNSC on IAEA 1 was characterized by Rich Hooper, an Iraq Action Team inspector, as a "typical safeguards inspection report, cryptic and without details."[47] Kay says the report would have been understated, affirming simply that the team had seen "everything and the Iraqis have nothing."[48]

Allegedly, it was only under pressure from the three U.S. experts seconded to the group that caveats about the robustness of the findings were added to the report.[49] Kelley states that the experts feared the IAEA, especially the "old guard," was ready to "fold" after one inspection, on the basis that all nuclear materials in Iraq had been located, verified, and placed under safeguards.[50] Even after it was released, the report was criticized for the innocuous way it revealed that during the inspection, Iraq had for the first time declared that it had about two grams of plutonium; it simply noted: "Presence of recovered plutonium (2.26 g) was confirmed."[51] This gave no hint that Iraq had violated its safeguards agreement and may have been hiding more undeclared plutonium as part of a plan to illicitly accumulate enough material for a plutonium bomb.[52]

In the Agency's defense, Olli Heinonen says that Kay's complaints about the first report not containing enough information were misguided; the first inspection was not looking for "a smoking gun," but was a mapping exercise to provide a basis for subsequent inspections—a traditional safeguards approach.[53] Perricos confirms that "we decided [in] the first instance, we needed to take control over all nuclear material in Iraq. As was in the declaration from Iraq."[54]

A difference in the approaches of the IAEA and UNSCOM toward declarations being obtained prior to inspections soon emerged: Agency inspectors "never went into a building without a declaration [from the Iraqis] in hand," whereas UNSCOM made no progress until "they realized that they had to go back and get a declaration and something to inspect against." According to IAEA officials, only after three years did UNSCOM realize that "declaration is the basis of everything."[55]

Meanwhile, the IAEA began to adapt to the unfamiliar circumstances of the new style of inspections in Iraq. Its reports after IAEA 1 became more "representative of what the Action Team was learning and the work which was done," and, along the way, began conflicting further with the prevailing Agency safeguards culture.[56] The detailed reporting that the Iraq Action Team was increasingly required to provide to the UNSC was reportedly "anathema" to most of the old-guard IAEA safeguards inspectors, whose "culture did not admit to such an approach."[57] Traditional reporting culture was based on the legal requirement that information provided by a state was "safeguards confidential" and public reports should be kept to a bare minimum. Information derived from inspections should not be shared with any other entity, even one such as UNSCOM, which had been established by the UNSC. Traditionally, safeguards reports also tended not to mention verification problems until the state concerned had had a chance to redress them. In this case redress was irrelevant, as the whole program was to be shut down.

Demonstrating further how deeply ingrained the traditional safeguards culture was, the Agency briefly contemplated running two systems in Iraq—a traditional one to manage the usual tasks and a separate one to satisfy the UNSC's additional requirements.[58] The aim, in part, was presumably to keep the traditional system firewalled from the alarming precedents being established by UNSCOM. Iraq Action Team member Perricos appeared to confirm such inclinations when he recalled that:

> Zifferero insisted that I move my office to the 28th floor [at the IAEA's Vienna headquarters, where the Director General's office was located] and I insisted that I would not, because at that time I felt that the basic principles of the inspection effort in Iraq would have to come out of the Department of Safeguards. Because [it] had trained inspectors, trained, ready, knew how to work on a team, how to assess results from an inspection, to put it in perspective, how to measure nuclear material quite well. We had already the spirit and structure of a team right inside the house.[59]

Blix even appeared to assume that the normal safeguards model would be sufficient for an Ongoing Monitoring and Verification (OMV) system to ensure that Iraq did not resume its nuclear weapon program after its existing capabilities had been verifiably destroyed. Trevan notes that the IAEA's plan, drafted in Vienna and "insulated from the strength of feeling in New York," reflected the original intent of nuclear safeguards "rather than the new reality."[60] UNSCOM's plan for ongoing monitoring, in stark contrast to the IAEA's, was "an extraordinary document, giving the Special Commission powers that no UN agency had ever before enjoyed."[61] It was impossible, he said, to allow these two plans to proceed together. The IAEA plan was thus modified to bring it more into line with UNSCOM's.

Blix also initially opposed conducting searches for documentary evidence of the Iraqi nuclear weapons program, because this would be an "intelligence gathering mission," something IAEA safeguards culture did not allow. He reportedly acquiesced only when UNSCOM threatened to conduct its own inspections without the IAEA.[62] Traditional safeguards inspectors were also skeptical about using satellite imagery—ironic now in view of the Agency's current extensive use of it. Kelley recalls that "an old guard safeguards hand at my level in the hierarchy used to ask me why I bothered looking at pretty pictures. What would it tell me?"[63]

Culture Clash in the Field: Cowboys versus Bunny Huggers

Compounding the cultural challenges faced by the IAEA in dealing with UNSCOM at the organizational level, another culture clash soon emerged in the field between long-time IAEA inspectors on the one hand and outside experts and UNSCOM inspectors on the other.

Beginning with IAEA 2, which took place from June 22 to July 3, 1991, increasing numbers of personnel from outside the IAEA—either experts on loan to the Agency or from UNSCOM—joined the Action Team.[64] They were mostly from Western states. Many had no inspector training but were weapons experts, analysts, or intelligence officers. Experts from the U.S. Department of Energy laboratories, France, the United Kingdom, and UK-French-Dutch uranium enrichment company URENCO soon made up most of the team. It was they who carried out the investigations of weaponization, procurement, and centrifuges, invariably following U.S. intelligence leads. The attitude of this group was "aggressive, inquisitive and 'take charge'."[65] In contrast, Kelley records, rather disparagingly, that IAEA personnel handled traditional matters like verifying nuclear materials and fuel cycle disabling activities "like filling glove boxes with concrete."[66] The seconded and UNSCOM personnel also clearly maintained close links with their national organizations, something also anathema to traditional IAEA safeguards culture, which emphasized loyalty to the Agency.

Because UNSCOM was created to seek out and destroy all of Iraq's weapons of mass destruction (WMD) capabilities as soon as possible, it "had to be highly intrusive, in direct contrast to the IAEA's more bureaucratic and passive policy and culture."[67] Not having its own inspection rulebook, at least at the outset, UNSCOM could improvise procedures and techniques by trial and error. It thus endowed itself from its inception with a much more expansive and intrusive verification culture than the IAEA's. New, aggressive inspection methods, UNSCOM felt, "had not only to be forced upon Iraq but also imprinted in some IAEA inspectors' minds."[68] Ekéus argued that UNSCOM's adoption of a more aggressive approach was a reaction to Iraqi behavior. Perricos speculated that this may have been because in the chemical, biological, and missile domains they were dealing with military people, but these tactics were not necessary in the nuclear field, where the IAEA dealt largely with scientists, engineers, and other nuclear experts.[69]

Harrer says that the more aggressive techniques used during the now-infamous joint IAEA-UNSCOM inspection IAEA 2 produced the first major conflict between the two different inspection cultures.[70] During this operation, on June 28, the Iraqis were discovered trying to remove calutrons (machines for enriching uranium) from the Fallujah Military Transportation facility. According to the official UNSC report, "In the course of this incident, the Iraqi military did not comply with a request by the Acting Chief Inspector that there should be no movement of transport or equipment pending inspection. The Iraqi military fired small arms into the air when members of the Team endeavoured to photograph loaded vehicles leaving the site."[71]

This was the first time in its history that the Agency had experienced such physical threats against its inspectors. It was also the Agency's first experience inspecting an undeclared site (identified by UNSCOM using U.S. intelligence information) and its first experience of an unannounced, no-notice inspection. It was the second exposure to a state's attempt to blatantly deceive and obstruct its inspectors, following earlier incidents during IAEA 2 on June 23 and 24 at Abu Ghraib, when the Iraqi authorities denied the nuclear inspection team access to certain facilities and removed suspect items.[72]

In reality, the episode in Fallujah was less of an organizational culture clash than a case of one individual, IAEA team leader David Kay, paradoxically applying the UNSCOM zeitgeist to an IAEA-led inspection.[73] It was Kay, Acting Chief Inspector in the absence of Zifferero, who decided that the operation would be a "no-notice" inspection; until then, a 12-hour-notice rule had been in force. Kay also misled the Iraqis by pretending that the inspection team was headed to Abu Ghraib, not Fallujah. Scott Ritter called him "the aggressive inspector from the IAEA."[74] An American with no previous inspector experience, Kay was an IAEA Technical Cooperation (TC) administrator who had been assigned to the Iraq Action Team by Blix because "he had a talent for writing and a reputation for getting things done."[75] Back in Vienna, there was shock at what had transpired in Fallujah. Kay recalls a safeguards inspector berating him in an elevator at IAEA headquarters, saying, "What you are doing in Iraq will get one of us killed in our future jobs."[76] Kay had allegedly put their lives in danger by setting a precedent for future unannounced inspections in any country under safeguards.[77] Contrary to Kay's approach, IAEA inspector culture enjoined them to play by a strict rulebook. Kay says that if his rule-breaking inspection had not been successful, he "probably would have been hung in the IAEA."[78] Blix says, "To UNSCOM, the IAEA inspectors seemed too much like proper civil servants; to the IAEA, some of the UNSCOM inspectors seemed to act Rambo-style."[79] Mohamed ElBaradei, the Agency's chief legal advisor at the time, criticized "UNSCOM's abrasive behavior."[80]

The Agency clearly regarded itself as having a superior inspector culture to that of UNSCOM's "cowboys"—some of whom, in turn, called the IAEA inspectors "bunny huggers."[81] Perricos joked about the physical difference between the "sedentary" IAEA inspectors and those from UNSCOM, who very often came from military programs and could "jump over a fence."[82] Gallucci claimed that the IAEA, with its traditional culture, could never have exposed the whole Iraqi nuclear program.[83] As for cultures clashing, Ekéus says diplomatically, "I wouldn't say 'clashing' but [they] were in friction, because I think in the early stages, I think no one in the IAEA believed, really, that Iraq had a nuclear weapons program."[84]

Although unannounced inspections did become normalized, most of the IAEA members of the Iraq Action Team, notwithstanding Kay's precedent, never reconciled themselves to UNSCOM's aggressive mode of operation. Harrer reports that a "constructive dialogue" between the Agency and Iraqi officials that evolved during IAEA 3, from July 7 to July 18, 1991, even convinced some UNSCOM team members that "the quiet methodological process of the disarmament of Iraq" would have had the same result as the "more confrontational process actually used."[85]

Another infamous example of Iraqi obstruction occurred during IAEA 6, from September 22 to September 30, 1991, when inspectors were detained at a site in Baghdad for four days, with CNN cameras rolling, in a stand-off that became known as the "parking lot incident."[86] Here the differences in IAEA and UNSCOM cultures revolved around two issues: whether attempts should be made to locate and seize documents detailing the Iraqi nuclear program; and the role of intelligence information in guiding such document searches. In this case, UNSCOM had received intelligence information indicating that a large cache of documents relating to the Iraqi nuclear weapons program was housed at the Nuclear Design Center in the Al-Nakabat building in Baghdad.[87] According to Kay, the personnel requirements of such a mission would be unusual: "people who could go through doors," break into safes, and "do other things outside the thinking of an ordinary safeguards inspector."[88] Gallucci did not want the IAEA to have a role in the inspection, and even threatened to label it a chemical weapons inspection in order to automatically exclude the Agency.[89] Blix had "expressed his discomfort" with the idea of his inspectors participating in an "intelligence gathering mission," which IAEA practice and culture did not abide.[90] He ultimately agreed that the Agency would lead the mission and participate in it in order to avoid being shut out of the operation altogether.

Led by Kay, the large, 45-person team included only two other IAEA officials.[91] Gallucci and 27 other Americans were involved, as well as members from other Western countries and three Arab states, many presumed to be from the "intelligence community." When the operation proved spectacularly successful, definitively revealing the existence of an extensive Iraqi clandestine nuclear weapons program, Blix was forced to reconsider the value of intelligence information and documentary searches, albeit not with the forceful methods employed during IAEA 6.

Despite these initial humiliations, the Agency soon rose to the occasion, performing the nuclear part of the Iraq WMD disarmament effort creditably. Agency safeguards personnel quickly became accustomed to using intelligence information to guide their inspections, conducting documentary searches and analysis, interviewing personnel to expose deliberately concealed activities, deploying environmental sampling and nuclear archaeology and forensics, and planning the destruction of facilities safely and reliably. The Secretariat also sought to acquire necessary new skills, including the ability to:

- identify installations for production of nuclear weapons–usable materials (HEU [highly-enriched uranium] and plutonium);
- assess industrial capacities for indigenous construction of uranium enrichment and extraction plants, including inspections in non–nuclear industry sectors;
- identify potential weaponization activities (including testing of non-nuclear components of nuclear weapons);
- search for evidence of plans for a nuclear weapons research and development program; and
- remove nuclear weapons–usable material and destroy facilities.[92]

A panel chaired by Brazilian diplomat Celso Amorim, established in 1999 by the UNSC to assess the performance of UNSCOM and the IAEA in Iraq, concluded that:

> in spite of well-known difficult circumstances, UNSCOM and IAEA have been effective in uncovering and destroying many elements of Iraq's proscribed weapons programs in accordance with the mandate provided by the Security Council. Both UNSCOM and IAEA

have . . . been adopting a pragmatic approach which assumes that 100 [percent] of verification may be an unattainable goal.[93]

IAEA safeguards culture took this for granted, but the inexperienced UNSCOM did not. Amorim did not refer to clashing cultures, but recommended that "unnecessary confrontation or disproportionate reactions should be avoided."[94] His only cultural references were to the needs for an enhanced security culture and for training to improve inspectors' sensitivity toward the national cultures of inspected states. Krasno and Sutterlin conclude of the IAEA-UNSCOM relationship that "the clash of [their] two agendas and cultures were frequently played out by the two directors, but to their credit, they managed to overcome these differences, for the most part, and together support the intent of Resolution 687."[95] Relations between the two bodies reportedly improved when Australian Richard Butler took over as head of UNSCOM in 1997.[96]

The relationship between the IAEA and UNSCOM's successor, the United Nations Monitoring, Verification and Inspection Commission (UNMOVIC), which operated in Iraq from 2003 to 2007, was much smoother.[97] First, the IAEA was quickly able to determine that Iraq had not attempted to reconstitute its nuclear weapons program, so prolonged investigative interaction with UNMOVIC was unnecessary. Second, planners for UNMOVIC had evaluated and modified some of the heavily criticized aspects of UNSCOM that had made its culture so different from that of the Agency. For example, its inspectors were now employed by the organization rather than seconded by states, making them more like IAEA and other UN personnel.[98] Moreover, elements of the new inspection techniques adopted by the IAEA in Iraq, whether learned from UNSCOM or from its own experiences, were incorporated into UNMOVIC's mandate.[99] Finally, and perhaps most important, any culture clash was ameliorated by Blix's appointment as Executive Chairman of UNMOVIC, while ElBaradei, his long-standing colleague at the Agency, became IAEA Director General.

Strengthened Safeguards: 93+2 and the Additional Protocol

The Iraq case had a profound impact on IAEA safeguards, with lasting implications for safeguards culture. The changes, the Safeguards Department said, were "of such magnitude that they can rightly be characterized as a revolution, rather than an evolution, in the way in which safeguards activities are planned and implemented, the results are analyzed, safeguards conclusions drawn and follow-up activities are carried out."[100] Perricos joked that before Iraq the IAEA was called the "sleeping dog," while afterward it was the "watchdog."[101] Former Australian Ambassador Michael Wilson argued that the Secretariat became "more alert and assertive in its approach to verification and more confident that it could rely on the positive support of the Board of Governors and the wider international community should there be another challenge."[102] ElBaradei boasted that "a decade of dealing with challenges such as Iraq and North Korea had made us more resourceful and more confident. We had considerably more legal and technological verification tools at our disposal."[103]

Veteran safeguards expert Tom Shea concluded by 2008 that "all things considered, the Agency's verification capabilities are today vastly superior to where they stood when the NPT came into force, or when Iraq and the DPRK [North Korea] first violated their nonproliferation undertakings."[104] In the same year, two other renowned safeguards analysts, James Tape and Joseph Pilat, concurred that "the new IAEA system that is emerging is more flexible and should be better suited than the old to allocating scarce resources where they are needed most in countering proliferation risk."[105] They cautioned, however, that the Agency still "faces limits on safeguards inspections inherent to the agreements that authorize them." Other limits to the effectiveness of safeguards in a given state, they said, "can also stem from *residual cultural issues* [emphasis added], gaps in available technology such as wide area environmental sampling, and cost issues."[106] It is not clear whether they were referring to national cultural issues or those of the IAEA.

Initially, the Secretariat appeared to regard the Iraq case as unique, involving the enforced nuclear disarmament of a single state, but with no lasting impact on the IAEA's traditional ways of doing things. This soon changed, however, as the Agency realized that Iraq had methodically gamed the system. This undermined the Agency's culturally-embedded assumptions about the "gentlemanly" fashion in which traditional safeguards had been implemented. The IAEA was finally forced to concede that states might misuse the safeguards system's focus on diversion to camouflage undeclared nuclear weapons programs. On the positive side, the Iraq case opened a political window of opportunity that the Secretariat could leap through to remove some of the constraints on safeguards that had long hindered it.

To his credit, Blix, after initial uncertainty, soon realized—under pressure from some member states and as a matter of organizational (and perhaps professional) survival—that the Agency would have to change the way it implemented safeguards. As early as July 1991, as Iraq's non-compliance was being revealed, Blix told the Board that:

> I conclude that the lesson to be learned from the present case is that a high degree of assurance can be obtained that the Agency can uncover clandestine nuclear activities if three major conditions are fulfilled: first that access is provided to information obtained, *inter alia* through national technical means, regarding sites that may require inspection; second, that access to any such sites, even at short notice, is an unequivocal right of the Agency; and third, that access to the Security Council is available for backing and support that may be necessary to perform the inspection.[107]

In 1991, he also told the U.S. Senate Foreign Relations Committee that "the IAEA safeguards inspection system should be given sharper teeth," with enhanced access to locations and additional information, including "data obtained through national technical means, satellite cameras and other intelligence gathering activities."[108]

Between 1991 and 1992, Blix managed to secure agreement on all three of the propositions he had put to the Board, including the Secretariat's right to use additional sources of information, not just from intelligence sources; and reaffirmation of its right to seek special inspections. With regard to UNSC support, on January 31, 1992, the president of the UNSC helpfully declared on behalf of the Council that it considered the proliferation of all weapons of mass destruction a threat to international peace and security and that its members would take appropriate measures in the case of any violation reported by the IAEA.[109] The IAEA Board also agreed that states should provide design information or information on changes to nuclear facilities as soon as they decided to construct, authorize, or modify such a facility.[110]

On October 8, 1991, Blix told the UNSC that the lessons of the Iraq episode "were almost written on the wall."[111] Inspections had achieved great success because member states had provided intelligence information to the Agency. He reported that at the request of the Board of Governors, he was preparing a proposal for "a mechanism and procedure in the IAEA through which member states may make information available to a special Secretariat unit regarding the possibility of undeclared nuclear facilities in NPT states."[112] If credible, such information would allow the Secretariat to request a special inspection in that country. If the inspection was refused, the matter could be put to the UNSC.[113] The planned unit was never created, but Blix did hire former intelligence officials who were to report only to him and the DDG for Safeguards.[114]

The North Korean non-compliance case, which emerged publicly in 1993, when the Agency discovered that the state had not declared all of its plutonium production, reinforced the need to deal with the Agency's blind spot regarding undeclared activities at declared facilities.[115] In the same year, South Africa's revelation that it had developed nuclear weapons, albeit outside the NPT and safeguards, provided additional lessons as the Agency sought to verify the state's claim that it had destroyed all of its nuclear weapons and repurposed or destroyed associated infrastructure.[116]

In 1993, the Secretariat codified the results of its experiences in Iraq, North Korea, and South Africa into a series of proposals for strengthening safeguards.[117] These became known as Program 93+2, which envisaged giving the IAEA a range of new rights and tools by the time the 1995 NPT Review and Extension Conference met in two years' time (hence 93+2).[118] Although this did not happen in time, the conference extended the life of the NPT in perpetuity and urged "support for Agency efforts to strengthen safeguards and to develop its capability to detect possible undeclared nuclear activities."[119] The international politics of the safeguards issue had clearly changed, opening the window for the Secretariat to act.

Bruno Pellaud, DDG for Safeguards from 1992 to 1999, is credited by former IAEA official John Tilemann with "trying to bring safeguards out of the dark silo they had created."[120] Tilemann also says Pellaud led the cultural changes that accompanied 93+2. Tilemann recalls that Blix and his team, including Pellaud, Laura Rockwood, and John Rames, spent endless hours considering how the safeguards regime might be strengthened: "It was exhilarating to observe, and sometimes contribute to, an international team striving for a radical re-configuration of non-proliferation standards."[121]

Program 93+2 was implemented in two parts. Part 1 came soon after the IAEA's legal office determined that existing legal authorities under CSAs already gave the Secretariat the necessary legal authority to undertake additional activities.[122] This included a reassertion of the rights to seek special inspections. IAEA legal expert Rockwood exclaims that following the "epiphany" of Iraq, "sure enough, when we went back and looked at [INFCIRC] 153, there was the authority staring us in the face, all the whole time."[123]

Part 2, which included the measures for which the Secretariat required additional legal authority, came in 1997, when the Additional Protocol was agreed by the Board and states began to gradually adopt and implement it, albeit only voluntarily.[124] The negotiation of the Additional Protocol illustrated once more the limits of states' tolerance of intrusive verification measures and the need for the Secretariat to proceed cautiously in testing those limits. The Additional Protocol is appended to and amends a state's CSAs only when that state has negotiated a text that is based on the model with the

Secretariat, and only after the state has signed (and ratified if their national procedures require it) what is essentially another bilateral agreement with the IAEA (like its original safeguards agreement).

The negotiations on the Additional Protocol were complex. Two of the major objectives of states with significant nuclear industries were "cost neutrality" and "safeguards neutrality."[125] There would be no increase in the cost of safeguards in the IAEA budget, and the implementation of new safeguards measures would be balanced by a reduction in other measures. The primary interest of the developing countries was ensuring that funds needed for safeguards did not take away from funds available to the Agency for technical assistance.[126] Among the nuclear weapon states, China, and to a lesser extent Russia, were "preoccupied with ensuring that whatever new authorities were given to the IAEA would not be applied to them."[127] France, the United Kingdom, and the United States were interested in ensuring that the Agency had as much authority as possible to detect undeclared nuclear activities, and were generally supportive of the Secretariat draft. A final group consisted of the four countries that were not parties to the NPT at that time: Cuba, India, Israel, and Pakistan, which sought to ensure that whatever was negotiated did not affect them.[128]

Nonetheless, the resulting protocol produced a new safeguards standard and significantly modified the existing comprehensive safeguards system for those states that adopted it. Such states agreed to permit the Agency to use new authorities and tools, including expanded inspection rights, known as "complementary access," and to provide it with essentially "cradle to grave" information about their nuclear activities. To implement the new system, the Secretariat would develop new concepts such as "correctness and completeness," "information driven–safeguards," and eventually the State-Level Concept (SLC) and State-Level Approach (SLA).[129] The State-Level Concept set out how the Secretariat would gather much more information about each state than previously, from a greater variety of sources, as it sought to conduct enhanced evaluation of such information for each state as a whole, rather than simply considering each nuclear facility or cache of nuclear material discretely, as under the old system. This innovation drew directly on the Iraq case, as the first State Evaluation Report (SER) had been done on Iraq in 1995.[130] The Safeguards Department would also begin preparing to deploy the new verification techniques and technologies now permitted, such as environmental sampling, which had been used in Iraq for the first time.

The intention of the State-Level Approach (SLA) was that the Safeguards Department would now produce an annual assessment for each state under comprehensive safeguards, drawing on all relevant information about the state, and develop custom-made safeguards for it. The SLA was first applied, beginning in 2001, to states that were considered eligible for integrated safeguards, then to all states with an Additional Protocol, and finally to all states with CSAs. Such significant adjustments portended cultural change. For states with an Additional Protocol, there would also be an attempt to draw what was called the Broader Conclusion, an annually-renewed assessment that all nuclear material that the state had ever acquired was accounted for and remained in peaceful activities.[131] For many countries, reaching the Broader Conclusion would take years of intensive investigation, but there would be a reward at the end of it.

Once the Broader Conclusion had been reached, states were then eligible for so-called integrated safeguards, designed to rationalize the accretion of many layers of safeguards over the years to ease the verification burden on both the state and the Agency.[132] Under integrated safeguards, considerable savings were expected from reducing the number of regular inspections and relying instead on unannounced random inspections; the operator's provision of advance notification of planned facility operations; the automated collection of data such as accounting declarations and operating records; and enhanced remote monitoring.[133] After 2002, integrated safeguards reduced overall field inspection efforts by approximately 40 percent.[134] The largest reduction resulted from changes in the safeguards approach that applied to nuclear power reactors, which were the majority of inspected facilities. This savings allowed the IAEA to increase information collection and analysis activity at headquarters.

While it is beyond the scope of this book to describe comprehensively the strengthened safeguards system, Table 4.1 summarizes the Agency's enhanced powers and responsibilities.

It is notable that, far from being cowed by what had occurred in Iraq, the Secretariat continued to assert its autonomy at several key points in the process of strengthening safeguards. In 1991, Director General Blix chose to "simply inform" the Board of his intention to implement the so-called Part 1 safeguards strengthening measures. The Board, in turn, simply "took note," in effect giving its approval.[135] When it came to drafting the Additional Protocol, differences arose between Blix and then–chief legal advisor ElBaradei over whether the Board should be involved in putting the finishing touches to the Secretariat's draft. Blix favored simply presenting it to the Board in the hope that it would be noted as approved, while ElBaradei proposed that the Board be invited to help craft it.[136] The issue became moot when Canada proposed

Table 4.1. Enhanced IAEA Powers and Responsibilities under Strengthened Safeguards.

Part 1: Existing Authorities under Comprehensive Safeguards.

- collect environmental samples at any place where inspectors have access
- use unattended and remote monitoring of movements of declared nuclear material in facilities
- transmit authenticated and encrypted safeguards-relevant data to Vienna
- use, to a greater extent than previously, unannounced inspections within the routine inspection regime

Part 2: The Model Additional Protocol.

- access all parts of a state's nuclear fuel cycle, from uranium mines to nuclear waste, and any other location where nuclear material that is intended for non-nuclear use is present
- access, at short notice, all buildings on a site
- access nuclear fuel cycle research and development activities not involving nuclear material
- access manufacturing and import locations
- collect environmental samples at locations beyond those provided under safeguards agreements
- gain readier entry to states' territory through streamlined designation and multiple entry visas valid for at least one year
- use internationally established communications systems, including satellites and other forms of communication
- use wide area environmental sampling after Board approval and consultation with the state under inspection

Source: "IAEA Safeguards: Staying Ahead of the Game," IAEA, September 2007, pp. 12–13.

the establishment of Committee 24 to conduct the Board's negotiations on the text. Even so, the Secretariat draft formed the basis of the negotiations, and its staff were intimately involved in shaping it until the very end. Rockwood's remarks about integrated safeguards captures the Secretariat's attitude perfectly:

> When we first introduced integrated safeguards to the Board of Governors, again, it was not to seek the permission or approval of the Board, but to put them on notice as to where the Secretariat was going. Certainly, to give the member states an opportunity to say, we don't like that. But not formally required to get approval.[137]

Beyond the strengthened safeguards system itself, comprising new documents, techniques, and technologies, the Secretariat also engaged in organizational change to implement the operational and personnel requirements of the new system. The Safeguards Department concluded early on that strengthened safeguards required "a whole new way of working, new organizational arrangements, new responsibilities and new infrastructure."[138] Major departmental reorganization did not occur, however, until 2011, which is quite late considering that the strengthening of safeguards had begun in 1993. The reorganization was driven by two major objectives, both of which had clear implications for safeguards culture, even if this was never articulated:

- Inspectors, in their field activities, would move away from their focus on the criteria approach and nuclear accountancy to a more wide-ranging view of safeguards implementation, essentially becoming as concerned with "completeness" as "correctness."
- The Safeguards Department, including analysts and inspectors, would move from an individualistic outlook toward a more collaborative approach, as they would need to work together to implement the State-Level Concept.

The focus of structural reorganization was the Operations Divisions (A, B, and C), which plan, organize, and carry out safeguards in different regions of the world. Traditionally, sectional units within the divisions were organized by the type of facility that they inspected; one unit in a section was responsible for inspecting reactors, while another would only inspect enrichment plants.[139] A separate section was devoted to analysis rather than inspections. Different aspects of a single country's inspections could be spread across several sections—Germany, for instance, was handled by three. In the reorganization, country units combined both analysis and inspection for a single large country or for several small ones. Unit heads were now to be titled Senior Inspectors for State-Level Evaluation. State Evaluation Groups (SEGs) were established, comprising managers, analysts, and inspectors, to bring multidisciplinary expertise on each country together to enhance state-level analysis and evaluation. They would also promote collaboration between the state evaluation process and inspection activities.

The expected outcomes from these changes included increased effectiveness; enhanced ability of the system to concentrate efforts on areas of greater safeguards significance ("differentiate, not discriminate"); increased responsiveness to changing circumstances on the ground; and better use of resources for increased efficiency.[140] The new approach stressed the need for vigilance about all possible "acquisition pathways" to a clandestine nuclear weapon. A longer-term view of possible challenges to the non-proliferation regime, compared with the traditional reactive approach, was also promoted. In sum, the new safeguards approach was meant to implement the long-heralded need to be more collaborative, cooperative, internally transparent, analytically rigorous, technically ambitious, anticipatory, and watchful.[141]

Strengthened safeguards also promised to enhance the relationship between the Agency and its member states that were under safeguards. States were now obliged to cooperate more closely with the IAEA through their state or regional authorities, the designated entities which managed their State Systems of Accounting and Control (SSACs) or Regional Systems of Accounting and Control (RSACs). This sent the message that safeguards were no longer something simply "applied" by the IAEA to states, but were to be a collaborative, cooperative endeavor. The old, passive, "going through the motions" aspect of safeguards culture, on the part of both the Agency and its member states, was no longer acceptable.

As for the role of inspectors, in the words of the *INMM Chronicle*, it "changed dramatically" after the post-Iraq strengthening of safeguards.[142] There was considerable expansion in the range, number, and complexity of the inspectors' duties. As a direct result of the Iraq experience, the Additional Protocol contained a new concept called a "site," which, "if in effect at the Tuwaitha facility would have permitted the IAEA to detect the undeclared nuclear activities conducted there."[143] A "site" includes the "area delimited by [the state] in the relevant design information" and "shall also include all installations, co-located with the facility or location, for the provision of essential services, including: hot cells for processing of irradiated materials not containing nuclear material; installations for treatment, storage and disposal of waste, and buildings associated with specified activities."[144] The state must provide a map of the site where a safeguarded facility is located, with a general description of each building, including its use and, "if not apparent from that description, its contents."[145] Inspectors may seek complementary access to any place on the site and the state must provide that access, with 24 hours' notice. When the Agency is conducting INFCIRC/153 inspections on a site, only two hours' notice is required, "but in exceptional circumstances it may be less than two hours."[146]

New skills were needed by the inspectors to implement these provisions, including expert knowledge of plant operations, strengthened investigatory abilities, an enhanced ability to use remote monitoring capabilities, and language proficiency. Inspectors were encouraged to be more inquisitive, willing to engage with nuclear regulators and operators, aware of their legal rights and obligations, and more willing to take the initiative in the field rather than, as a first reaction, seeking advice from headquarters. Inspectors now had the freedom, for example, to use an equivalent piece of evidence as a substitute for another requirement that they were unable to check off their list. Furthermore, the Additional Protocol enhanced the ability of inspectors to communicate with IAEA headquarters and to transmit to Vienna information generated by its safeguards equipment.[147]

Inspectors' judgments would be increasingly involved in drawing safeguards conclusions. Australian safeguards expert John Carlson commented that "the revolutionary aspect of strengthened safeguards" was that "judgment is coming to the fore."[148] This was to be encouraged even though "conclusions about the *absence* of something—undeclared activities—can never be as definitive as conclusions based on quantitative methods applied to a finite problem—the verification of a declared inventory."[149] Former IAEA safeguards official Myron Kratzer notes that under strengthened safeguards, inspectors were expected to seek two types of information: technical information, and that which is more "subjective, maybe even impressionistic" or "observational."[150] He also described the latter task as a revolution in inspector behavior. Jill Cooley, then head of IAEA Concepts and Planning, remarked that with the advent of the Additional Protocol, the notion of the "inquisitive inspector" had emerged.[151] Training in the use of proliferation indicators and other analytical skills was provided to inspectors to enhance the state evaluation process. As a result, they were exposed to "eye-opening experiences, for instance, when U.S. personnel showed them how to do competing hypotheses analyses."[152] Michael Rosenthal and his colleagues note that as well as clandestine nuclear programs in a number of states, another factor, the emergence of illicit trafficking in nuclear material, equipment, and technology, "led to a generally held view that the Agency should shift emphasis so that inspectors should be 'less like accountants and more like detectives.' They would need to ferret out undeclared activities."[153] This would require not only putting more effort into field activities such as complementary access, but also placing more emphasis on headquarters activities, especially the gathering of information and its analysis.

The change with the greatest immediate implications for inspectors on the ground was "complementary access." Inspectors could now request such access for three purposes: to assure themselves of the absence of undeclared material and activities at a declared site; to resolve questions, including inconsistencies, relating to correctness and completeness of information provided by a state; and to confirm the status of decommissioned facilities or locations outside facilities where nuclear material is customarily used.[154] While this did not rise to the level of a special inspection or a challenge inspection, as permitted in other arms control regimes, it did expand the options that inspectors had to resolve questions and anomalies on the spot. The protocol itself says that "the IAEA shall not mechanistically or systematically seek to verify" information provided by the state, an implied criticism of the old safeguards system and culture.[155]

Strengthening safeguards since the Iraq case has been an iterative process involving organizational learning and adaptation, now stretching over more than thirty years. States took some time after 1997 to negotiate, sign, and begin implementing their Additional Protocols. The development of new concepts, such as integrated safeguards, often proceeded simultaneously with their provisional application. Sometimes new concepts were developed by the Secretariat that generated concern among member states. "Risk-informed safeguards" was a term used initially in internal discussions, ultimately replaced publicly by "information-driven" safeguards.[156] Some member states found both of these terms threatening or at least too all-encompassing, perhaps implying a reliance on intelligence information. Some developments, such as the State-Level Concept and State-Level Approach, were gradually introduced, tested, and refined, with some initial opposition by some states. Consequently, there has not been a neat linear path from the old system to the new one. Hence, cultural change wrought by strengthened safeguards has also been slow to emerge and solidify.

Resistance to Change—Cultural and Otherwise

By its very nature, organizational change is disruptive. Resistance to it may be motivated by a wide variety of factors, especially as it usually implies "winners" and "losers." Personal ambition, "office politics" (known as "games" in some of the organizational culture literature), and potential loss of status and benefits are usually involved.[157] A key element is often cultural: change and the need to relearn one's job is perceived as a threat to "the way we do things around here." As organizational culture guru Edgar Schein notes, "learning anxiety is a combination of several specific fears, all of which may be active at any time as you contemplate having to unlearn something and learn something new."[158] These include: fear of temporary incompetence; fear of punishment for incompetence; fear of loss of personal identity; and fear of loss of group membership. Defensive responses include: denial; scapegoating, passing the buck, and dodging; and maneuvering and bargaining. In many cases, unless cultural change accompanies organizational change, the affected staff simply proceed to behave much as they did before. Jim Casterton warned as late as 2005 that "vigilance must be maintained in order to identify those that may be reluctant to change and ensure that they are not placed in positions of influence."[159]

In the IAEA's Safeguards Department, three identifiable groups—inspectors, analysts, and managers—were affected in different ways by the change that accompanied strengthened safeguards. Their reactions may, at least in part, be explained in cultural terms.

The Inspectors

Within the Safeguards Department, the inspectorate—which had once been treated as "gods"—was a key barometer of cultural change.[160] It was among this group, at the "coalface" of the system, where change portended its greatest impact, where the threat to the existing safeguards culture was highest, and where resistance was most likely. Naturally there was some hankering by some individuals for "the good old days" when "inspectors ruled."

A major point of contention was the change from a strictly quantitative approach to a more qualitative one. Especially troubling to many inspectors was the decision to move away from reliance on the safeguards criteria. These criteria became increasingly formalized over time until, in the late 1980s and early 1990s, documents known as the "Safeguards Criteria" or "91–95 Criteria" were produced.[161] These were meant both to guide the inspections and to assist in evaluating them. The advantages included: the exclusion of personal judgment; the certainty gained by states and facility operators; the simplicity and uniformity of the approach; the efficiency in inspection planning and preparation; the production of unambiguous results; and the uniformity in application of safeguards to counter any claims of discrimination.[162] The traditional inspector culture had, as a result, come to favor a mechanistic, checklist approach, based on the criteria and with detailed instructions on how to fulfill them.

Conversely, the disadvantages of the criteria-based approach were that personal judgment could not be used; facility operators might take advantage of certainty about inspection measures; uniformity in reports made it hard to identify verification problems resolved in the field; inspection effort was concentrated on states with the greatest number of facilities rather than those of proliferation concern; and uniformity prevented the Agency from differentiating between different facilities and inspection situations.[163] Great attention was paid by inspectors to avoiding mistakes in gathering and recording data to fulfill the criteria, but there was no assessment made of the overall contribution of the inspection to detecting the diversion of material to illicit purposes, much less to uncovering undeclared nuclear material or activities.

Under the new safeguards system, inspectors were now expected to be more interested in "flow and process," drawing more on tools such as Acquisition Path Analysis, which investigates hypothetical schemes that a state could consider to acquire nuclear materials for weapons purposes, than in strict materials accountancy.[164] In view of what had happened in Iraq, they were also now supposed to be more concerned with the possibility of undeclared materials and facilities, not just with their list of declared items and facilities. Although the safeguards criteria were not abandoned altogether, remaining the principal verification tool in states without integrated safeguards, they were no longer to be the centerpiece of inspections but just another tool in the inspectors' toolbox. A report for the Australian Safeguards and Non-Proliferation Office noted that replacing the safeguards criteria with methodologies derived from the State-Level Approach, State Evaluation Report (SER) and an Annual Implementation Plan (AIP) for each state, while "incorporating all of the positive aspects of the criteria, has the potential to substantially improve both the effectiveness and the efficiency of the safeguards system and to result in significant resource savings for the IAEA."[165]

It was not, however, always clear to inspectors, especially in the early days, how they were expected to carry out their new role. Carlson says that fear of the unknown partly drove inspectors' early unease about the new system.[166] While some inspectors liked the idea of being more like investigators, even if they were unsure of the boundaries, others felt "at sea," not quite sure of their new roles. Fear of making a mistake and over-stepping the boundaries produced confusion and hesitation. Kay alleged that some safeguards personnel reportedly reacted "hostilely" to these changes, but Shirley Johnson says it was more a matter of confusion.[167] She bluntly says that in the beginning, without the criteria, and in the absence of anything to replace it, "it was a mess."[168] Now, she says, "almost twenty-five years later, they are realizing that some sort of 'criteria' are needed, not just for the inspectors but to explain to the member states what the hell they are doing."[169] The confusion is settling down, she says, but "they still have some work to do." Concerns among some staff that downplaying the criteria approach went too far persist to this day.[170]

Inspectors' concerns were reportedly heightened by the scrapping of the Inspectors' Manual, a detailed guide to every aspect of the inspectors' tasks that had been assembled over many decades. The aim in getting rid of it was to "deconstruct" the inspection process and force inspectors to take more initiative.[171] It was replaced by Instructions and Guidelines, comprising thousands of documents that inspectors could access online.[172] Inspectors, especially new ones, complained that they did not know where to look for relevant information and advice. Heinonen says that the manual had contained much valuable information beyond the mechanics of inspections, including how to handle anomalies.[173] Some section heads continue to this day to demand that the manual be reinstated.[174] Some began compiling their own informal handbooks, a typical culturally-driven "work around."

One of the biggest challenges was the integration of inspectors into the newly broadened verification enterprise at headquarters. The Institute of Nuclear Materials Management (INMM) noted that inspectors had gone from doing mostly verification activities at facilities to also conducting complementary access, reviewing additional state declarations, and preparing SERs.[175] Some inspectors felt outside their "comfort zone" in the new system, despite training and other efforts at acculturation. The attitude of some was: "So you want me to do inspections and all of this other stuff?"[176] Those who had become inspectors because they enjoyed the adventure of field work resented the change in their job description: they simply felt that desk work was "not what I signed up for."[177] Not all inspectors relished the opportunity to have a broader role or to be forced to collaborate so closely with non-inspectors in analytical work. Training in analytical skills provided to IAEA inspectors by the MITRE Corporation was resisted so much that a senior inspector was heard to mutter after leaving a training session: "Now back to my real job."[178] The situation was reminiscent of attempts by the U.S. Federal Bureau of Investigation to integrate its traditional field agents with a new analyst group: "That involves the delicate job of meshing the very different cultures of the streetwise agent and the brainy analyst, who reads secret dispatches, pores over intercepted communications, absorbs news media accounts, and digests it all."[179]

Other inspectors felt that their status and expertise were threatened by the new arrangements. They were once the most important element in the system. They would plan, manage, and conduct their own inspections. On their return from the field, they would normally simply file their reports. Occasionally the reports would be randomly checked to ensure that all expected tasks had been completed and that the data appeared accurate. Debriefing, if it took place at all, was cursory, and there was no intermediary "second-guessing" of the inspectors' assessments. If there were reportable anomalies or other troubling compliance issues, the inspectors themselves could, at least in theory, trigger a high-level response by the Secretariat or even member states (as provided for in the IAEA Statute).[180] No information beyond their reports and state declarations was considered when questions about anomalies arose. Now, inspectors' reports were to be just one element in assessing safeguards compliance. Their work would be scrutinized by SEGs in which they participated but which they did not necessarily dominate. The groups were even supposed to plan future inspections, which used to be the inspectors' job, albeit often informally and somewhat mechanistically.

The SLC, the SLA, and the Broader Conclusion themselves also created confusion. As late as 2014, it was not yet certain that most inspectors were comfortable with the SLC, implying that the Safeguards Department did not do a good job explaining it internally (as it had also failed to do for member states).[181] Some inspectors did not feel comfortable about drawing the Broader Conclusion for some states, such as Japan, which operate complex nuclear facilities.

There were other threats to the inspectors' domain. Initially there were significant cuts to the number of inspections conducted, as it was decided that there was more important work to be done at headquarters than in the field. Former DDG for Safeguards Jon Jennekens recalls that achieving "the level of acceptance required" on revised safeguards approaches for 1991–1995, including unified planning, implementation, and evaluation criteria, was "a major factor in the reduction of our person-days of inspection effort."[182] Despite the subsequent significant growth in the number of states with safeguards agreements, the quantities of nuclear material, and the number of facilities under safeguards, the size of the inspectorate has remained relatively constant for several years at fewer than four hundred.[183] Technological advancements, declining unit costs of key equipment, and budgetary constraints have combined to make remote monitoring devices more attractive than employing additional humans.[184]

Integrated safeguards, meanwhile, promised to lower the number of inspections conducted in states with an Additional Protocol and further encourage the use of remote monitoring and automatic data transmission to Vienna, rather than employing more inspectors. Then–Director General Yukiya Amano explained in 2019 that "partly as a result of these efforts, the number of days spent by our inspectors in the field has barely changed since 2010, increasing by just 0.4 percent."[185] Despite these developments, however, there has remained a strong lingering belief in the Safeguards Department's culture about the enduring primacy of human inspectors—sustained by the fact that humans can still outperform machines in many ways. This was demonstrated in Iraq once IAEA inspectors were given virtually unhindered access to Iraq's territory. Since the advent of strengthened safeguards, the inspectors' activities in the field have, in fact, become more extensive and intrusive, employing more sophisticated technology during their inspections and giving greater responsibility to the inspection team. In 2017, the Iran nuclear deal, the Joint Comprehensive Plan of Action (JCPOA), required 24-hour monitoring by IAEA personnel at key Iranian facilities, highlighting the importance of on-site inspectors once more.[186]

The Analysts

Strengthened safeguards have led to greater demand for analytical work at IAEA headquarters. The sources of information requiring analysis has expanded beyond safeguards-derived data to encompass open sources, satellite imagery, remote monitoring products, intelligence information, trade analysis and procurement activities, the level of international cooperation exhibited by a state, and its national political situation. New types of experts who had joined the Agency for the Iraq inspections stayed on in some cases, but many more were required, including intelligence specialists, environmental sampling experts, information analysts (notably trained in using open sources), and even nuclear weapons experts.[187] The Agency estimated in 2008 that it would need to increase its "desk evaluation" activities at headquarters by up to 50 percent by 2030.[188] The IAEA budget for 2016–2017 noted that "the increasing complexity and volume of information analysis requires an enhanced profile of expertise and explains the increased budget in this area."[189]

Analysts bring with them their own sub-culture because they have different professional backgrounds and skills than the inspectors and have stronger links with their epistemic communities than with the inspectorate. Some are from their country's national intelligence services—often without admitting as much to the Agency, which has no way of formally checking. They have markedly distinct expertise from the typical inspector, such as political science, non-proliferation studies, intelligence studies, information technology, and various technical fields like satellite imagery. They are likely to be enamored of fact-based analysis, skilled in utilizing social media, and experienced in operating in data-rich environments. Unlike inspectors, they are focused on desk work and may have had little or no field experience. When the department attempted to rectify this by giving analysts the rank of inspector and permitting them to go on inspections without doing inspection work, regular inspectors resented it as "eating up" so-called person-day inspection quotas. There was also resentment that political science majors and international relations specialists, rather than engineers or other technical experts, were being designated as inspectors.

One can immediately see the possibilities for cultural discord between the traditional inspectors' culture and that of the analysts. As one interlocutor put it, "[analysts] are always asking *about* things and inspectors are always *verifying* things."[190] Heinonen has argued that "for the IAEA to keep ahead [of] the game, it has to develop further its safeguards analytical culture," presumably meaning that analysis should be more prominent.[191] The "balance of power" in safeguards has obviously changed for good, but in the meantime, the cultural unease about the changing balance has had to be managed.

The Managers

The effort to strengthen safeguards did not initially encompass attempts to improve the management of safeguards, important though that might be. In general, UN-type organizations were the last to adopt the modern management practices that were sweeping through the business world at the time.[192] Indeed, the laudable goal of introducing the State-Level Concept to safeguards resulted in changes to the role of managers that were initially sub-optimal.

The attempt to turn middle-level safeguards managers into analysts by making them Senior Inspectors for State-Level Evaluation resulted in confusion and some resentment. Kelley points out that most of the IAEA's mid-level safeguards managers had backgrounds in nuclear material accounting and verification, but little to no experience or skills required to do state-level analysis, such as exploitation of open sources and analysis of satellite data. They viewed the staff who carried out such analytical functions as support staff, subordinate to safeguards inspectors.[193] As one account has it,

> overnight and without warning a group of people called "unit heads" were told they were now "senior inspectors for State Level evaluation" and henceforth they would direct analysis. These people who were possibly ten years into their career . . . had been inspectors and had become schedulers and bean counters for ensuring that criteria were met. They did not have any analysis skills unless by chance. They were not asked if they wanted the jobs. They did not have clear job descriptions and assurance they would meet the job requirements. They were just suddenly dropped into largely undefined new roles. So, these unhappy leaders were given jobs they were not trained to do and would not receive training to do. They were

guiding people who did not know how to do analysis and they were working for "section heads" who also didn't know about analysis, only safeguards.[194]

Ironically, the senior manager who devised the changes kept referring to the new "senior inspectors" as "unit heads," sending a strong cultural signal that nothing much had really changed.[195] Tariq Rauf reports that the SEGs themselves developed some dysfunctional practices. An unstated assumption arose after a while that the teams would not challenge other teams' conclusions, lest their own be challenged. He says that there developed an unspoken agreement between the three divisions to ensure that "no one division was seen racing ahead (or being more successful) especially with the Broader Conclusion."[196] There were exceptions, but mostly the teams began accepting other teams' conclusions at face value, which was not how their peer review was supposed to work and which amounted to a cultural work-around.[197]

Eventually, as the strengthened safeguards system was being progressively implemented, the Agency turned to improving the overall management of safeguards. This promised not only to enhance the management of the Safeguards Department as a bureaucratic entity, but also to make the new safeguards system more effective and efficient. It too would involve cultural change, as effectiveness and efficiency, as in all UN-type organizations, had not been a prominent traditional value at the Agency outside of the Department of Management.

In 2004, more than a decade after the 93+2 program was launched, and following recommendations from, among others, the Special Advisory Group on Safeguards Implementation (SAGSI) and an internal review, the Safeguards Department introduced a Quality Management System (QMS).[198] The QMS would comply with the ISO 9001:2000 standards promulgated by the International Organization for Standardization (ISO).[199] Such standards had become an internationally recognized management tool used by corporations and other organizations, including those in the UN system.[200] In a QMS Statement of November 23, 2004, the Safeguards Department said that the aim was to "enhance the efficiency and effectiveness of the Department's verification and evaluation activities through continual improvement."[201] Introducing a QMS, it said, "will force the Department to look carefully at how it does its work, how decisions are taken and to understand what it knows—and what it doesn't know. It will lead the Department to concentrate not on the conclusion itself but on how the conclusion was drawn. It will require more analysis and assessment of the 'what ifs.'"[202]

The ISO standards the department would use were based on eight quality management principles: customer focus; leadership; involvement of people; process approach; a system approach to management; continual improvement; a factual approach to decision-making; and mutually beneficial supplier relationships. Although the department claimed that the standards were sufficiently flexible to suit the IAEA's needs, some seem particularly unsuited to an international organization. The reference to "customers," as if a mere commercial transaction were involved, appeared to miss the point about the spirit of collaboration between the Agency and its member states that was supposed to characterize the newly strengthened safeguards system, not to mention the IAEA's wider responsibility to the international community and the cause of peace and security. The QMS Statement's identification of the 35-member Board of Governors as the sole "customers," rather than all IAEA member states and other stakeholders in the safeguards enterprise, could only heighten concern that the QMS was not going to be culturally transformative. The department's promise that "staff will consider the needs of their customers and will understand better where they fit into the overall process" could even be interpreted as undermining the autonomy of the Secretariat if the "customers" were limited to the Board.[203] Culture itself did get a mention in the statement but, in a familiar refrain, it was again restricted to recognizing that "if the Department is to improve how it works a culture of transparency and knowledge sharing must be instilled."[204]

At the time of the introduction of the QMS, John Patten noted that:

> The international safeguards system has changed, and is still changing as the IAEA responds to the need to draw conclusions regarding the absence of undeclared nuclear material and activities. Compared to verifying the absence of diversion of declared material this task is intrinsically more challenging and at the same time Member States require increasing confidence in the conclusions drawn. A change of culture is needed to deliver this. The Agency must move out of the comfort zone. It must move away from the familiar evaluation methods and apply different thinking to the problem.[205]

He explained that the work of the department would henceforth be "managed as a system and a set of inter-related processes, rather than as activities of staff in different units, sections or divisions," an undoubtedly laudable goal.[206]

Patten did, however, identify a logical conundrum of safeguards that QMS would find difficult to resolve. If the result of an inspection was "zero, i.e., no evidence of undeclared nuclear material or activities was found, then it is not possible to apply quality control to the result itself." The Secretariat could only be confident in the result "if it is able to confirm that the necessary and sufficient processes are in place, they are carried out correctly, adequately monitored and that appropriate feedback loops—the *check* phase in plan, do, check, act—are in place."[207] Only then could any weaknesses or shortcomings be handled correctly.

Safeguards Culture Overlooked

Change management experts recommend that significant organizational change should encompass planning to manage the impact on an organization's culture. In its 2014 report, "Managing Organizational Change in Nuclear Organizations," the IAEA's own Department of Nuclear Energy recommends that "achieving and sustaining the goals of organizational change requires that all changes be clearly connected to an organization's culture. Making this connection not only enables effective implementation, but also embeds change into the daily life of an organization."[208] The report recommends that ten elements be used to "screen" which aspects of organizational change require cultural change:

- Organizational structure
- Customs and norms
- Goals and measurement
- Rewards and recognition
- Rules and policies
- Communications
- Leadership and management behavior
- Ceremonies and events
- Training
- Physical environment.[209]

Once these elements have been identified, deliberate attention should be paid to cultural change:

> Intended culture changes call for creativity, insight, coherence and considerable persistence. Cultural change calls for receptiveness among employees to new ideas, values and meanings. Without such openness—which may be facilitated by cultural changes in society, or

growing awareness of fundamental problems in the organization—
radical, intentional cultural change is very difficult.[210]

It is unlikely that the document was even noticed, much less consulted and
implemented, by either the Safeguards Department or the Secretariat as a
whole. Because it was targeted at national nuclear organizations, it was in any
case unlikely to have been considered applicable either to the IAEA itself or
to its Safeguards Department—despite its obvious relevance to any "nuclear
organization" undertaking organizational change.

Indeed, in ignorance of the report's sensible admonitions, little thought
was apparently given by any of the Agency's stakeholders to the impact of
organizational change resulting from the "safeguards revolution" on safeguards culture and how such developments might be managed. In crafting
its organizational change strategy, the Secretariat appears to have paid almost
exclusive attention to organizational structure, goals and measurement, rules
and policies, and training—rather than more directly cultural items such as
customs and norms, communications, rewards and recognition, and ceremonies and events.

Even then, not all of the elements in the IAEA Department of Nuclear
Energy's list that were attended to, such as rule-making and training, were
dealt with comprehensively or with cultural implications in mind. Contrary
to organizational theory, revolutionary changes were made to the way safeguards operated even before new guidelines, procedures, and training were
in place. Staff had not been consulted about these changes and were unsure
about what was expected of them. Many were reportedly confused and resistant. As one insider put it, "It had been assumed that everyone understood the
general direction of change; but the devil is in the details."[211] In 2005, more
than ten years after Program 93+2 began, Cooley conceded that the Agency
was "still struggling with changing the culture."[212]

Following the Iraq incident, some outside observers had identified the
need for the IAEA to develop a new "safeguards culture."[213] Under the heading "IAEA Organizational Culture and 'Mindset," the U.S. Office of Technical Assessment (OTA) called for change in April 1995, claiming that "many
feel that the IAEA is more conservative and cautious than it should be or
needs to be, and that it cannot adapt to a new, more ambitious agenda."[214] The
report of a Commission of Eminent Persons (the 20/20 Commission), established by the IAEA in 2007 to consider the future of the Agency through 2020,
recommended that "the IAEA should take additional steps to strengthen its
safeguards culture," but left it at that.[215] Two years later, the International

Commission on Nuclear Non-Proliferation and Disarmament (ICNND) called for "cultural change in the way that [safeguards relevant] information is used," as well as recommending change "going to the whole organizational culture of the institution."[216]

Those recommending cultural change or "strengthening" of the culture, however, provided no specifics on how this might be done or in what direction cultural change should occur—apart from the idea that IAEA inspectors should act more like investigators or detectives.[217] Indeed, on the rare occasions that safeguards culture was mentioned, it appeared to refer only to the "working culture" of IAEA inspectors at nuclear facilities. Use of the word "culture" often seemed to be an attempt to add a certain sophistication to organizational analysis, but lacked any real understanding of what it implies.

Despite periodic use of the term and recommendations that it be considered, the concept of safeguards culture was not widely employed and was even contested in the move to strengthen safeguards, both inside and outside the Agency. Directions on changing safeguards culture did not emanate from the top leadership, which culture theory establishes should take the lead. Director General Blix did reportedly attempt to introduce the term "safeguards culture" after Iraq and even spoke internally in cultural terms, but he encountered resistance from the "old guard," who were puzzled by it.[218] Veteran inspector Perricos called it a "candy concept."[219] Blix made no mention of it in his public statements or writings. He retained a traditional perspective on key aspects of inspector culture. Yet he was also aware that in a regime like Saddam Hussein's, "nobody gives away information because inspectors are soft-mannered."[220]

ElBaradei, Blix's immediate successor, also did not typically speak of safeguards culture or of cultural change within the organization more widely, the one exception being his "One House" campaign to break down the Secretariat's culture of programmatic information silos, which potentially had profound implications for safeguards culture.[221] The Agency's background paper for the 20/20 Commission, initiated by ElBaradei in 2007, did contain the heading "Changing the Culture of Verification," but the word "culture" never reappeared in the text and seemed to refer mostly to the department's move toward "information driven" safeguards and the need for states to improve their SSACs, rather than to safeguards culture itself.[222]

The one area where both Blix and ElBaradei made extensive comments that touched on safeguards culture concerned the behavior of inspectors in the field, but their remarks on this tended to be reactions to what had happened in Iraq rather than efforts to recast inspector behavior for a new era. Blix concluded from the Iraq experience that:

> Inspectors should avoid humiliating the inspected. I think a Rambo-style attitude on the part of inspectors antagonizes more than it intimidates. Inspection is not the pursuit of war by other means. Inspectors are not occupiers and should neither shoot nor shout their way in. Many inspectors have told me that Iraqi scientists and technicians provided more information when they were talked to calmly than when they were bullied.[223]

Blix remained wedded to the traditional inspector values.[224] His successor, ElBaradei, offered a similar characterization:

> IAEA inspectors are not detectives, nor are they security officers or police. They are accustomed to looking for and pointing out quantitative and qualitative discrepancies—including deliberate cover-ups—and they do not shrink from confronting the party under inspection with the evidence. But their style is respectful, whether the country being inspected is Canada or South Africa, Japan or the Netherlands—or, in this case, Iraq. For my part, I firmly believe that this respectfulness, a hallmark of IAEA inspections, has repeatedly proven to be a key Agency asset.[225]

For ElBaradei, the "basics of the IAEA approach" remained "professionalism marked by tenacity and respect."[226]

Meanwhile, the Safeguards Department itself did not publicly deploy the language of culture, whether traditional or otherwise, except for the briefest of mentions. In 2005, INMM and the European Safeguards Research and Development Association (ESARDA) held a joint workshop on safeguards culture, apparently the first such event ever devoted entirely to the subject.[227] Despite its title, it mostly skated over culture in favor of the more familiar subjects of organization, process, and technology.[228] DDG for Safeguards Olli Heinonen's address to the opening session, "Views on and Expectations for the Workshop," did not mention culture once.[229] Representatives of two U.S. nuclear research laboratories urged the workshop to "clarify the vision for the future of safeguards and what 'safeguards culture' means, and to drive the dialogue between policy experts and technical leaders" (implying that this had not yet happened).[230] John Carlson, then Director General of the Australian Safeguards Office and former Chair of the Special Advisory Group on Safeguards Implementation (SAGSI), was more direct, emphasizing the need to consciously direct cultural change in safeguards and at the Agency.[231] In their summary of Working Group 1 on the Evolution of Safeguards, Cooley, head of Concepts and Planning at the Agency, and Roger Howsley, then of British Nuclear Fuels Limited, concluded that:

> "Safeguards culture" needs to be addressed if the efficiency and effectiveness [of safeguards] are to continue to be improved. This will require commitment and change at all levels, from States to facility operators. Cultural change has to come from good leadership, doing the right thing and "beliefs" are not sufficient—behavior is what counts. We are optimistic that with sufficient effort and the right incentives, change can be accomplished quickly.[232]

Rather pointedly, this conclusion did not mention the Agency, but once more appeared to be directed entirely at member states and their nuclear enterprises.[233]

Even as of 2021, there has still been no IAEA study or documentation on safeguards culture, whether initiated by member states or by the Secretariat. The Secretariat confirms that there has never been any in-house research devoted to it. SAGSI has not tackled the issue and has never been asked to do so. Safeguards culture has rated only the briefest of mentions, if at all, in IAEA promotional publications. A widely-distributed glossy 1998 publication designed to laud strengthened safeguards, "The Evolution of Safeguards," gave no hint that the new system had any cultural implications.[234] In 2007, "IAEA Safeguards: Staying Ahead of the Game" finally recognized that "major changes in safeguards implementation since the early 1990s have brought about the need for new ways of thinking, new behavior and a new culture."[235] Training and recruitment were recognized as playing key roles in such a transformation, but no further details were given.

A literature survey for this book revealed no expert or academic studies, at least in the public domain, on IAEA safeguards culture; this assessment is confirmed informally by several Agency officials. Some research papers presented to IAEA conferences do deal with the broad organizational culture of the Agency, while some work on safeguards touches on cultural issues in passing, usually without employing the term.[236] References to safeguards culture have appeared periodically in papers presented at IAEA and other conferences by national technical experts, but they deal primarily with national safe-guards culture rather than the Agency's.[237] The sole example of in-depth analysis touching on safeguards culture is Tom Shea and David Thompson's "Human Factors Affecting IAEA Inspector Performance," an unpublished report for the U.S. Arms Control and Disarmament Agency (ACDA) in the early 1980s—well before the Iraq case—which may never have been seen by the Agency. It also eschews the term "culture," although it clearly deals with aspects of it; its focus was inspector productivity rather than culture *per se*.[238]

It is telling that, unlike safety and security cultures, no attempt has ever been made to produce an Agency definition of safeguards culture, much less an internationally agreed description. IAEA planning documents have also neglected safeguards culture. While the Agency's Medium-Term Strategy for 2006–2011, prepared jointly by the Secretariat and member states in 2005, insisted that "a nuclear safety culture must be strengthened and a nuclear security culture fostered," any mention of safeguards culture was absent.[239] Culture in any realm of the IAEA's work made no appearance whatsoever in the Medium Term Strategy for 2012–2017 or in the most recent Medium Term Strategy for 2018–2023.[240]

In contrast, the Agency held a conference in 2016 that was entirely devoted to "Human and Organizational Aspects of Assuring Nuclear Safety—Exploring 30 Years of Safety Culture."[241] In 2020, it issued a Harmonized Model for Enhanced Safety Culture in Nuclear Organizations.[242] In the nuclear security field, the IAEA published an Implementing Guide to Nuclear Security Culture as part of its Nuclear Security Series as long ago as 2008.[243] An article in the IAEA Bulletin in 2016 titled, "Security Culture: One for All, and All for One," sent the appropriate cultural message.[244] While all of these efforts are directed at member states rather than the Agency itself, they indicate the degree to which the concepts of safety and security culture have permeated the IAEA's thinking in those fields.

In the safeguards area, contrary to the Agency's general organizational culture recommendations for national nuclear organizations, no one was put in charge of cultural change management for the extensive 93+2 program to strengthen safeguards. Some management experts contend that culture is so entwined with organizational change that it cannot be delegated to a single person. They recommend instead that all change managers be required to adopt a cultural perspective. In practice, so many people are uncomfortable with cultural thinking and try to avoid it that it is more effective to put a dedicated person or group in charge of inserting the cultural perspective into management deliberations. This person or persons can also devise appropriate strategies in cooperation with other change managers and monitor the impact of change on organizational culture.

Former SAGSI chair Jim Casterton of Canada was engaged by the Agency in 2011 to assist the DDG for Safeguards and the Safeguards Department in managing "the continuing evolution of safeguards implementation," but his "change management" assignment did not come with an explicit cultural change remit.[245] Nonetheless, Casterton was one of the few IAEA officials encountered in this research who spoke in cultural terms about his work. He adopted several guiding principles:

- Opportunities for staff input are essential.
- Management decisions must be implemented rather than simply promulgated.
- There is a need to recognize that there will be "casualties" in the change process.
- Fundamental changes to work practices will take time to implement consistently.[246]

In line with this philosophy, staff working groups were created to produce a work plan for change management. These groups reportedly became "significant change agents."[247] Departmental oversight mechanisms were also improved. Guidance documents were developed for internal use, especially regarding the conceptual approach to safeguards implementation. Regular briefings for staff were arranged, along with communications with external stakeholders, primarily the safeguards authorities in member states. During this consultation process, however, explicit references to changing organizational culture were restricted to modifying the persistent Agency-wide culture of "stove-piping" information.[248]

Cultural Murmurings in Strategic Planning

Even though the term "culture" was still not widely used in its initial iteration or in its subsequent revisions, the Safeguards Department's Long-Term Strategic Plan (2012–2023) is one of the most heartening recent developments in the evolution of safeguards culture.[249] The full version is not available publicly, so the following analysis is based on a summary posted on the Agency's website and discussions with Secretariat personnel. The plan is intended to be a "living" document, periodically reviewed and updated. Its methodology "provides for the data, forecasts and analysis on which the plan is based—in particular those related to the external environment—to be reviewed every two years in order to identify any developments which would require the Department to make adjustments."[250] There was a single reference to culture in the public version of the plan, a call to "review and make the necessary adjustments" to safeguards "working culture."[251] No further details were provided about what this culture was or should be and how it should be adjusted. In setting out in detail the goals and aspirations of the Safeguards Department, however, the plan systematically established new espoused values for safeguards culture for the first time. (These will be considered in Chapter 5 of this volume.) In subsequent iterations, this could be an important vehicle for signaling changes in espoused values.

In the meantime, in preparing its Long-Term Strategic Plan, the department probed the existing underlying assumptions of safeguards personnel. One of the goals was to "clarify" the "view and understanding" of safeguards personnel about how their activities related to the department's strategic objectives.[252] This process, which is recommended by organizational change experts, was said to have greatly helped staff feel consulted about pending changes. To ensure that all staff views were considered, a Strategic Planning Team (SPT) was formed, representing the whole department, not just management.[253] The group began with a Strengths, Weaknesses, Opportunities, and Constraints (SWOC) exercise. The results were culturally revealing. Among the department's identified strengths were its "expertise, know-how and safeguards experience," "multicultural, dedicated staff," and "improvements through quality management." Among its weakness were: "expertise sometimes inadequate," "recruiting difficulties," and "room for improvements in interdepartmental cooperation, coordination, communication, and culture." The SPT recommended an Overall Strategy, including "Changing Organizational Culture (More Collaborative)," which seems to indicate that culture was given some prominence in the group's "brainstorming." In addition, nine strategies were proposed, including three especially relevant to culture: "Coping with the Workload," "Changing the Way the Department Works," and "Managing the Workforce."

A presentation on the Strategic Plan by Jill Cooley at the IAEA's International Safeguards Symposium in 2010 mentioned the need to improve "departmental communication and collaboration, addressing structural, technical and cultural aspects."[254] The latter was at last a recognition that problems with intra-departmental communication and collaboration have a cultural aspect. Under the heading of "Workforce and Knowledge," she noted the need to make the department "a more attractive workplace," and that training and developing the department's current and future workforce involved "aligning associated activities with evolving needs."[255] Again, both items have cultural implications. Because the Strategic Plan is confidential, however, it is not known what steps were envisaged or taken as part of the planned strategies.

Nonetheless, the fact that the department had engaged in intensive consultations on the future of safeguards with its entire staff was unique in the Agency's history, and augured well for an approach to organizational change that is sensitive to the role of safeguards culture. As Shea and Thompson note, "it is widely accepted that the process of involving employees in determining their working arrangements in itself has a beneficial impact on morale, and improved morale is one of the commonly accepted means to improve productivity."[256] On the negative side, however, they caution that such surveys "may lead to exaggerated importance on relatively insignificant issues, and inevitably foster expectations that once employee opinions are measured, some corrective action will be taken."[257]

As envisioned, the Long-Term Strategic Plan has been updated every two years. In 2016, the Safeguards Department adapted the strategic planning processes to "make them more responsive to changes in the operating environment."[258] A 2018 revision apparently promotes elements of organizational culture relating to work practices, performance, and communication in the department; however, a public version is not yet available to help determine whether this is safeguards-specific or is more generically organizational.

A Foray into Leadership Culture

In 2013, then–DDG for Safeguards Tero Varjoranta launched several initiatives under the banner "Management Matters" in order to tackle issues of leadership culture in the Safeguards Department.[259] While privileging the term "climate" over "culture," the aim was to "diagnose the existing Departmental climate and leadership capacity." "Interventions" would then be designed "to help create a more conducive climate and enhance staff engagement."[260] To create an atmosphere that would be open to giving and receiving constructive feedback, Varjoranta bravely offered himself as the first to undergo an assessment of his leadership qualities and performance. A brief survey was sent to the colleagues who worked most closely with him; the results are not publicly known.[261]

In 2014, a wider survey was conducted in the department to gather information on staff perceptions of various issues that affected their "engagement, motivation and morale" and to compare the changes over the decade since the previous survey in 2004.[262] Over 80 percent of staff responded, with more than 175 "free format" comments. While the overall employee engagement score was "extremely positive," at 87.9 percent, concerns were expressed about a wide range of issues, such as career prospects, long-term well-being, and certain management practices. To delve deeper into these areas of dissatisfaction, six focus group discussions were held, facilitated by an external consultant and on a non-attributable basis, to create a "safe space" for staff to share their opinions openly.

As expected, some raised issues related to policies and practices affecting the entire IAEA, or, in some cases, the wider UN system. The department therefore decided to focus on areas where it could make a difference on its own, among them the strengthening of management and leadership, improving accountability, and enhancing multicultural sensitivity. Following a Leadership Competency Assessment of each departmental leader down to section-head level, a series of "developmental workshops" was held and executive coaching sessions were organized. The "Management Matters" project was certainly worthwhile in seeking to raise awareness of the importance of the "tangible and intangible" aspects of organizational culture among both leaders and staff. Yet the focus was on management and leadership, with an emphasis on individual leadership skills and competence, not on safeguards culture *per se*. Moreover, it seems that the activity was funded by extra-budgetary contributions from a member state (or states), indicating that it was not regarded by the Agency itself as worthy of regular budgetary funding and indeed had probably been instigated by a concerned member state, not the Secretariat itself. Finally, the program was explicitly framed as dealing with "organizational climate" rather than culture. According to one of the consultants involved in it:

> Changing organizational culture at the invisible levels is significantly challenging. It is relatively more feasible to address the here and now or the more visible aspects of culture that would also be referred to as "organizational climate." If culture is the personality of the organization, then climate has more to do with the prevailing atmosphere within the organization. The climate is prone to more short-term fluctuations and is determined by many factors, including leadership, structure, rewards and recognition.[263]

Apart from the fact that these seem like distinctions without a difference, there is a hint that the Safeguards Department was more comfortable discussing "climate" than "culture."

Why the Aversion to Culture?

The reasons why the IAEA Secretariat has almost entirely avoided both the language and the reality of cultural change, either for the whole organization or for safeguards specifically, are complex. Like many institutions facing organizational change, the IAEA has tended to view culture as too difficult to tackle. Instead, it has focused on "the more tangible aspects of change"—legal documents, operations, equipment, systems, and procedures—seemingly in the belief that positive cultural change will automatically follow.[264] Paradoxically, it could be that the Agency's predominant organizational culture—scientific, technical, bureaucratic, legalistic, mostly male, Western, and UN-derived—conspires to keep a "fluffy" concept like culture out of the change conversation.

Casterton speculates that one reason for the Secretariat's reluctance to talk openly about cultural change is a fear that it may be misunderstood by the Agency's culturally diverse member states and the multinational Secretariat staff as being a threat to their national cultures.[265] An example was provided by Stephen Dahunsi, a Nigerian participant in the INMM Safeguards Culture Workshop in 2016, who noted that "it is the general belief in Nigeria that cultural identity, political affiliation and personal relationships help to access and enhance a quality livelihood from one generation to another."[266] The principle of "man know man," he said, was common in hiring and promotion decisions; in Western culture, this practice is known as nepotism. While some aspects of national cultures may be a barrier to an optimal, unified safeguards culture, those in charge of organizational change at least need to be aware of them, drawing on values that will enhance the collective culture while disincentivizing those that do not. The existence of national cultures has not, in any case, stopped the IAEA from debating and promoting nuclear safety and security cultures, although again, this is directed at member states, not at the Agency itself.

Five

Contemporary Safeguards Culture

D ESPITE THE IAEA's reluctance to consider safeguards culture and devise a strategy to handle cultural change, it was inevitable that simply by undertaking organizational change to implement strengthened safeguards, the Agency brought about *de facto* cultural change. As might be expected, change has occurred in all three areas of Edgar Schein's cultural schema: artefacts, espoused values, and underlying assumptions.[1]

New or Modified Artefacts and Espoused Values

As illustrated in Table 5.1, the safeguards revolution has produced new artefacts in the legal, planning, and process categories. Collectively, these new artefacts have reinforced perceptions of the uniqueness and centrality of safeguards in the overall IAEA mission, their continuing legal and technical basis, and the enhanced powers, responsibilities, and autonomy of the Safeguards Department in running the system. The most important legal change,

Table 5.1. Safeguards Artefacts, Pre- and Post-1991.

	Pre-1991	Post-1991
Legal artefacts	IAEA Statute; Nuclear Non-Proliferation Treaty (NPT); INFCIRC/66 agreements; INFCIRC/153 agreements; Voluntary Offer Agreements; Small Quantities Protocols	Strengthened safeguards; Additional Protocols; revised Small Quantities Protocols
Planning artefacts	Annual Program and Budget; Medium-Term Plans	Safeguards Department's Long-Term Strategic Plan
Process artefacts	The "Inspectors' Document"; nuclear materials accounting; containment and surveillance; inspector reports; safeguards conclusions; Safeguards Implementation Report (SIR); special inspections	State-Level Concept; State-Level Approach; State Evaluation Groups; country officers; integrated safeguards; Broader Conclusion; open-source information; intelligence

in extending the inspection, information-gathering, and analytical powers of the Secretariat, is the Additional Protocol.[2] As for new planning artefacts, the most significant in cultural terms is the Safeguards Department's Long-Term Strategic Plan (2012–2023).[3] Finally, new process artefacts have emerged, like the State-Level Concept (SLC), the State-Level Approach (SLA), integrated safeguards, "weaponization" indicators, and means for detecting illicit nuclear trafficking.

The new artefacts of strengthened safeguards have signaled new or modified espoused values. Table 5.2 sets out the new or reworked espoused values of safeguards culture that have emerged since 1991, alongside a description of the change from previous values. Most of the traditional espoused values of IAEA safeguards culture remain intact after the buffeting occasioned by Iraq, North Korea, Iran, Syria, and other lesser-known non-compliance experiences.[4] The principles of non-discrimination and respect for state sovereignty are key examples. Both remain holy grails for member states. The

Table 5.2. Cultural Change in Espoused Values.

New or reworked since 1991	Changes from previous values
"IAEA verification contributes to a more secure world by helping to deter the proliferation of nuclear weapons and to advance States' aspirations for a nuclear weapon[s] free world."[1]	An implied role for the IAEA in nuclear disarmament was not accentuated in the early decades of the Agency—notwithstanding the promise of the Statute.
The Safeguards Department, "backed by the necessary legal authority, the required technical capabilities (expertise, technology and infrastructure), as well as adequate resources . . . works as a team to effectively and efficiently carry out the IAEA's nuclear verification mission."[2]	Inspectors had long been considered the most important component of IAEA verification. The emphasis on teamwork is new, reflecting the fact that the state-level evaluations require an integrated approach to information-gathering and evaluation.
The IAEA should be "recognized as the pre-eminent international nuclear verification agency" and have the "confidence and support of the international community."[3]	Traditionally this was implied but is now much more explicitly and assertively stated.

New or reworked since 1991	Changes from previous values
The Agency's safeguards personnel are professional, technically proficient, well trained, and dedicated to the task.	This is newly emphasized and more explicit than in the past, reflecting criticism of IAEA capabilities both before and after the Iraq case.
It is "vital . . . to detect and report early any *potential* [emphasis added] misuse of nuclear material and activities."[4]	Safeguards were traditionally designed to detect non-compliance after it occurred, not before. The Agency repeatedly emphasized that it could not verify "intentions." It still asserts this, but assessing proliferation "potential" inevitably leans in the direction of assessing "intentions."
Safeguards are concerned with verifying correctness *and* completeness; undeclared materials/activities/facilities must be pursued; all non-compliance scenarios must be considered.	Previously the emphasis was on declared materials, activities, and facilities. Diversion was considered the most likely non-compliance scenario.
All validated sources of information, including open sources, trade analysis, satellite imagery, and intelligence information from member states, should be used for drawing safeguards conclusions.	Previously, only information provided by the state and that derived from Agency inspections should be used.[5]
Continuous improvement in safeguards is necessary.[6]	The need for *continuous* improvement was underplayed, if mentioned at all, presumably to avoid alarming member states about the existing difficulties with safeguards and the cost of fixing them. In practice, improvements were occurring continuously in response to both technical and political developments.
Strategically, the Safeguards Department should be proactive, using risk assessment techniques to identify potential threats to the non-proliferation regime and to safeguards.	The old underlying assumption appeared to be that because safeguards were designed to detect non-compliance only after it occurred, the department should be reactive and not try to predict where future threats might come from.[7]

(continued)

173

Table 5.2. *(continued)*

New or reworked since 1991	Changes from previous values
Inspectors should be more inquisitive, investigatory, and innovative.	Inspections were primarily focused on nuclear material accounting and strategic points and guided by a prescriptive, criteria-driven approach which concentrated on the quantity and type of nuclear material at declared nuclear facilities and other locations.[8]
A collaborative, team-oriented approach involving inspectors, analysts, and other experts is necessary for effective verification, especially in implementing the State-Level Concept.	The old, individualistic culture, especially among inspectors, is no longer seen as appropriate in delivering effective nuclear safeguards.

Notes:

[1] Department of Safeguards, Long-Term Strategic Plan (2012–2023) Summary, 2011, www.iaea.org/safeguards/documents/LongTerm_Strategic_Plan_%2820122023%29 -Summary.pdf (hereafter Strategic Plan), p. 2.

[2] Ibid.

[3] This is an aspiration that the Safeguards Department should work toward. Ibid.

[4] Ibid., p. 4.

[5] As Fischer and Szasz put it, "it is, perhaps rightly, an article of faith for the IAEA that it cannot rely upon any verification except its own." David Fischer and Paul Szasz, *Safeguarding the Atom: A Critical Appraisal* (London: Taylor and Francis for the Stockholm International Peace Research Institute [SIPRI], 1985), p. 36.

[6] "It is vital that the Department continually improves its capabilities to draw soundly-based safeguards conclusions." Strategic Plan, p. 4. The NPT Review and Extension Conference in 1995 endorsed this, calling for IAEA safeguards to be "regularly assessed and evaluated." See 1995 Review and Extension Conference of the Parties to the Treaty on the Non-Proliferation of Nuclear Weapons [NPT] Final Document, Decision 2: Principles and Objectives for Nuclear Non-Proliferation and Disarmament, para. 11, http://www.un.org/Depts/ddar/nptconf/2142.htm. This was new, and implied the need for continuous improvement.

[7] The department's strategic planning methodology "provides for the data, forecasts and analysis on which the plan is based—in particular those related to the external environment—to be reviewed every two years in order to identify any developments which would require the Department to make adjustments." Strategic Plan, p. 9.

[8] IAEA, The Conceptualization and Development of Safeguards Implementation at the State Level, Report by the Director General, GOV/2013/38, Vienna, August 12, 2013, available at: http://armscontrollaw.files.wordpress.com/2012/06/state-level-safeguards-concept-report-august-2013.

Secretariat has made a brave attempt to promote "differentiation" in the application of safeguards to different states as being different from "discrimination," but this is still viewed by some states with suspicion, as they sound like much the same thing.[5]

The most significant and universally accepted change in espoused values is the move to "correctness and completeness," rather than just correctness. As Jim Casterton notes, the change in the safeguards approach from a uniform one that depended on the size of a state's nuclear program to one that looked at all states' nuclear programs holistically implied "significant changes to safeguards culture."[6] The steady adoption of Additional Protocols by member states has reached the point where it is becoming the norm and is therefore firmly embedded as a cultural artefact. As of December 16, 2021, Additional Protocols were in force for 137 States and EURATOM.[7] Another 14 states had signed Additional Protocols but have yet to bring them into force.

Notwithstanding the attacks by Russia and a few supporters beginning in 2013 (considered in detail below), there has also been growing acceptance by member states of the Secretariat's application of an SLA to their nuclear activities, with all the benefits that can bring in terms of the Broader Conclusion and integrated safeguards.[8] The Broader Conclusion has also become a prized artefact of safeguards culture, with approximately 70 states having received it so far. In addition, an espoused value has emerged that says states should all strive to acquire and keep the Broader Conclusion. The Secretariat added immeasurably to the credibility of the Broader Conclusion in 2019 by, for the first time, withdrawing it from a state.[9] The Australian governor on the Board noted the "important precedent and reminder that the Broader Conclusion is neither automatic nor permanent. It is based on an annual evaluation of all safeguards-relevant information available."[10]

One striking but little noted change in espoused safeguards values is the admonition in the Strategic Plan that it is "vital . . . to detect and report early any *potential* [emphasis added] misuse of nuclear material and activities."[11] Safeguards were traditionally designed to detect non-compliance after it occurred, not before. Historically, the Agency has repeatedly emphasized that it could not verify "intentions." It still makes this claim, but the move to assessing proliferation "potential" inevitably leans in the direction of assessing intentions. Strategically, the department is now enjoined to be proactive, using risk assessment techniques to identify potential threats to the non-proliferation regime and to safeguards. The old underlying assumption appeared to be that because safeguards were designed to detect non-compliance only after it occurred, the department had to be reactive and not try to predict where future threats might come from.[12]

Another newly voiced espoused value in the Strategic Plan is the need for continuous improvement in safeguards: "It is vital that the Department continually improves its capabilities to draw soundly-based safeguards conclusions."[13] Although safeguards staff have from the outset been keen to improve safeguards, and sought to do so where it was politically and technically possible, the need for continuous improvement was underplayed. This was presumably done to avoid alarming member states about the existing weaknesses of the system and the political and financial costs of fixing them. Aspiring for continuous improvement is now supported by two additional espoused values that appeared explicitly for the first time in the Strategic Plan. One is the need for effectiveness and efficiency.[14] The other is that the IAEA should be "recognized as the pre-eminent international nuclear verification agency" and have the "confidence and support of the international community."[15]

Unfortunately, member states often stand in the way of continuous improvements in safeguards. In 2005, when the Board established a Special Committee on Safeguards and Verification (otherwise known as Committee 25) to consider strengthening safeguards further, the Secretariat was able to produce overnight a list of at least twenty items for consideration. These included enhancing the Agency's satellite imagery capabilities and encouraging states to provide additional information on specific equipment and non-nuclear material to enable the Agency to better address clandestine nuclear trade. The Safeguards Department had clearly been thinking about such reforms for some time. None of them were adopted, however, as the committee descended into acrimony and, after six meetings, was concluded two years later.[16]

One legacy that the Iraq experience and subsequent non-compliance issues—including the most recent, continuing case of Iran that began in 2003—have not imparted to the Agency, either in terms of new artefacts or espoused values, is acceptance of the aggressive inspection techniques used in Iraq. Key aspects of the traditional inspector culture, notably the way inspectors comport themselves, have proved enduring. The IAEA simply concluded that aggressive inspection techniques were unprofessional, unnecessary, and even counterproductive. As Tom Shea and David Thompson put it: "for safeguards to be effective, the inspector must be sufficiently assertive to accomplish his appointed tasks and to resolve any anomalies identified. At the opposite extreme, the inspector must not be so arrogant that cooperation is inhibited."[17] Getting the balance right is in part a matter of the prevailing culture. For Robert Kelley, "In practice, it boiled down to respect and professionalism."[18] Since the Iraq episode, the Secretariat has reinforced these elements of traditional inspector culture. In any case, for normal safeguards activities, states would not abide a more aggressive approach by inspectors.[19]

Another abiding taboo in safeguards culture is the use of special inspections. Here the Secretariat continues to balance its perceived need to act deferentially toward states while seeking to use its autonomy to strengthen the safeguards regime. In February 1992, the Board of Governors helpfully reasserted the right of the Agency, under both the IAEA Statute and comprehensive safeguards agreements (CSAs), to request such inspections "when necessary and appropriate"—though naturally they could still only be conducted with the consent of the given state. The Board, however, unhelpfully added a caveat that previously had not been made explicit: that special inspections should be used "only on rare occasions."[20] This seemed to reaffirm the way that standard practice had developed, and should be read by the Secretariat as a signal that they should not suddenly regard special inspections as any more feasible than in the past.

John Carlson says one interpretation of this is that the Board simply assumed the majority of states would cooperate fully with the Agency, rendering the need for special inspections, in practice, rare.[21] The record also indicates some concern that substantial additional costs would be incurred if special inspections became more commonplace. Rosenthal and his colleagues agree that the Board's anticipation that special inspections would be rare could simply be because non-compliance itself is rare. Alternatively, it might have been "guidance to the Director General that he should not seek to undertake a special inspection except in the most egregious circumstances."[22] As Board proceedings are confidential, it is impossible to be certain, but the ambiguous language most likely reflected a compromise among Board members with starkly different views.

Faced with this ambiguous signaling, the Secretariat has never called for a special inspection since it requested one in North Korea in 1993, even when it seemed warranted on several other occasions, as in the cases of Iran and Syria. As one report understatedly puts it, despite the "affirmations of the legitimacy of special inspections when required, the IAEA has shown some reluctance to invoke the relevant provisions," leading to their "atrophy as a potential verification tool."[23] James Tape and Joseph Pilat commented in 2008 that the shortcomings of the new safeguards system were "exacerbated by the fact that the Agency does not fully use all its authorities, especially the authority to conduct special inspections."[24]

Instead, the Secretariat has drawn on its assumption of autonomy to proceed with what are sometimes called "information visits," which are special in all but name. There have been many cases where, to clarify a matter arising from a routine inspection, or to complete a task which could not be completed in such an inspection, the Secretariat, with the agreement of the member state, has undertaken additional activity in excess of that specified in CSAs.[25] The best-known instances have occurred in Egypt, Iran, South Korea, and Syria.[26] In Iran's case, more intrusive inspections have been permitted by the Joint Comprehensive Plan of Action (JCPOA), but it has not been necessary to label them "special."[27] The IAEA simply calls them "access," while Iran stresses that it is "voluntarily providing" such access, undoubtedly to save face, but also to permit it to decline such visits at any time.[28]

The right of the Agency to request a special inspection is, of course, quite apart from the right granted to the Agency by member states with Additional Protocols to conduct "complementary access" outside declared facilities and locations, which may, for all practical purposes, at least partly displace the need for special inspections. As increasing numbers of member states acquire Additional Protocols, special inspections may continue to be the rarity that the Board envisaged in its statement in 1992. This may be a case of practice and culture aligning.

The controversy over whether the Secretariat should concern itself with weaponization rather than simply with nuclear materials and facilities has largely been resolved in favor of the Secretariat proceeding to explore such cases, especially as there are several precedents relating to Iraq, Iran, and South Africa. This does represent a change in the culture. According to Rockwood and her colleagues, however, there still remains "residual skepticism about that authority, especially in connection with activities not involving the use of nuclear material."[29] They argue, though, that the IAEA need not justify its request for access to locations only on the basis that there is undeclared activity at such locations or that activity being carried out there might involve nuclear material. The Agency, they state, "may seek such access if it believes that access will contribute to fulfilling its mandate."[30] Director General Yukiya Amano was uncharacteristically forceful in insisting in April 2019 that:

> For me it is clear that the Agency has the authority to look into weaponization because our mandate is to prevent the diversion and misuse of nuclear material. If the Agency becomes aware of possible weaponization activities in a country, and these could be related to nuclear material or the nuclear fuel cycle, then the Agency seeks clarification

from the State under the comprehensive safeguards agreement and/ or additional protocol of the country concerned. Our authority in such cases is clear.[31]

Underlying Assumptions

Cultural theory contends that significant changes to an organization's existing artefacts and espoused values will bring changes in underlying assumptions, whether intended or not. Cultural theory also holds that employees will take their cues from leaders and managers as to whether new artefacts and espoused values are mere window-dressing or whether the culture is really supposed to change. A strong indication of whether cultural change is really occurring is usually found at the lower levels of an organization, where passive resistance and "work-arounds" in response to change will be most evident.

At the IAEA, the way in which the radical changes involved in strengthened safeguards were implemented at the "working level" would largely determine their impact on safeguards culture. In 2005, an INMM/ESARDA workshop stated that a "new IAEA safeguards inspector is emerging."[32] It found "the change in safeguards culture to be underway," and was "optimistic that with sufficient efforts and the right incentives the change can be accomplished quickly."[33] Certainly this accords with the new, much repeated Agency mantra, mentioned at the workshop by John Patten, the Safeguards Department's Quality Manager, that "the absence of evidence is not evidence of absence."[34]

Today, thirty years after the post-Iraq safeguards "revolution," there is unlikely to be any significant overhang of the old culture, not least because of retirements and generational change. Most of those involved in the Iraq case have retired, and most current safeguards personnel have not known anything other than "correctness and completeness." Yet, because culture is constantly evolving, new issues are emerging in safeguards culture. One is the danger that the culture will ultimately tip toward an obsession with completeness, which is a more challenging and interesting task, and downplay the importance of correctness, which is now perceived as a mundane, routine matter. Another is the possibility that mutual peer reviews by State Evaluation Groups (SEGs) will become tainted by the desire of each to have their conclusions accepted. Not "rocking the boat" can lead to counterproductive "group think." Such issues can only be dealt with if there is proper appreciation of the role of culture.

Research for this study nonetheless indicates continuing skepticism about the importance of the cultural perspective among safeguards staff. Reactions within the Agency to mentions of safeguards culture continue to range from puzzlement to apprehension to derision. A senior safeguards official remarked, after a version of this paper was presented at the IAEA Safeguards Symposium in 2014: "Is this really culture? Isn't this just what we do?"[35] In 2016, a second INMM workshop devoted to safeguards culture put emphasis almost entirely on national safeguards culture, with no recognition that the IAEA itself has a safeguards culture.[36] The IAEA representatives, like the other participants, seemed at a loss to define safeguards culture and retreated to the safer ground of safeguards techniques and technologies rather than the trickier issue of culture.[37] Tellingly, a report on the workshop concluded that, regarding culture, "there is no international organization that champions the concept."[38]

Without validated data from surveys that canvass the opinions of Agency personnel, it is impossible to confirm the current underlying assumptions of IAEA safeguards culture. Preferably such surveys should be part of an independent organizational culture study by qualified professionals with access to confidential documentation. This could confirm the veracity of the presumed underlying cultural assumptions considered in this book. Anecdotes and stories, personal biographies, and clues in official pronouncements are insufficient. The persistence of deep cultural assumptions and anecdotal evidence of their effects on behavior point to the utility of a systematic survey and study of the Agency's culture. As Director General Mohamed ElBaradei's submission to the 2008 Commission of Eminent Persons on the Future of the Agency to 2020 (known colloquially as the 20/20 Report) noted, "The IAEA should perform regular in-depth assessments of its safeguards culture—just as it recommends that nuclear facilities regularly assess their safety culture—and take steps to correct any weaknesses identified."[39] Such surveys are particularly valuable when organizational change is being contemplated, but are also a useful way to determine whether cultural change is occurring that is contrary to the espoused values of the organization. Various methodologies are available, including "functional analysis," structured interviews, focus groups, behavioral observations, and use of the Behavioral Anchored Rating Scales (BARS).[40]

Issues in Contemporary Safeguards Culture

Even without a scientific survey of attitudes and values among safeguards personnel, it is possible to identify several challenges and dilemmas facing contemporary safeguards culture. Some aspects of organizational culture and practice exist across the IAEA Secretariat and cannot be tackled by the Safeguards Department alone, no matter how well its strategic planning and change management approaches work. Others reside below the departmental level, within divisions or sub-cultures of the department. Others operate at the level of individuals. The following discussion will necessarily range across these different levels.

Organizational change throughout the Agency, preferably accompanied by a cultural perspective at the highest levels of the organization, supported by the Board of Governors and the general membership, is necessary if long-standing safeguards practices are to change. Unfortunately, neither the general membership nor the Board of Governors yet seems cognizant of the safeguards culture issue. The Medium-Term Strategy 2018–2023, which was adopted by the Board and the General Conference, does not mention cultural issues.[41] This document does recognize the need to attend to safety and security cultures, demonstrating that it is not the concept of culture *per se* that is at issue, but rather the lack of recognition that safeguards culture exists and must be systematically considered in the same way as safety and security cultures. For this to change, member states that take safeguards culture seriously will need to press for Board recognition of the benefits that can come from greater attentiveness to it.

Agency Morale: Pride and a Certain Defensiveness

One of the ways to build a unified safeguards culture is to cultivate a dedicated, permanent safeguards workforce that is valued for its skills, sees the possibility of career advancement and promotion, and has a long-term commitment to the organization. Such an environment is obtained "through efforts to understand employees' needs and desires" and then "adjusting the organization, management practice, rewards, opportunities, and the nature of the work itself to satisfy those needs."[42]

Today the Safeguards Department is especially proud of the way it quickly rebounded after 1991 from the shock of Iraq's duplicity and adapted itself to handling the new style of verification regime imposed there.[43] Safeguards personnel brim with satisfaction over how the Agency deftly demonstrated U.S. intelligence information to be wrong in the lead-up to the 2003 coalition invasion of Iraq, especially over allegations that Iraq had illicitly imported yellowcake, outside of safeguards, from Niger.[44] The IAEA's quick discovery that the Niger documents were fake is portrayed as the triumph of an international organization over an over-bearing superpower. Safeguards personnel are also pleased with the way they successfully learned from their experiences with Iraq, as well as those with South Africa, North Korea, Libya, and, more recently, Iran, in strengthening safeguards.

The award of the 2005 Nobel Peace Prize to the Agency and Director General Mohamed ElBaradei boosted the Secretariat's confidence, especially that of the Safeguards Department and its inspectors, vindicating their professionalism and defiance of those who sought to misuse or disparage safeguards.[45] David Waller, Deputy Director General (DDG) for Management and Budget from 1993 to 2011, notes the "enormously important impact" of the Nobel Prize on staff morale. When "things got a little rugged," he would remind staff of it.[46] The award has also become a cultural artefact, visibly affirming the hitherto somewhat nebulous link between safeguards and peace, as noted in the citation from the Nobel Peace Prize Committee. It also signifies the Agency's unique status *vis-à-vis* other UN organizations at the Vienna International Center (VIC). The medal now sits on proud display, strategically placed in the IAEA office's atrium on the way to the cafeteria.

Behind this pride, though, there remains considerable defensiveness among safeguards personnel when questioned about their culture. This is no doubt due to past criticisms of safeguards, continued skepticism about their efficacy in some quarters, and the sensitivity of dealing with a series of complex non-compliance cases. Myron Kratzer contends that the Agency's initial implementation of comprehensive safeguards was "very defensive" due to states' apprehensions about the new system.[47] Having to deal with sensitive, security-related issues involving member states, along with the higher diplomatic profile and frenetic media attention these episodes bring, is bound to produce caution and a certain defensiveness in any organization.[48]

Tariq Rauf speculates that the Secretariat's defensiveness also comes from an unwillingness to have its past decisions and practices examined too closely and a certain embarrassment about the Agency's lack of expertise at critical junctures. Most safeguards personnel are from non–nuclear weapon states (NNWS) and therefore have no experience with the weaponization aspect of nuclear proliferation. The Secretariat has thus had to rely on experts from nuclear weapon states with precisely such expertise in nuclear weapons research, design, and production.[49] These experts are clearly in a more exalted category than the regular inspectors and have a certain mystique about them, even though regular inspectors are usually nuclear engineers with a basic understanding of how nuclear weapons work.

The Iran Case   A certain defensiveness within the Agency seems to have been resurrected by the Iran case, which has now dragged on for nearly two decades. Attacks on the Agency by Iran after its non-compliance was first revealed in 2003 have only reinforced a culture of caution. Iranian delegates badgered safeguards personnel at meetings, questioned their impartiality and technical competence, and attempted to inject complications into areas of the Agency's work that had nothing to do with its case. Iran also helped railroad the Board of Governor's Special Committee on Safeguards and Verification, which, as mentioned above, survived for just two years (2005–2007), thereby precluding consideration of potential safeguards improvements that the Secretariat had long sought.[50]

Although direct Iranian attacks on the Secretariat ceased with the negotiation of the JCPOA in 2015, withdrawal from the agreement by the Trump administration in 2017 led to renewed tensions between the Agency and Iran.[51] As Iran began to renege on its undertakings and its compliance began to fray from mid-2019 in response to the U.S. withdrawal, the Secretariat has had to adjust its safeguards activities in the country, with the potential for disagreement about which should continue and which should end.

In February 2021, the Iranians announced that, in accordance with a law passed by the Iranian parliament, they would stop implementing "voluntary" transparency measures under the JCPOA, including short-notice inspections outside of declared sites, until the United States dropped sanctions imposed against Iran by the Trump administration and rejoined the JCPOA.[52] IAEA Director General Rafael Grossi felt compelled to go to Tehran immediately to salvage whatever parts of the agreement that he could. He succeeded in negotiating a "temporary understanding" to enable the Agency to continue with its verification and monitoring activities for three months.[53]

As with its piecemeal approach to violating the JCPOA constraints on its nuclear activities, Iran is attempting to ramp up pressure gradually by whittling away IAEA verification modalities piece by piece, while insisting that all changes can be quickly reversed. The Iranians are being careful not to do lasting damage to their relationship with the Secretariat, which would be bad for their international image, but they clearly see it as a pawn to be used to pressure the United States. As of late 2021, President Joe Biden's administration is grappling with how the United States should re-enter the JCPOA in return for Iran resuming full compliance with its obligations under the agreement. Once again, the IAEA Director General and his Secretariat are involved in a delicate dance as they attempt to satisfy the divergent demands of member states while protecting the Agency's autonomy and ensuring impartiality.

Russia and the State-Level Concept   The most striking recent example of Secretariat-led change is the State-Level Concept (SLC).[54] This gave rise, in turn, to the most serious challenge to the autonomy of the Secretariat in years. In 2013, Russia, accompanied by other states including Iran, Brazil, South Africa, and, most surprisingly, Argentina, inexplicably began questioning the right of the Secretariat to develop the SLC, which it had begun in 2004, without, they alleged, keeping them informed.[55] This was unexpected because Russia itself is not subject to the same comprehensive safeguards as NNWS. Russia sought not just increasingly detailed explanations bordering on pedantry, but also tried to obtain the release of working documents the Secretariat had prepared for itself while devising the new concept. Part of a wider drive by President Vladimir Putin to assert Russian power and influence in international organizations and elsewhere, the campaign by Russia to rein in some aspects of strengthened safeguards put the Safeguards Department on the defensive. Russia and other states were able to take advantage of the Secretariat's unveiling of its new approach without sufficient consultation with member states. Safeguards personnel subsequently acquitted themselves well in providing exhaustive briefings and explanations and in satisfying demands for greater transparency, but this nonetheless rattled the Secretariat's confidence.[56]

Concerned with rebuilding the "Vienna Spirit" of cooperation and compromise through U.S.-Russian expert dialogue, a 2020 joint report on the future of safeguards by the Washington-based Nuclear Threat Initiative (NTI) and the Moscow-based Center for Energy and Security Studies (CNESS) has called for greater openness and transparency from the Secretariat in reporting to the Board of Governors. In what might be taken as a challenge to long-standing Secretariat prerogatives, however, it also suggested that in further developing the SLC, the Secretariat "should be encouraged to provide more detailed status reports to the board on the work as it develops, as opposed to presenting it as a *fait accompli* after implementation begins."[57]

The clash with Russia over the SLC speaks to the Secretariat's traditional culture of opacity, as well as its broader culture of autonomy in managing the safeguards system. Ironically, it is argued by some observers that the Secretariat brought the entire incident on itself by seeking to promote what it was already doing as a newsworthy new approach. If the idea was to highlight its creativity in further strengthening safeguards implementation and demonstrating its autonomy, it backfired badly. The outcome can only have added to the perception in the Safeguards Department that a constant battle must be waged to protect its hard-won achievements and the autonomy that it has carved out in managing and improving the system.

In other areas, the Safeguards Department has quietly reasserted its autonomy. Its Strategic Plan was devised and has been periodically reviewed at its own initiative without consulting member states; the department simply informed them of its existence. The plan itself has not been released to member states, and is available to them and to the public only in summary form. The Safeguards Department, like other parts of the Agency, routinely reorganizes itself without seeking prior approval of member states. While portrayed as merely bureaucratic tinkering, such changes can have potentially significant implications for the way safeguards operate and for safeguards culture.

As for the sensitive issue of non-compliance, the Secretariat has traditionally not brought safeguards implementation problems to its governing organs at an early stage, but rather has approached the state in question in order to try to resolve the matter quietly, using "soft" approaches, like requests for further information or "transparency visits."[58] Ultimately it is the prerogative of the Director General whether to report a case of non-compliance to the Board of Governors, giving him considerable power and latitude. As Berhanykun Andemicael and John Mathiason put it:

> the statutory delegation of authority regarding verification from sovereign States to the Agency's governing bodies has been followed by a tendency in the Board to allow considerable leeway on compliance issues to the Director General, [enabling] the latter to emerge as a strong chief executive and inspection manager who acts with considerable autonomy.[59]

Transparency about Safeguards Effectiveness

The Secretariat has traditionally been reluctant to debate the effectiveness of safeguards, despite persistent prodding by critics to do so.[60] "Defending the IAEA's role and reputation," Olli Heinonen says, "is a natural reaction by its management but this can also have its limitations—resulting in oversensitivity [about] weaknesses exposed."[61] The Secretariat clearly recognizes, for instance, that information about bulk-handling facilities, such as plutonium reprocessing plants, is difficult to verify because of the large amounts of material moving through them, but it will not be drawn into specifying exactly how verifiable they are. The Secretariat will also not enter into debates on revising downward the decades-old definitions of "significant quantity" of nuclear material required for a nuclear device, although the Agency did briefly look into the question after the Iraq episode and concluded that there should be no change.[62] Moreover, despite renewed emphasis on "correctness and completeness" since the Iraq case, the Agency will not publicly characterize, much less seek to quantify, its abilities in detecting undeclared materials and activities. Carlson calls this "a potentially serious problem: avoiding unrealistic expectations of what safeguards can deliver."[63] Over-expectation has long been an element of safeguards culture.

No organization likes to admit its limitations, but there are particular additional considerations operating here. The Agency's stance is partly driven by the fact that verification can never be 100 percent effective, because it is impossible to prove a negative. Beyond that is the eternal verification dilemma posed so well by Allan Krass: "How much is enough?"[64] Barring a hyper-intrusive, lavishly funded verification regime that no state would countenance, the Secretariat is unable to guarantee that all undeclared materials and activities will be detected. All it can offer are "credible assurances." Hans Blix was adamant:

> 100 percent assurance cannot be achieved. Accordingly a so-called "clean bill of health" can never be issued by the Agency. Member states will have to do with reports that after such and such extensive verification measures, the Agency has found no evidence suggesting the existence of undeclared material or activities which should have been declared.[65]

Luckily, effective verification benefits from a degree of uncertainty on the part of potential violators about whether their illicit activities will be detected. While a verification agency may be unable to precisely calculate the power of its verification system, neither can potential violators. Deterrence is thus strengthened.

On a more practical level, delving into the issue of how much fissile material is needed for a nuclear weapon—the estimated quantities have generally been assumed to have dropped over the years—could involve the Agency getting close to discussing classified information. Officially lowering the quantities would also imply the need for a more intensive and expensive safeguards regime.[66]

A reluctance to discuss verifiability and adverse safeguards findings is fed by a Secretariat-wide predisposition toward unnecessary secrecy and non-transparency. Although many, if not all, bureaucracies exhibit such tendencies—after all, knowledge is power—in the case of safeguards, this is heightened by the legal requirements of safeguards confidentiality and by deference toward its member states. The Safeguards Department's Strategic Plan seeks to address such challenges, which is a clear recognition that a problem exists.[67] Indeed, the goal of greater transparency is given a whole section in the Strategic Plan.[68] The department admits that its failure to satisfactorily and fully explain the SLC in recent years contributed to the dissatisfaction of certain member states with the Agency's openness and transparency. An external auditor has notably criticized the lack of transparency regarding the touted cost savings resulting from integrated safeguards.[69]

Rockwood and her colleagues suggest that "challenges to the IAEA's authority stemming from States' mistrust of the Secretariat can be ameliorated with transparency, consultations and messaging that underscores a safeguards relationship characterized by partnership rather than contestation."[70] Carlson advocates a safeguards audit function—"a small group of trusted experts" reporting to the Director General, who could review the way safeguards decisions are taken in specific cases and, where appropriate, make recommendations on process.[71] Carlson understands that such an arrangement operated in the 1980s, "and it could have a useful role today where states are looking for assurance on the directions in which safeguards practice is evolving."[72]

An internal audit body for the Agency as a whole was formed in 2011 to provide "objective, independent and systematic assessments, helping the IAEA to efficiently and effectively carry out its mandate and ensure accountability and good governance."[73] As its remit covers all Agency programs and activities, however, as well as investigating alleged fraud, corruption, and misconduct, it is unlikely to be able to focus much attention on safeguards. In any event, its findings, which are shared annually with member states through the Board of Governors, are otherwise confidential, so it is unable to contribute to Agency-wide or external transparency.

While not exactly what Carlson had in mind, the department also established a State-Level Effectiveness Evaluation Review Team (SEERT) in 2018; the SEERT is a dedicated departmental team performing a review in the areas of planning, developing, implementing, and concluding safeguards activities.[74] Consisting of cross-cutting expertise in the Safeguards Department, it is "ultimately responsible for providing the DDG with an additional level of assurance on the correctness of safeguards conclusions."[75]

Atoms for Peace and Development: Mixed Messages about Safeguards

It is a cultural paradox that despite the obvious importance—some would even say preeminence—of safeguards among the IAEA's various missions due to its singular contribution to international security, the Agency does not generally portray safeguards as its most important role. Although safeguards is the best-funded Agency program (38 percent of the regular budget) and has the greatest number of staff (approximately 30 percent), the IAEA does not, in its external publicity, accord notable attention to the enterprise.[76] The IAEA website treats all IAEA missions equally. The fiftieth anniversary history of the Agency that was published in 2007 devoted much more attention to the promotion of the peaceful uses of nuclear technology than to safeguards, which were relegated to the back of the book.[77] In the Agency's annual reports and program and budget documents, safeguards are not given any special attention. Nuclear verification is traditionally Major Program 4, not 1.[78] Even the word "verification," which is much more readily understandable by non-experts, appears to be of relatively recent origin in the Agency's lexicon: the more obscure "safeguards" has been traditionally preferred. The Agency's Medium Term Strategy 2018–2023 lists safeguards as Part E, just before management issues. It blandly calls the section "Delivering Effective and Efficient Agency Safeguards," rather than "Verifying Compliance with Non-proliferation Objectives," or, even more appropriately, "Contributing to the Non-proliferation of Nuclear Weapons."[79]

Accolades are generally not showered on safeguards personnel. The 2005 Nobel Peace Prize ceremonies in Oslo and Vienna were a notable exception, but the focus on the Agency's verification role was likely due to the way the Nobel Committee framed the award.[80] Public speeches by Directors General lauding IAEA safeguards inspectors, despite the often difficult and sometimes dangerous work they do, were rare.[81] This has changed with the latest Director General, Rafael Grossi, who has lauded the inspectors and support staff for their "incredible resilience" during the COVID pandemic and reported that "we worked tirelessly to help all our staff to do their jobs and care for their families, and to get the respite they needed when they could."[82]

The proverbial visitor from Mars who landed at the Vienna International Center would have no idea that safeguards are so important. Too much media attention to inspection activities in Iraq, North Korea, or Iran is almost seen as unseemly, perhaps because it risks detracting from the serious, impartial, technical work required. There also seems to be a reluctance to privilege the work of inspectors over that of other Agency employees. Although the Agency has on occasion used the term "nuclear watchdog," it dislikes it, presumably because it gives too much weight to the safeguards and inspections part of its mandate, as well as implying that the Agency is engaged in enforcement rather than verification.[83] The public relations activities of the Agency emphasize the peaceful uses of nuclear energy rather than safeguards, and seem targeted as much at member states that are developing countries as the broader public.

Downplaying the prominence of the safeguards mission is ultimately political in intent—designed to avoid highlighting an activity that some influential member states, such as Brazil, Cuba, India, Iran, and Pakistan, do not see as a priority and are reluctant to pay for. Some of these states still regard safeguards either as a pointless Western preoccupation or as a plot to stymie their pursuit of peaceful uses of nuclear energy—or ultimately, in a few cases, the acquisition of nuclear weapons. More appealing to the developing-country membership is the Agency's technical assistance, such as the use of nuclear techniques to eliminate the tsetse fly in Africa, improve clean water accessibility, and increase crop production. Safeguards are still regarded by some as a problem only in developed countries. This in turn is tied up with budgetary struggles about where to allocate the Agency's finite resources.

As Carlson puts it, IAEA Directors General must constantly be "looking over their shoulders" at the developing countries.[84] Today, with the growth of their membership within the Agency, no Director General can be elected without their support. Former Director General Amano was particularly attuned to this dynamic, beginning his tenure with a pledge to redouble the Agency's contribution to using nuclear techniques in the fight against cancer and advocating an active Agency role in attaining the UN's 2015 Sustainable Development Goals.[85] He said openly that he wanted "to change the widespread perception of the Agency as simply the world's 'nuclear watchdog' because it does not do justice to our extensive activities in other areas, especially in nuclear energy, nuclear applications, and technical cooperation."[86] In 2017, he announced that the Agency's "Atoms for Peace" slogan would be changed to "Atoms for Peace and Development."[87] In fact, "Atoms for Peace" itself had only recently been resurrected as the Agency's slogan, after being retired when nuclear power generation lost its gloss in the 1980s and the proliferation effects of the Atoms for Peace program became increasingly evident. The revival of that slogan was designed to reassure developing countries that the Agency was giving greater prominence to what they perceive as its primary mandate—assisting them with nuclear technology. Adding "Development" drives the point home further.

Downplaying the importance of safeguards, however, comes at a cost in terms of engendering international public support for the Agency through its safeguards and disarmament work. A senior IAEA official has noted that "if you ask the average person if they have heard of the IAEA they shrug, but if you ask them have they heard of the nuclear watchdog they tend to know what you are talking about."[88] Failing to take advantage of this public awareness of the Agency's inspection activities also does not comport with the Strategic Plan's ambition for the IAEA to be the "pre-eminent international nuclear verification agency."[89]

Paradoxically, as mentioned by several interlocutors during the research for this book, the Safeguards Department is widely regarded within the Agency as receiving favorable treatment in the allocation of political, bureau-cratic, and financial resources from both the regular budget and the states' voluntary contributions. While the Agency portrays itself to the outside world as being mostly concerned with peaceful uses of nuclear technology, safeguards is considered favored internally, in terms of both funding and the attention given to it by the media.[90] The view is that "what the Safeguards Department wants, it gets"—presumably because the Agency's major funders, the United States and its allies in the Geneva Group and the Vienna Group of Ten, want it that way.[91] They can do this through the regular budget and extra voluntary contributions and by providing cost-free experts. Developing countries perceive a danger of extra-budgetary funding not only driving the direction of safeguards but also dictating it.

Leadership and the Cultural Tone of the IAEA

Although cultural formation is both a top-down and a bottom-up phenom-enon, an organization's leadership plays a critical role in culture formation. Cultural signals from leaders reflect not only organizational imperatives but also their own personal, professional, and national proclivities. The Director General of the IAEA thus plays a critical role in setting the general tone of organizational culture and in framing safeguards culture, whether deliberately or inadvertently. The role of the Director General has become increasingly important since the Iraq case, as the Agency has been buffeted by highly vis-ible international crises involving safeguards non-compliance controversies and nuclear accidents. The 24-hour news cycle has even elevated the Director General to the erroneous status of "chief weapons inspector" or the exagger-ated one of chief multilateral nuclear negotiator.[92] This personalization of the Agency began with the Iraq issue under Blix and ElBaradei and has continued with the Iran case under ElBaradei, Amano, and Grossi.[93] For example, Direc-tor General Grossi's urgent trip to Tehran in February 2021 in an effort to salvage parts of the IAEA's verification activities under the JCPOA propelled him instantly into world headlines just over a year after taking office. Setting the tone at the top is consequently more important than ever.

Blix was a strong supporter of safeguards and succeeded in ensuring that the Agency reacted creatively to the cases of Iraq, North Korea, and other states that challenged the safeguards regime. He also supported an expanded role for the Agency in other nuclear matters, making the ultimately unsuccess-ful case for the IAEA to host the verification system for the Comprehensive

Nuclear Test Ban Treaty and for applying safeguards to "surplus" nuclear material released from nuclear arms reductions.[94]

ElBaradei was a highly capable lawyer with a long career at the Agency before he became Director General. He joined the Agency in 1974 and became its legal advisor three years later.[95] He knew its shortcomings well and was avowedly interested in reform. He was especially determined, he says, to defend the rule of international law and to ensure that procedures and policies were not ruled by exceptions.[96] He is said by some observers to have "followed through on Blix's plan to fundamentally change the IAEA's culture," although the public evidence for this is slim.[97] He promoted the "One House" idea, designed to promote better inter-Agency communication and information-sharing, and established the Commission of Eminent Persons on the Future of the Agency (the 20/20 Commission), but his personal predilection for secrecy was at odds with One House, while the 20/20 report was filed away without appreciable follow-up action.[98] ElBaradei was regarded in some quarters as somewhat "Pharaonic" in his management style.

While supportive of the Agency's safeguards mission, ElBaradei was allegedly more remote from the safeguards enterprise than Blix. According to one account, he attended a meeting of the Standing Advisory Group on Safeguards Implementation (SAGSI) early in his tenure, gave an inspiring speech on safeguards, and never came back.[99] He is, however, likely to have been represented by someone from his office and would have received briefings on the meetings.[100] During the tenure of one of his DDGs for Safeguards, there was reportedly a "virtual war" between the Safeguards Department and the Director General's office. ElBaradei apparently regarded that particular DDG as a "technical boffin with no political nous."[101] He tried to avoid meeting or working with him and forced him to send reports through his office. This is not the way to establish an optimal safeguards culture characterized by dedication, collaboration, transparency, effectiveness, and efficiency. It is also ironic given the importance that the Secretariat places on being a technical agency that abjures politics. Furthermore, it does not square with the assessment of Andemicael and Mathiason in 2007 that "the IAEA, consciously through its management training, as well as through its culture, has sought to inculcate a participative leadership style and has largely achieved it."[102]

ElBaradei did, however, have an expansive vision of the IAEA's future global verification and compliance mandate, but his view was not widely shared by member states. Despite long-standing self-characterization of the Secretariat as a purely technical body, he saw a wider role for the Agency in nuclear diplomacy: "If dialogue is used as a tool to resolve nuclear proliferation tensions, it cannot be limited to a conversation between the inspectors and the accused country."[103] The cultural dissonance that this produced was evident during the 2003 crisis over the alleged continued existence of weapons of mass destruction in Iraq. U.S. Secretary of State Condoleezza Rice objected to ElBaradei's diplomatic efforts to resolve the crisis, declaring that "the IAEA is not in the business of diplomacy. [It] is a technical agency." He responded: "I don't sit here and feel I'm only a technician. I owe it to the international community to give them my advice. They don't have to take it."[104] He added, "for a war to be fought over unsubstantiated WMD [weapons of mass destruction] charges—and for the IAEA's nuclear diplomacy role to be pushed to the side, serving merely [as] a fig leaf of due process—was for me a grotesque distortion of everything we stood for."[105]

ElBaradei also displeased the United States over allegedly exceeding his mandate in the Iran case, again by seeking to conduct high-level nuclear diplomacy separately from that of member states. Critics of ElBaradei accused him of refusing to report Iran's violations as non-compliance, "taking the view that it was for the Board of Governors—in other words governments—to make this determination."[106] This could be interpreted as a derogation from both his own responsibilities and from the hard-won prerogatives, autonomy, and impartiality of the Director General and the Secretariat. It was not until the end of 2005 that the Board was able to bring itself to declare Iran in non-compliance, a decision, in Carlson's view, "that had been obvious for over three years."

Finally, ElBaradei advocated openly for nuclear disarmament and asserted the Agency's unrivaled qualifications to verify it.[107] He even apparently had ambitions for the IAEA to move beyond nuclear energy to encompass all types of energy by scrapping the word "Atomic" from the Agency's name.[108]

ElBaradei's tenure as Director General coincided with the widely noted "politicization" of the Board of Governors and the dissipation of the cooperation and compromise of the "Spirit of Vienna."[109] This had begun in the early 1970s, when the Group of 77 developing countries started to "flex its muscles" in pursuit of increased Technical Cooperation (TC) funding.[110] Dissension escalated after the arrival of the more politically attuned and formal Non-Aligned Movement (NAM) presence in Vienna in 2003.[111] Such politicization of the Board challenged the self-proclaimed scientific and technical culture of the Secretariat, which has fought determinedly to preserve this reputation; its autonomy, and even its very existence, depend on it. In what were obvious references to the cases of Iraq and Iran, in which he was accused of being too lenient, ElBaradei contends that:

Despite my attempts to define the scope of the IAEA's jurisdiction consistently and to draw a clear distinction between what the Agency can and cannot judge, the pressure was often extreme for us to behave in a partisan way. When we held fast to our objective evaluation of the facts, when we refused to lend our voice to someone else's interpretation of a country's intentions, we were sometimes accused of playing favourites, of ignoring evidence, or conversely, of speaking "outside the box."[112]

Amano, ElBaradei's successor, was a Japanese career civil servant and diplomat. Previously Japan's representative on the IAEA Board of Governors, he too was familiar with the Agency's organizational culture and challenges. Amano had a long-standing interest in non-proliferation matters. He regularly attended SAGSI meetings and was active in seeking its advice. He also sought to lower the political profile of the Agency and reassert its technical orientation after the ElBaradei era, despite the difficulties arising from the Agency's highly sensitive role in verifying Iran's compliance with the JCPOA. He was tougher, as he promised prior to his appointment, in drawing safeguards conclusions about Iran.[113] Overall, however, he avoided the political spotlight, stuck tightly to his script when communicating, and was not a commanding presence. He was considered by many as "ElBaradei's opposite, as he adopted a very reserved, cautious, and careful working style."[114]

Amano notably put greater emphasis on the peaceful uses of nuclear technology than on safeguards, changing the Agency's slogan to "Atoms for Peace and Development," as noted above. He also led a highly successful campaign to raise funds for the Renovation of the Nuclear Applications Laboratories (ReNuAL) project to refurbish the Agency's decrepit facilities at Seibersdorf. The Yukiya Amano Laboratories were named after him in tribute.[115] After the 2011 Fukushima disaster, he necessarily focused on nuclear safety after a halting start in which he seemed determined to lower expectations of the Agency's role in responding to the worst nuclear catastrophe since Chernobyl in 1986.[116]

As for management style, one interlocutor told the author that the Agency's culture under Amano's watch had noticeably "moved East." Amano, however, seemed to offer the United States assurances that he would not seek to impose a Japanese management style on the Agency by undertaking to add only one Japanese national to his team.[117] The Japanese *ringi* system stresses consensus decision-making, in which all parts of an organization are expected to sign onto a proposed course of action.[118] This style seems

compatible with the management style of international organizations generally, where decision-making is achieved through persuasion and the pursuit
of consensus.[119] Other features of Japanese management, such as deference
to seniority and hierarchy and aversion to loss of face, can result in opaque
decision-making and aversion to open criticism. Given his long exposure to
international diplomacy and multilateral organizations, though, it is not clear
to what extent Amano was representative of Japanese organizational traditions. In some respects, he adopted a surprisingly more casual style than his
predecessor, reportedly sitting anywhere in the cafeteria and being willing to
talk to anyone. He even reportedly dialed out of his office without using his
secretary.[120]

Director General Grossi, an Argentinian, has already put his own
national stamp on the Secretariat even though he has only held the position
since December 2019. He is more relaxed and open than his predecessor, and
more willing to speak extemporaneously rather than sticking to a rigid script.
He has more than thirty-five years of experience in the disarmament and non-
proliferation field, both as an Argentinian diplomat and as an international
civil servant. He was previously Argentina's representative to the IAEA and
served as Assistant Director General for Policy and Chief of Cabinet at the
Agency under Director General Amano from 2010 to 2013.[121] Grossi, therefore, is thoroughly grounded in the work and, undoubtedly, the culture of the
IAEA and the broader UN system.

It is not clear to what extent Grossi has been able to change the Agency's
culture at this early stage in his term, but clearly the media appreciate his
loquaciousness compared to the reserve of his predecessor. Of all the candidates for the Director General position, he ran the most public campaign,
and he called for the IAEA to make its voice heard more in international
debates.[122] After his election, he pledged to "give *my* [emphasis added] member states and the international community the guarantee that I am absolutely
independent and impermeable to pressure."[123] As for safeguards culture, he
rather quirkily said at his first press conference that "an inspector is not a
friend. He's someone who comes and needs to ascertain the facts without bias,
without agenda, in an objective and impartial way."[124] Not only did this appear
to ignore the increasing numbers of female inspectors, but there was also no
hint that the inspection process had been transformed in recent years.

The differing management styles exhibited by successive Directors General illustrate the challenges faced by international organizations like the
IAEA, which comprises multiple nationalities, in forging a common, enduring
organizational culture. They also, however, highlight the need for the Director

General and other top leadership of the Agency to be conscious that they play a fundamental role, for better or for worse, in shaping the culture of the organization. From the perspective of safeguards culture, if they are armed with such an awareness, they can more readily cultivate a culture amenable to achieving the IAEA's safeguards goals.

In 2002, the Agency commissioned Swiss management consultancy company MANNET to assess "the extent to which the Agency [had] institutionalized a culture of continuous improvement at all management levels to facilitate change management, accountability, quality assurance and cost effectiveness."[125] This project was intended to study the Agency in its entirety, not just safeguards, but its conclusions seem particularly pertinent to the Safeguards Department. In its study, MANNET identified various symptoms of "system distress," including many deemed to be cultural, such as "competition between departments"; "reliance on personal networks to get things done"; "unclear lines of authority"; and "evolution of differing and sometimes conflicting systems."[126] The authors used cultural terminology to explain their critique: "the main issues affecting the efficiency and effectiveness of the Agency are 'below the surface' relating to the beliefs, systemic forces and culture of the organization."[127] The report's primary recommendation to then–Director General ElBaradei was that he "develop a comprehensive and integrated change management strategy as a matter of some urgency to overcome the systems stress and, above all, to embed the change in the organizational culture."[128]

Regrettably, there is no evidence that the Agency took the MANNET Report's cultural recommendations to heart at that time, or even began to think about organizational change as suggested. By all accounts, the reaction of Agency personnel to the report was negative and even derogatory. It was widely and cynically seen as an attempt by the Director General to secure his reappointment and believed that it would likely be filed away once this had happened. Such views, whether true or not, are an indicator of a culture resistant to change. In fact, the report was commissioned at the insistence of the United Kingdom and other member states. ElBaradei reportedly "hated" it, but DDG for Management and Budget David Waller felt obliged to at least try to implement it.

Tariq Rauf, who worked in the External Relations and Policy Office (ExPo) at the time, recalls that a change management strategy was put in place across the Agency following the report, but it proved to be "inefficient and disruptive." He comments that "management consultants' recommendations such as change management, open floors, etc., work well in theory [but] not in practice when real work and deliverables are at stake."[129] This is, of course, a common rationale for resisting organizational change. As detailed below, in more recent years, as modern management methods have seeped into the Agency, the Safeguards Department has adopted some of the concepts mentioned in the MANNET Report that were geared toward "continuous improvement at all management levels to facilitate change management, accountability, quality assurance and cost effectiveness."

Handling Safeguards Findings

A sensitive issue that goes to the heart of inspector culture and safeguards culture generally is the way that management, whether at the level of the Director General or the departmental level, is perceived as dealing with the findings of inspectors, especially anomalies and other indicators of potential non-compliance. In 2015, a former IAEA official noted a recurring cultural issue with inspectors concerning the value placed by the Agency on their work: sometimes they feel their findings are ignored, misinterpreted, or glossed over.[130] Given the confidentiality that necessarily surrounds such cases, it is difficult for an outside observer to gauge the extent of this problem. Since the Iraq case, there have been several instances—involving Egypt, Iran, Japan, and South Korea—where there were suspicions that recommendations from safeguards staff had been ignored. Former DDG for Safeguards Pierre Goldschmidt has called on the IAEA to acknowledge where it has acted inconsistently in the past, notably in the cases of South Korea and Egypt, both in 2005.[131]

Rauf, who was Head of the Verification and Security Policy Coordination Office from 2002 to 2011, says that there was "no demonstrable case where technically based safeguards assessments were disregarded by the 28th/27th floors [the Director General's offices]." "Innuendos" about political influence playing into safeguards decisions by the Director General, Rauf says, "hide . . . the occurrence of incompetence, mistakes and sloppiness by inspectors." He notes that "several such incidents were picked up in the meetings of the high-level committee by inspectors from other divisions." There was, Rauf claims, a "healthy practice of inter-divisional questioning of findings/conclusions at the high-level committee and low-level committee meetings on reaching safeguards conclusions."[132] Heinonen confirms that in his experience, Directors General have usually made no attempt to substantively change safeguards compliance reports drafted by the Safeguards Department.[133] On the other side of the coin are insinuations that the Safeguards Department sometimes failed to consult the Director General or his office in making safeguards decisions. Rauf has spoken of "some fantasy of the 19th floor" (where the Safeguards Department is located) about its presumed preeminence in making safeguards decisions.[134]

Clearly it is ultimately the statutory right and responsibility of the Director General to decide on the veracity of alleged safeguards violations. The Secretariat and Director General cannot simply present evidence of non-compliance to the Board of Governors without a substantive case being made on legal and political as well as technical grounds. False or unsubstantiated accusations against a member state would undermine the credibility of safeguards. "Political" factors must necessarily be considered in the sense that the Director General needs to be aware of what the political situation will bear in taking a possible case of non-compliance forward. Political factors are a part of life in all international organizations, not just the IAEA.

Differences between management and operational levels can reflect a cultural divide between the scientific, technocratic, fact-based perspective that predominates among safeguards inspectors, analysts, and managers on the one hand and, on the other, the non-technical culture of personnel at the political, legal, and higher management levels. Fostering awareness of the cultural differences between management and operational staff and between personnel of varying professional backgrounds may help alleviate some of the negative outcomes. The more collaborative approach recommended for drawing safeguards conclusions within the Safeguards Department should also be applied to the interactions between top leadership and safeguards managers in dealing with significant allegations of safeguards violations. The department has made attempts to ensure that all divisions work on the same basis to minimize the possibility of individual personalities playing a large role, but there are reports of continuing differences between the operational and planning divisions, with mixed cultural cues coming from divisional heads. Personality clashes cannot be dealt with easily, although top management should lead by example and human resources specialists should be able to help. Warning that "every bureaucracy has the potential for pathology," Barnett and Finnemore caution that "organizational coherence is an accomplishment rather than a given."[135]

The Continuing Struggle to Build One House

One element of IAEA policy that was praised by the MANNET consultants was ElBaradei's One House campaign, which was designed to have the Agency speak with one voice and to be more open, transparent, and collaborative internally.[136] The policy sought to break down the culture of informational "stove-piping" for which the Agency was notorious. It was reportedly triggered by ElBaradei's experience in asking the Secretariat for the number of nuclear research reactors worldwide and getting three different numbers from three different departments. In another instance, the Safeguards Department did not find out about a research reactor being built in Morocco with assistance from the TC program until it was about to start up.[137] Carlson says that some of the TC staff were suspicious of inspectors, regarding them as "spies."[138]

There was, however, widespread skepticism, even cynicism, within the Agency about One House. Most saw it as a lofty goal that would not be achieved without greater incentives, and removal of disincentives, for sharing of information and collaboration between departments. The IAEA's leadership itself, including the Director General, had not given consistent cultural cues in support of One House. A member of the 20/20 Commission remarked that:

> It is ironic, ElBaradei was able to make an impact [at the Agency] in part because he operated very much like the Bush White House— there was a lot of secrecy; he didn't feel there were many people he could trust. There was relatively little consultation with others, including with some deputies and department heads. There was a lot of talk at the agency about the need for transparency, but in some areas transparency was in short supply.[139]

The One House initiative eventually did lead to new information management systems that at least partly replaced the old, separate filing systems maintained by each department, division, and section, and sometimes even by individuals. An annual conference of senior managers was also instituted by ElBaradei and considered worthwhile by those involved.[140] A Director General's Meeting (DGM) was usually convened weekly.[141] In the view of at least one critic, however, enduring "structural obstacles" meant that "little progress" had been made toward One House.[142]

In 2011, Director General Amano created the Director General's Office for Policy (DGOP), which absorbed functions of the former Office of the Director General (DGO) and ExPo. According to the UN's Joint Inspection Unit (JIU), the DGOP reportedly played "a significant role in furthering and ensuring close coordination and information-sharing both within the office and across departments."[143] The JIU found that management and staff "generally feel informed adequately and in a timely manner about the main issues relevant to their work as discussed in the DGM and the respective senior management committees."[144] A few years later, Amano changed the name of this office to the Director General's Office for Coordination (DGOC), apparently on the grounds that only he made policy. Under Director General Grossi, it has simply become the Director General's Office. Presumably it continues to play a coordinating role.

The changes made in management and coordination have, in the view of other observers, including diplomatic missions in Vienna, been designed to assert greater control by the executive, a common trend in both international organizations and national governments, rather than achieving greater transparency and consultation. The Agency's culture, they contend, has become even less transparent in recent years, both internally and externally—in relations with member states, the media, and non-governmental organizations. The JIU report concluded that, notwithstanding the Amano reforms, the IAEA would "benefit from a change in culture toward more and open communication and information-sharing, not so much vertically, but . . . across the different departments."[145] This could be done through existing formal mechanisms or more informal channels, but would benefit both the Agency and the [Safeguards] department and "constitute an important step toward a 'one-house' approach."[146]

Given that the Safeguards Department is the repository of most of the confidential information held by the Agency, it has understandably struggled more than other departments with embracing One House over the years. Indeed, "safeguards confidentiality" is explicitly designed to guard against information-sharing by the Safeguards Department. A provision in the IAEA Statute, embodied in subsequent documents, requires that IAEA staff "shall not disclose any industrial secret or other confidential information coming to their knowledge by reason of their official duties."[147] Safeguards information is classified as either restricted, confidential, or highly confidential, and may only be used for safeguards purposes. Outside the Safeguards Department, only the Director General's office, including the Office of Legal Affairs, has access to some but not all confidential information. In addition, both the Board and the Directors General have ruled that information collected by other parts of the Agency, such as Technical Cooperation and Nuclear Sciences and Applications, should not be used for safeguards purposes. The confidentiality mantra, unusual for an international organization, is now so firmly embedded in IAEA safeguards culture that it has contributed to a general culture of opacity about all safeguards matters and a lack of transparency on the part of the Agency generally.

A genuine shift from "need to know" to "need to share" without breaching safeguards confidentiality is required to remedy this.[148] Change in this cultural norm can only come from the top, at the levels of both the Director General and all other department heads. Appropriate incentives and disincentives need to be put in place to achieve this, but nothing would work as well as leading by example, with more transparency from senior management.

The work on the "Department-specific internal communication strategy to enhance senior leadership and departmental staff member communication capabilities," which was initiated by Deputy Director General Tero Varjoranta in 2013, has continued into 2021 with the support of Member State Support Programs.[149] According to project manager Frédéric Claude, it aimed at "nurturing a *culture* [emphasis added] of collaboration, teamwork, and information/knowledge sharing, as well as increasing staff trust in leadership and co-workers."[150] The work includes revising and implementing an existing Strategic Internal Communication Plan, which identifies four factors in the department's efforts to enhance internal communications:

- The strategic use of internal communication tools, events, and channels;
- Facilitation of vertical and horizontal communication through meetings and reports;
- Communication capacity-building; and
- The establishment of feedback and survey mechanisms.[151]

Specific initiatives include a departmental communication working group, convened regularly to promote further coordination and collaboration across divisions, as well as other combinations of meetings to "ensure effective information/knowledge sharing at appropriate levels." Capacity-building programs were identified as still being needed on topics such as strategic communication, risk mitigation and crisis communication, communication working group skills, mentoring, and improving accountability. This long list, combined with the intensity of the meetings schedule, suggests that the Safeguards Department had an internal communications problem.

Public-Facing Transparency

The external transparency of safeguards culture can also only be changed with an Agency-wide cultural shift. Greater openness by the Agency about its strategic goals, budget and finances, organizational restructuring, and performance measurement would embolden the Safeguards Department to be more open about the effectiveness of safeguards, emerging proliferation challenges, and generic concerns about non-compliance by member states.

The Secretariat's refusal to publicly answer its critics on these matters leaves safeguards culture always on the defensive. The most recent example followed the Israeli government's contention in 2018 that there were two sites in Tehran containing documents and equipment related to the Iranian nuclear program that had not been declared to or visited by the IAEA. Instead of a clear and robust response, the Agency issued a statement that was so obscurely worded that even safeguards cognoscenti had trouble discerning its meaning.[152]

In its 2019 Safeguards Statement, as mentioned above, the Secretariat revealed that one state had lost its Broader Conclusion. In typical IAEA fashion, however, this was not announced directly, and the state—Libya—was not explicitly named but was only apparent from the drop in the number of states with the Broader Conclusion from 71 to 70. Observers had to work out the answer for themselves. This would have been the perfect opportunity to nudge safeguards culture toward increased transparency, because Libya was in the midst of a civil war and could hardly be blamed for a safeguards lapse. Similar considerations applied to Japan during and immediately after the 2011 Fukushima disaster, when the Agency's inspectors clearly could not verify that all nuclear material in the reactors was accounted for. In past years, the Secretariat has mentioned states that, through no fault of their own, could not comply with their safeguards obligations (such as Ukraine after Russia's seizure of Crimea), so the struggle for greater transparency appears to continue.

Amano reasoned that "there is a limit to what we can say publicly . . . because we are legally obliged to protect confidential information entrusted to us. This can sometimes make it difficult for Member States and the public to understand what the Agency is doing."[153] He also conceded that "it is frustrating for us when we see inaccurate information under discussion in the public domain. However, public silence on our part on a particular issue should never be taken to mean inaction."[154] This seems to be confusing two issues: a statutory obligation to protect confidential information and a need to challenge "fake news" that has already been aired publicly. In 2019, a study commissioned by the Swedish Radiation Safety Authority and produced by the Vienna Center for Disarmament and Non-Proliferation (VCDNP) recommended that "false assertions regarding the IAEA's legal authority should be challenged by Member States and by the Secretariat."[155] Even more pointedly, the authors of this study noted that "independence and transparency need not be viewed as mutually exclusive qualities, but rather as complementary ones."[156] The situation appears to be changing with Director General Grossi, who is more open about sharing information and more outspoken in his public pronouncements, but it remains to be seen if this will last.

Ideas for increasing external IAEA transparency include publicly releasing the SIR on a routine basis, as the Secretariat has proposed on several occasions. Such a document should, "when relevant, identify when problems are attributable to the IAEA, whether due to equipment failures, staff issues or administrative challenges."[157] At present, member states and the public are left guessing, which contributes to a type of cultural amnesia about the true problems facing safeguards: no one appears to be fully responsible for their implementation. The Secretariat is loath to blame member states, and member states are reluctant to blame the Secretariat, presumably on the grounds that the outcome would be a justifiable demand for greater resources for safeguards implementation. Rockwood and her co-authors propose giving SAGSI a public-facing role "to help challenge false statements about safeguards, [and to] offer independent opinions on safeguards issues to the public and to the Board."[158]

The Status of the Inspectorate

An issue closely related to the prominence of the safeguards mission is the status of IAEA inspectors, who remain a key element of the safeguards system and whose cultural cohesion is vital to the health of safeguards culture overall. As noted, the inspectors were traditionally the "heroes" of safeguards. With the rise of analysts and their increasing importance to safeguards, the inspectors have faced the risk of being relegated to a supporting role. The static number of inspectors and the rising significance of technological alternatives to human inspections are also a challenge to the status of inspectors.

The way in which inspectors are characterized as an identifiable group is also problematic for building safeguards culture. Although the term "Inspectorate" was used in some of the earliest IAEA documents, it fell out of favor early on, probably because of its regulatory and enforcement overtones (like the term "Inspector-General," which was abolished early on by Director General Sigvard Eklund).[159] The IAEA Statute refers blandly to a "staff of inspectors," and the Nuclear Non-Proliferation Treaty (NPT) does not mention inspectors at all.[160] The vaguer, less threatening collective term "inspectors" has become the Agency's preferred nomenclature.

Moreover, the Agency does not treat inspectors as a special category of staff but rolls them into the general category of "professionals," in part to allow flexibility in moving personnel between desk positions and field activities, but also to avoid drawing attention to them as an "inspectorate." It is difficult to discover, for instance, how many inspectors the Agency has at any one time and what their demographic, national, and educational characteristics are;

in this author's experience, even Agency employees have trouble finding this information out. This is no doubt due to the system of "designating" inspectors, which depends on the agreement of the state receiving the inspector(s) and the Board's approval of the roster. The numbers vary somewhat year from year. Moreover, in theory, any employee of the Safeguards Department may be designated as an inspector, subject to member state and Board approval. This practice has generated some resentment among inspectors with science and engineering backgrounds, who believe that political scientists, intelligence experts, and other types of analysts are diluting the qualifications of the inspector corps. In sum, the system seems almost designed to minimize the notion of an identifiable group of dedicated, professional investigators fanning out across the world to monitor and verify member states' nuclear activities—although this is what IAEA inspectors are and do.

Inspectors are still vital to the Agency's work and need to be recognized accordingly. The task of doing so is now more complicated with the rise of an analyst sub-group in the department. The Agency should consider re-adopting the term "inspectorate" and begin releasing the numbers and types of inspectors it fields each year, which would give the public a better idea of the scale of operations than the unfathomable figure of "person-hours" spent on inspections. Appropriate rewards like "Inspector of the Year" should be considered, although this would probably now have to be matched with "Analyst of the Year" to avoid deepening animosity between the two equally vital categories of safeguards employees.

Rotation: A Complicating UN Artefact

The greatest incentive for IAEA personnel to commit themselves wholeheartedly to the safeguards mission would be the reassurance of a lifetime career in Vienna—subject, of course, to satisfactory performance. This would mean discarding the present system in favor of a permanent inspectorate. Currently the Agency operates a UN-style rotation system designed to fulfill the IAEA Statute's stricture that "permanent staff shall be kept to a minimum."[161] This system was instituted by the United Nations to supersede the League of Nations experiment with a permanent international civil service, which allegedly produced an entitled, self-satisfied, and ultimately calcified culture. Under the rotation system, personnel are recruited on renewable short-term contracts of three years, which may be twice extended by two years. Most professional staff are thus limited to a maximum of seven years. An extension of employment beyond that is considered "long-term"—in effect, permanent. Extensions are apparently subject to a rigorous review process before the final decision is made by the Director General. The Secretariat has established an informal 40-percent benchmark for staff on long-term contracts beyond the seven-year limit, although the percentage differs widely among departments.[162] The Safeguards Department fares better than most, probably due to the convincing case it makes that its heavy investments in training and providing staff with professional experience should not be lost.[163]

The official aim of the IAEA's policy is to "keep the collective knowledge of the staff up to date and at a high level, especially in scientific and engineering fields, and to ensure regular introduction of new ideas."[164] It also gives the nationals of more member states opportunities to work at the Agency, for which developing countries are constantly pressing the Secretariat. This also benefits the Agency in being able to send experienced safeguards personnel back to member states to enhance their national safeguards implementation. Furthermore, the system may have additional benefits for propagating good national safeguards cultures. If returning personnel have absorbed the Agency's safeguards culture properly, they may help embed it in the safeguards culture of their own national authorities. Finally, an implied but real benefit of the system is that it allows the Agency to let go of under-performing staff, contrary to complaints that most international organizations find it impossible to dismiss anyone.

The rotation system does, however, have what the Agency's external auditor has called "some avoidable negative effects."[165] One is the loss of experienced, trained staff whose short-term contracts are not renewed or who are not given long-term contracts and never return. The reasons may be departmental funding constraints or pressure to keep long-term contracts within established Agency targets. As the external auditor notes, this can produce "permanent loss of institutional memory."[166]

A second dysfunction is the gaming of the system by the Agency and by employees. Robert Kelley points to several alarming cases:

> The worst being people who were told to leave, find other work, wait a year and then come back for 7 more years. . . . Will H. took this advice and came back after his year of exile. A person I know was recruited, sent back to Canada with all his moving expenses etc., then brought back again for the same treatment. There are more. The young lady in Safeguards information systems re-engineering who waited until the last day to find out if she got a new contract, and when she got it on the last day she told them what to do with it![167]

From the perspective of a healthy safeguards culture, such practices produce negative attitudes among staff, including short-term career outlooks and cynicism about the Agency's commitment to excellence and to its employees. Valuable employees are sometimes deterred from staying, reapplying, or even joining the Agency in the first place. Again, Kelley says on the subject:

> If you take a reasonably competent person and give them one or two years notice that their contract is ending they will find another job. If they cannot, and they can also afford to sit out a year or more without pay and believe you will hire them back into the same job at the same pay with one-year loss of salary and promotion. Who will you get? You will get the losers who cannot compete in a three-year job search and come back to you out of comfort.[168]

Such cases are detrimental to a healthy safeguards culture because they discourage loyalty to and identification with the Agency and its safeguards mission and encourage a short-term outlook toward employment at the organization. At worst, they may deter staff, including inspectors, from revealing anomalies in safeguards compliance, criticizing safeguards implementation, or challenging safeguards conclusions. As Kelley bluntly puts it: "Everyone wants [long-term contracts] after seven years, so they don't rock the boat." The U.S. dollar pension is "much sought after."[169]

The JIU contended in its 2012 study that the rotation policy "poses a significant issue" for IAEA staff. Managers were accused of using the policy to remove poor performers instead of using the performance appraisal system, which might involve personal confrontation. Such a work-around negates the whole point of the appraisal system. The external auditor recommended that the rotation system not be used for this purpose because of its negative effects, particularly on staff motivation.[170] A further deleterious effect on safeguards culture is that the constant process of contract renewals and a lack of transparency about the process leads to unending anticipation, speculation, and gossip among the staff.

The JIU's survey further revealed that most respondents disliked the rotation policy and favored its reform.[171] On the other hand, most IAEA officials and representatives of member states favored keeping the current policy, at least in principle, "bearing in mind the benefits it offers." They also, however, saw room for improvement in its implementation, especially in making the process for granting long-term appointments as transparent as possible. The claim that the JIU inspectors were "informed" about the views of "a majority of member states" must be viewed skeptically, as they did not manage to canvass the views of a majority.[172] Nonetheless, anecdotal evidence suggests that member states and senior management have a different attitude toward the rotation system than staff, which is one reason why it remains in place. The Agency would also have to break with long-standing UN tradition to replace it.

Short of abolishing the rotation system, the Agency could implement measures to alleviate its dysfunctional aspects and thereby minimize the effects on safeguards culture. Increased transparency about the process seems warranted, as do better procedures to handle under-performing staff and to ensure that staff are informed well in advance of decisions about terminating or renewing their employment. Rockwood and her co-authors warn, however, that "basing extensions on performance . . . means that managers have to be willing to document both good and bad performance by staff, which, by most accounts, is not a prevailing culture at the IAEA."[173] They suggest that given the competing interests in preserving or abolishing the rotation system, the Secretariat should consider a study of staffing practices. This could include examining how other international inspectorates, such as the International Civil Aviation Organization (ICAO) and the International Maritime Organization (IMO) deal with the challenge of creating a loyal, permanent inspectorate within the confines of UN employment practices, notably how they maintain a healthy regulatory culture. Rockwood and her colleagues also propose offering long-term leave or sabbaticals to enable staff to acquire fresh perspectives on scientific and safeguards issues at a national level while ensuring that the IAEA retains the institutional memory of qualified staff.[174]

In addition to continuing concerns about the rotation policy, a widespread assumption persists in the Agency's organizational culture that the requirement for geographical balance in recruitment results in a certain number of appointments and reappointments that are due not to proven competence but to politics. Furthermore, Barnett and Finnemore explain that many international organizations do not face strong external performance pressures and that successful performance may be difficult to measure. Staff may not compete on the basis of output, but rather are "valued more for what they represent than what they do."[175] An inability to evaluate performance, "either within the organization, so it can self-correct and learn in a rational fashion, or from outside, increases the potential for pathologies."[176]

An associated, enduring cultural assumption is that "the way we do things around here" is to work around the resulting "dead wood."[177] The MANNET Report urged the IAEA to "deal with the problems of chronic under-performance" produced by the UN's recruitment practices, although it is not clear to what extent this applied to the Safeguards Department.[178] The 2020 report by the Nuclear Threat Initiative (NTI) and the Center for Energy and Security Studies (CNESS) noted that "certain aspects of safeguards staff resource management have improved over time (e.g., job interviews have become mandatory, the hiring process has become more transparent, and training is intensive and ongoing so that today's inspectors and analysts now have better skills)."[179] Yet there remain "disparities" both in selecting new staff and in terminating marginally performing staff.

The Secretariat has been making valiant efforts to recruit better-qualified candidates from developing countries and other under-represented countries in all departments, including safeguards, but the requirement for geographical balance continues to prove problematic. There was a global surge in nuclear science and engineering studies in the 1970s, with large numbers of national atomic energy agencies established and funded. But over the decades since then, this "expertise bubble" has burst, and nuclear science knowledge in developing countries is now a rare commodity. There has been something of an upsurge since the beginning of the 2000s and the emergence of the so-called nuclear renaissance, but the geographical scope remains patchy and most new graduates with nuclear science proficiency are being swept up by industry. It may be some time before they become available to the IAEA.

Reliance on External Consultants

Another difficulty facing all parts of the IAEA, but safeguards in particular, is its dependence on outside consultants, even in core positions. As the Secretariat puts it, "Consultants are not staff members and are engaged to supplement the staff of the Secretariat for a limited time, normally providing expertise, skills or knowledge for the performance of a specific task or piece of work of a professional nature that has a defined end date."[180] In the Safeguards Department, such personnel only work in the concepts and planning area, not in operations and not as inspectors. They do not have access to confidential safeguards material.

Member states, however, have nonetheless long complained about both the Agency's reliance on external consultants and the predominance of non–developing country personnel among them. The Secretariat reported in 2019 that regular budget spending on consultants had increased by 5.1 percent compared to the first year of the biennium 2016–2017, which had seen a decrease of 6.1 percent.[181] The percentage of consultants from developing countries has been stable at approximately 30 percent for several years.[182] These figures do not include cost-free experts provided by some member states, who are also mostly from Western countries. Although the Safeguards Department spends less from its regular budget on consultants than any other major Agency program, it is still heavily dependent on them.[183]

The Agency's auditor noted in 2013 that the IAEA did not have a strategy for developing knowledge in areas where it regularly employs consultants, nor did it have a well-defined succession plan to fill in the skills gap left by the departure of staff members who do not complete their assigned tasks before retiring.[184] In many cases, former staff members have been rehired as consultants for significant periods to carry forward the work they were originally employed to do. Even SAGSI is not immune from this "revolving door" phenomenon. Kelley contends that: "Just as I was retiring two [safeguards] directors retired. Both were immediately appointed to SAGSI and both were well-known for their resistance to any change. Consultancies like SAGSI pay very well and many meetings are held in very desirable world spots."[185]

Such employment practices nullify some of the benefits of the rotation system and leave a greater cultural overhang than might otherwise be expected. This also means that a career path in IAEA safeguards is not being developed, and that commitment and loyalty to the organization cannot be assumed. Consultancy contracts do not provide for knowledge transfer and do not even guarantee completion of projects by the end of the consultancy. The Agency made several changes to consultancy policy in response to criticisms from the external auditor, including restricting consultancies to two years. After the new external auditor reminded the Secretariat in 2017 of the previous auditor's recommendations for reform, the response was that implementation "was in progress."[186] Almost ten years after the recommendations were made, the external auditor was being told by IAEA management that it "will" establish an effective knowledge transfer process for departing staff and that it considers several of the other recommendations to be still "in progress."[187]

Cultural Implications of Demographic Change

One of the ways that culture changes is through the influx of new blood into an organization. Staff turnover and generational change help to ensure that a new culture develops over time. Unlike other bureaucracies, the turnover of staff at the IAEA is relatively high due to the rotation policy, although those who rise to the managerial level tend to be long-term employees and key bearers of the culture. In addition, there is currently an unusually high turnover of experienced staff on long-term contracts due to a generational change triggered by the compulsory retirement age (previously 63, now 65). In 2017, the Secretariat predicted that over the seven years leading up to 2024, more than half of the Agency's current 1,114 regular professional staff would retire, an astonishing turnover.[188] There is also continuous churn in the international mix of the Agency. The Secretariat reported in 2020 that over the next two years, three member states would see 50 percent or more of their nationals retire, while five other states would see 25 to 50 percent of their nationals retire.[189]

Responding to political and societal pressures and in accordance with current managerial practice, the Secretariat is admirably seeking to broaden the recruitment pool for safeguards personnel to encompass a wider variety of professions and backgrounds, achieve greater multinationality, and redress its gender imbalance. Increased diversity is now viewed as a strength in regulatory-type organizations, where "group think" can be disadvantageous. But it also poses challenges for attempts to deliberately fashion a common organizational culture.

Presumably the expectation is that increased diversity will lead to more effective safeguards by expanding the range of skills, talents, intuitions, and even national cultural perspectives that are brought to bear on the Agency's verification tasks. In 2015, the Secretariat reported that it had augmented its talent acquisition program to include: a more focused sourcing strategy, including even more active outreach to potential candidates (such as hard-to-reach niche professionals); a research-based candidate search; and the formation of talent pools for future vacancies.[190] The Secretariat also reports that it has launched "talent pipelines" in multiple areas to facilitate recruitment of qualified experts for short-term assignments. Remarkably, only in 2019 did the Secretariat begin to design specific outreach activities for the needs of the hiring department.[191] In 2019, the Agency undertook recruitment missions to eight member states with their assistance; education and training on the recruitment process as well as information on working conditions and related topics was provided.[192]

A Junior Professional Officers (JPO) program now allows member states to fund temporary work for junior officers, some of whom may later apply for regular positions.[193] Such assignments are being advertised online via the Agency's official social media channels and among relevant professional organizations in member states. Recruitment through such modern means should result in a younger cohort of applicants, with obvious implications for culture. It is also likely to broaden the range of qualifications away from science and engineering. One of the best inspectors recently recruited is reportedly a non-proliferation specialist, not an engineer.

The Secretariat is still clearly struggling, under continuing pressure from member states, to internationalize its personnel. Figures on the percentage of developing country inspectors are not available publicly, but the percentage of staff members from developing countries in the entire Secretariat decreased from 31 percent in 2015 to 28.5 percent in 2017, and has hovered around the latter percentage ever since.[194] In the Safeguards Department, the extensive assistance given to improve member states' State Systems of Accounting and Control (SSAC) and Regional Systems of Accounting and Control (RSAC) will, incidentally, help expand the pool of potential IAEA recruits from a wider variety of member states.[195] Ultimately change will come, and with it, the need to consider the impact on safeguards culture.

The Agency's gender balance is also relevant for cultural change. One indisputable aspect of the traditional IAEA culture has been its maleness. As noted, all of the Agency's Directors General to date have been men, as have all Deputy Directors General for Safeguards. Since 2007, the Secretariat has sought to implement what it calls a comprehensive Gender Equality Policy, which "addresses gender balance and enhances work-life balance measures in the Secretariat, as well as mainstreaming a gender perspective in the Secretariat's programmes and activities."[196] Under pressure from member states and the external auditor, the Agency has been repositioning itself to become a leader in this area among international organizations.[197] Director General Amano became a Gender Champion in June 2017 at the launch of the Vienna chapter of the International Gender Champions initiative.[198] Grossi, his successor, is also a Gender Champion, and has been named to the Board of International Gender Champions.[199] Grossi has pledged that the IAEA will reach gender parity among its professional staff by 2025. In addition, he launched the ambitious annual Maria Sklodowska Curie Fellowship program, under which the first one hundred female winners in November 2020 were awarded fellowships to finance their master's degrees in nuclear subjects.

The Agency's progress toward gender equality has been slow but steady. In 2007, the percentage of female staff in the Secretariat was 41.9 percent; in the General Service category, it was 60.6 percent, while in the professional and higher categories it was just 22.5 percent.[200] By July 2019 (the latest year for which figures are available), the percentage of women in the Secretariat had risen to 44.5 percent; in the General Service category, it had increased to 63.8 percent, and in the professional and higher categories, the number grew to 30.2 percent.[201] The representation of women at senior levels remained higher in the administrative field than in the scientific/engineering field.

The Secretariat reports that it maintains "a proactive dialogue" with member states on the recruitment of women to encourage qualified female candidates to apply for vacant positions.[202] One continuing difficulty that the Secretariat acknowledges is that its efforts to appoint well-qualified women sometimes conflict with competing demands by member states for increased representation from developing countries and from unrepresented and under-represented member states.[203] The Agency has also long struggled with the fact that the overall number of graduate students, particularly women, in nuclear-related areas remains small, and "consequently, the resource pool of women in the nuclear field is and will remain limited."[204] This trend has continued despite the increased interest in nuclear science as a result of the growing importance of nuclear energy in some countries. A related problem is that increasing numbers of graduates are being recruited by the nuclear industry itself following the surge of building new nuclear power plants in various countries, leaving the Agency unable to compete. The advent of the World Nuclear University and increasing numbers of nuclear engineering courses at universities worldwide may help address this challenge.[205]

As for safeguards personnel specifically, there were no female inspectors until 1982. In 2007, the Agency reported that of its professional safeguards staff, only 12 percent were women with qualifications in science and engineering, while an additional 3 percent were women in safeguards administration.[206] By 2015, the percentage of women inspectors had risen significantly to 21.8 percent.[207] Most were from developed countries, with the significant exception of the Philippines.[208] By December 2019, 35 percent of all regular staff members in the Safeguards Department were female.[209] In the professional and higher categories, women represented 23 percent of the regular staff and 20 percent of the safeguards inspectors in the Divisions of Operations and the Office for Verification in Iran. Non-governmental organizations like Women in Nuclear (WiN) are helping to raise the profile of women in the nuclear industry, including at the IAEA.[210]

These changes in recruitment, although only slowly producing results, will inevitably change the IAEA's culture, at least by changing the mix of personnel who participate in the culture. Greater awareness of gender inequality and different attitudes toward work-life balance are already apparent at the Agency. When he joined in the early 1980s, Tom Shea says, there was "a lot of drinking," but now there are stress-related issues, presumably due to increased work pressures.[211] Work-life balance policies have already been implemented by the Agency, including flexible working hours; part-time arrangements; the possibility of working from home; maternity, paternity, and adoption leaves; nursing breaks for new mothers; family emergency leave; and a child-care facility located at the Vienna International Center.[212]

The impact of such changes on safeguards culture remains to be seen, but they could further complicate the scheduling of inspections, which is already hindered by logistical and budgetary factors. These changes may also deepen the trend toward the use of technology to supplant or supplement human safeguards activity.

The Lure of Vienna

One of the cultural tensions at the IAEA is between the attraction of Vienna as a comfortable place to live and work and the need for inspectors to be risk-takers. If too strong an emphasis is placed on the delights of Vienna (as Shirley Johnson worried she did in her recruitment efforts), rather than on the challenges of being an inspector in the field, the recruitment of a new generation of safeguards personnel could be counterproductive to fostering a new safeguards culture. As Kelley puts it, "when it comes to inspectors from less-developed countries the Vienna Euro salaries are princely, as is the retirement system, so there is a huge incentive to conform for seven years until achieving [long-term contracts] and then a life of comfort compared to 'back home.'"[213] Former IAEA inspector Mark Schanfein concurs:

> As I became friends with inspectors from around the world, some from [developing] countries . . . told me something that, while at first shocking, made complete sense. They said that they would never go back to their own country under any circumstances nor would they ever put their current job as IAEA inspectors in jeopardy. Unlike their home country, here in Vienna they had security, great housing, safe food, health care, and a future for themselves and especially their children who can obtain a good education and find excellent opportunities: unlike what they found at home.[214]

This elicits the question of whether inspectors would risk their IAEA jobs by raising concerns about the effectiveness of safeguards, member state non-compliance, or attempts within the Secretariat to downplay or cover up possible violations. Potential whistle-blowers in any field are disincentivized by their desire for job security, but at the IAEA this may be exacerbated by the high quality of life in Vienna.

Schanfein notes that no amount of training will induce inspectors, analysts, or other Agency personnel to take risks, essentially by becoming whistle-blowers about safeguards non-compliance, if the personal incentives outweigh the organizational incentives.[215] The Agency needs to come to grips with this problem, presumably by putting in place stronger incentives to encourage risk-taking and to promote, reward, and protect those who handle safeguards non-compliance cases best. There is a significant organizational management literature and precedent to draw on.[216] While lessons are mostly taken from commercial or industrial corporate experience, there are certain elements that are applicable to all types of organizations, including international ones.

Cultivating an Optimal Safeguards Culture

Fostering an optimal safeguards culture, one that is unified in meeting the challenge of nuclear proliferation through safeguards, is a constant work-in-progress. This section considers some of the best tools available to the IAEA in pursuing an optimal safeguards culture: defining what it is; fostering a personal stake in the success of safeguards; training; and incentives and rewards.

Officially Defining and Codifying Safeguards Culture

A major difficulty in promoting an "optimal" safeguards culture is that there is still no agreed international definition, official or unofficial, nor have its elements been systematically codified in IAEA documents. The first step in fostering an optimal IAEA safeguards culture would be to define and codify it. The 2016 INMM Workshop on Safeguards Culture could not agree on a definition, but concluded that, while "not a solution," an agreed definition "might contribute to a better common understanding and result in improved safeguards effectiveness and efficiency."[217]

The elements of an optimal IAEA safeguards culture have become increasingly apparent since the advent of strengthened safeguards. Some of these are boiler-plate aspirations that all organizations should aspire to: organizational excellence; a sense of service and loyalty; and a commitment to effectiveness and efficiency. Other values are more specific to the Agency as an international organization dedicated to a higher cause than its own well-being. The IAEA, along with other organizations in the UN common system, now promulgates a statement of its core ethical values of integrity, professionalism, and respect for diversity.[218] Of special relevance to safeguards are the integrity standards requiring staff, *inter alia*, to demonstrate in practice the values of the Agency in activities and behavior; to resist political pressure in decision-making; to ensure that official power or authority is not abused; and to seek to promote the Agency's interests.

Values more specific to safeguards are set out in various documentary artefacts emanating from the Safeguards Department as detailed above, most importantly the Strategic Plan. They are also embedded in the Safeguards Department's Quality Management System (QMS), which has been operating since 2004. In August 2018, Deputy Director General for Safeguards Massimo Aparo issued the latest iteration of the Safeguards Department's "Quality Policy," including the following admonitions:

> Each one of us has the power and responsibility to ensure quality through our personal actions and dedication.
>
> Quality is about building trust and confidence in our safeguards conclusions. Through our Quality Management System, we provide assurances to stakeholders and ourselves that our safeguards activities are implemented in an efficient and effective manner.
>
> In alignment with our strategic objectives to continually improve performance and productivity, we are committed to ensuring the quality of our work and strengthening the capabilities necessary to fulfil our legal obligations in a professional and impartial manner.[219]

The document then lists the department's "Quality Management Principles"—essentially the same ones identified in 2004: leadership; engagement of people; process approach; evidence-based decision-making; improvement (no longer "continuous"); customer focus; and relationship management. Two "Quality Objectives" were specified in support of this policy: "promoting a quality culture and encouraging ownership of quality responsibilities and accountabilities" and "implementing our quality policy and following our quality management principles in the way we work."[220]

As of this writing seventeen years later, this document still has not been specifically adapted to the safeguards enterprise (it still refers to "the Department's customers"), except for the reference to "building trust and confidence in our safeguards conclusions." Although it is one of the few official IAEA documents that uses the term "culture," it still does not delve into any specifics. The Agency is not alone in this. The Quality Management movement, while generally giving a nod to organizational culture and encompassing culturally-relevant elements, tends to ignore the need for analyzing underlying assumptions while emphasizing espoused values, as if they will be self-implementing. While the Quality Policy statement is a useful, continuing reminder to safeguards personnel about the need for attention to "quality management," from a cultural perspective it gives every impression of simply going through the motions.

The Director General includes a section on Quality Management in his annual report on safeguards implementation, which records the department's continuing efforts to improve the system.[221] The Safeguards Department, however, has struggled with implementing its QMS. To be fair, this is because, by its very nature, Quality Management involves a never-ending process of review, evaluation, and reform. But it is also, perhaps, because the implications for the department's organizational culture, including safeguards culture, are ignored.

In 2019, the external auditor sought to answer the question: "To what extent does the Agency have an Effective Quality Control [process] to support the implementation of [safeguards]?"[222] It concluded that:

> We observed that there have been various responses to the evaluation activities, particularly to the SEERT since it was first implemented in 2018. Some support it and consider it an important aspect of effectiveness evaluation, and some others find it as a duplication of work and an extra burden to their currrent job, particularly when it comes to resources. Since SEERT is a departmental team and involves experts and analysts from other divisions, it potentially causes conflicting responsibilties for the experts/analysts.[223]

Furthermore, the auditor observed that recommendations arising from the evaluation process were not detailed in the form of proper action plans that would "capture key activities, dates and milestones . . . in implementing the recommendation."[224] Although those responsible for implementing each recommendation had been identified by the department, an action plan would provide "assurance and commitment of the responsible parties to accept the recommendation."[225] The Agency agreed with the auditor's proposals. The problems identified by the auditor may well be grounded in culture rather than process, although assessing this would require further investigation.

Member states recognize the continuing challenges of Quality Management. Their Support Program for 2020–2021 includes a project to "strengthen and mature [sic] the Department's Quality Management System and monitor and report on its effectiveness," drawing on a 2017 internal self-assessment of the "maturity" of the QMS.[226] Although Quality Management can only ever be one element in transforming organizational culture, it should be carried out with potential cultural barriers in mind, along with an awareness of the potential cultural ramifications, often unintended, when significant organizational changes are contemplated.[227]

Although obviously influenced by the latest management concepts and jargon, the Quality Policy is not only a cultural artefact in itself, but also a deliberate attempt to shape the culture openly. For one of the few times in an official IAEA document, it even used the term "culture."

Drawing on the safeguards experiences and developments of the past thirty years since the Iraq case, I have identified the key values of an optimal IAEA safeguards culture and set them out below:

- The IAEA strives to be the pre-eminent international nuclear verification body deserving of the support of the international community for its safeguards mission.
- IAEA safeguards aim to contribute to international peace and security by helping to deter the proliferation of nuclear weapons.
- While all elements of the IAEA's mandate are important, nuclear safeguards, along with nuclear security, make the greatest contribution to international peace and security.
- IAEA safeguards require the necessary resources—human, legal, technical, and financial—to ensure their success.
- The Director General, guided by member states and subject to advice from the Board of Governors, is responsible for oversight of the nuclear safeguards system.
- The Safeguards Department has principal responsibility for the effective and efficient implementation and continuous improvement of IAEA safeguards.
- The Safeguards Department is proactive in identifying at the earliest possible stage any potential threats to the non-proliferation regime and to the integrity of safeguards.
- Safeguards are concerned not just with nuclear materials accountancy, but also with verifying correctness and completeness of member states' declarations; pursuit of undeclared materials/activities/facilities; indications of weaponization; and consideration of all potential proliferation pathways.

- All validated sources of information, including open sources, trade analysis, satellite imagery, and information from member states are used in drawing safeguards conclusions.
- A collaborative, team-oriented approach involving inspectors, analysts, and other experts is necessary for effective verification.
- All safeguards personnel—managers, analysts, and inspectors—are equally important in their contribution to safeguards.
- Safeguards personnel individually aspire to be professional, technically proficient, and dedicated to the safeguards mission.
- Inspectors are inquisitive, investigatory, and innovative, while at the same time respecting states' rights.

Some of these values are already alive in the Safeguards Department's culture, while others are a work-in-progress. Some may still be controversial in some circles, especially among some member states. Like a definition of safeguards culture, these core values would at least need to be acknowledged by all stakeholders, including member states, if a consistent safeguards culture is to be embedded at the Agency. The critical question thereafter is how such agreed values might be inculcated, encouraged, and strengthened. Some ideas on how this might be done are explored below.

Fostering a Personal Stake in the Success of Safeguards

At the level of the individual safeguards employee, the culture already exhibits a strong personal commitment to safeguards as a means of preventing the proliferation of nuclear weapons. This needs to be sustained and promoted, as it is easy in the day-to-day pursuit of technical goals to lose sight of the ultimate aim of the safeguards enterprise. There are, however, major differences between how this is achieved in safety and security cultures and how it might be done in safeguards culture.

Unlike the nuclear safeguards field, the nuclear safety and security realms raise the distinct possibility that the individual may be personally harmed by a nuclear accident or terrorist incident while on the job, which is a powerful incentive in embedding commitment to a robust culture among personnel. The campaigns to develop nuclear safety and security cultures thus have strongly personalized elements, both normative and prescriptive, designed to avoid nuclear catastrophe that may harm the employee. "Good" and "bad" cultures have been identified for emulation and rectification respectively. Modern safety and security cultures foster constant vigilance to avoid back-sliding and to respond to perceived new challenges. Leadership, awareness-raising, and a rewards system are key tools in changing safety and security cultures.[228]

In the case of safeguards, it is much more difficult to draw a direct line of personal consequences that might arise from a poor safeguards culture. The illicit acquisition of a nuclear weapon by a state under safeguards, or worse, the use of such a weapon is, from an individual IAEA employee's point of view, disturbing and a great professional disappointment, but he or she is unlikely to fear personal physical harm. Such an episode is likely to be a slowly unfolding scenario with many contributing factors, the least of which may be a failure of safeguards. In fact, safeguards are not designed to prevent nuclear weapons proliferation in the way that security culture is meant to prevent nuclear terrorism; rather, the role of safeguards is to alert the international community to the possibility that there has been a diversion of nuclear materials from a peaceful program to a weapons program or that there are suspected undeclared materials and activities. Therefore, the motivational element of safeguards culture is likely to be less compelling than in the case of either safety or security cultures, both individually and organizationally. The difference is illustrated by the fact that while IAEA inspectors do not fear the use of a nuclear weapon during their inspections, they have feared for their own safety at poorly maintained nuclear facilities or may even fear a nuclear security–related incident involving terrorists.[229]

Fostering a strong safeguards culture must thus rely on incentives other than those operating in safety and security cultures. This is one reason for stressing the contribution of safeguards to international peace and security. Jim Casterton says that the most challenging cultural issue is the need to maintain strong, united management oversight within the Safeguards Department to ensure the continued evolution of safeguards implementation, which features collaborative approaches rather than individual preferences.[230] The old individualistic culture of the department identified by the MANNET Report is no longer appropriate.

Training

Although the role of training in changing culture tends to be exaggerated, as if a simple retraining course will alter deeply ingrained habits and attitudes, it can be a significant tool in the organizational change toolbox if done with cultural awareness. Training can, of course, have both intended and unintended consequences for culture.

Among the initial moves taken to strengthen safeguards, training was reportedly the last to be implemented, after institutional, documentary, and procedural changes were put into place. Heinonen says that the old training courses were too long and involved too much theory, accountancy, and diplomatic concepts from the 1980s.[231] A 2005 U.S. Government Accountability Office (GAO) report was critical of some aspects of IAEA training, noting that recruits with poor English skills had difficulty fully absorbing the materials and presentations, which were all done in English.[232] In the same year, the Agency introduced, remarkably for the first time, a qualification process to ensure that all inspectors had the knowledge and skills to perform at an acceptable level.[233]

The Agency has since taken significant steps to improve training and has used it to embed new cultural values, although it has avoided using the word "culture." The overall IAEA training budget is modest, however, compared to those of other international organizations, which typically invest 1 percent of their annual budgets in training and management; the IAEA's training budget hovers around 0.35 percent.[234] Amazingly, extra-budgetary support from member states is required for most training courses and related travel, as well as cost-free experts to teach some courses.[235] Without support from mostly Western member states, the IAEA's training would shrink significantly. The downside of this reliance is that it perpetuates the idea that nuclear safeguards are a Western project far removed from the priorities of the developing world. From a safeguards culture perspective, it would be preferable that the sources of funding and cost-free training staff be broadened to signal that safeguards are a universal concern.

The Safeguards Department's Training Section is responsible for designing and delivering safeguards training courses for both Agency personnel and for staff from State or Regional Systems of Accounting and Control. In 2018–2019, in cooperation with member states, 212 courses were delivered to 2,340 departmental staff participants.[236] Courses ranged from one-day seminars to two-week courses. In addition, more than thirty training courses were delivered to member states for over five hundred of their personnel. This dual role presumably helps impart IAEA safeguards culture to national nuclear bodies, in addition to allowing the Secretariat to detect possible dysfunctional safeguards practices (and cultures) in such bodies. The Training Section concedes, however, that its work is complicated by budget constraints, staff turnover, reliance on external trainers (60 percent), increased restrictions on facility access for on-site training, and the need to update management and training tools.[237]

Currently, training for new inspectors begins with an Introductory Course on Agency Safeguards (ICAS) that lasts from three to four months.

ICAS modules cover the necessary technical topics, including non-destructive assay techniques, containment and surveillance, radiation protection, and design information verification. The values being inculcated include the need to facilitate a good flow of information, to understand the verification "tool-box," and to move from the broad picture to the details rather than the other way around. Inspectors are expected not just to check materials but also to determine if the facility in question is performing as indicated by the design information verification submitted by the member state to the Agency.

Training increasingly also involves "soft" skills that impart cultural values: observation, negotiation, communication, and interviewing techniques. Different communication styles are conveyed, notably the difference between interviewing and negotiation. The development and delivery of a new course on writing skills for safeguards is envisaged. Training also seeks to familiarize inspectors with the history of safeguards (including past non-compliance cases) and the background to safeguards treaties and agreements. As Heinonen has warned, too much legal detail can be overwhelming for non-lawyers, but the inspectors need to be familiar enough with safeguards agreements to know what tools are legally available to them. The introductory course concludes with a comprehensive inspection exercise at a light-water reactor and the presentation of a case study.

Courses for continuing safeguards staff are also offered by the Training Section. These cover the full range of safeguards activities conducted at facilities and headquarters and aim to develop both "technical and behavioral skills."[238] The language used indicates that there has been a cultural shift at least in the way that the necessary skills are described.[239] Quality Management training for all staff, not just inspectors, is now mandatory.[240]

The new "completeness and correctness" mantra is being embedded in the culture through training. Inspectors are now being trained and encouraged to be more inquisitive, more investigatory, more questioning of their facility or government hosts, and more willing to take initiative in the field rather than automatically requesting permission from Vienna. An experienced inspector who conducts part of the introductory training course for new inspectors claims that the training is working: "In addition to measuring nuclear material, reviewing accountancy and auditing the books, we're always looking for signs or indications of potentially undeclared nuclear materials and activities."[241] The new motto is: "Be prepared: know what you want and how you think you might be able to get it."[242] A level of confidence needs to be imparted for inspectors to do this.

One challenge in a multicultural environment is that some national cultures are individualistic and assertive, while others are collaborative and deferential.[243] Some trainee inspectors say that their native culture does not allow them to be as assertive as the training requires. Others, with a more forceful national culture, want to call out a state as "lying" or "guilty," but need to be told to stick to the facts and only discuss what is provable. Such differences must be handled creatively in forging an inspector culture that is both assertive and collaborative. A former director of inspector training cautions that the issue is more one of personality and approach than national culture.[244] The Safeguards Department told the auditor in 2019 that a Learning Management System is being developed to track training and link suitable courses to individual staff members' competencies.[245]

The new training system also seeks to tackle the diversity of professional backgrounds among inspectors, who may bring their own sub-cultures with them. The aim is to teach them to think not like physicists, chemists, or engineers, which the majority still are, but as investigators.[246] They must be aware of "red flags" and inconsistencies and not be intimidated by the technology they are seeing; an uncooperative state may try to engage inspectors in technical details to distract them from non-compliance. While engineers are trained to focus on detail, they must also learn to look for the "bigger picture." As Casterton has pointed out, because inspectors like to verify, they tend to try to verify all information provided by a state, when in fact the more information a state provides, the less verification may be required.[247] Essentially, inspectors must learn to be whistle-blowers. This involves not just being prepared to uncover evidence of non-compliance, but also being confident enough to make the case for a violation to a potentially skeptical senior Agency supervisor.

Given the prominence of the State-Level Approach, it is especially important for all safeguards personnel to be trained in the systematic use of new analytical techniques, including critical thinking and "structure analysis."[248] Analytical skills training is designed to help analysts and inspectors avoid "group think" and employ competing hypothesis analysis, which can reportedly be remarkably effective, and to remove individual bias as much as possible. Participants are taught that there are three levels of analysis: objective analysis, with which they are all comfortable; subjective analysis, where some degree of subjectivity is required to draw a conclusion on the evidence; and the political level, where they should not venture. The Training Section is conducting a series of one-day workshops to teach participants in State

Evaluation Groups how to work as teams.[249] Joint training for inspectors and analysts should help inculcate a common culture. Inspectors are also being trained to utilize information derived from state-level analysis at headquarters during their on-site inspections.

Analysis is difficult to teach, as it often does not come naturally. As Charles Duelfer notes of his Iraq Survey Group experience:

> One of the things I kept reminding the analysts (and myself) of was to look for what was not there. A photograph shows some things, but it may also be important to notice what is not there (e.g., an image of a facility that had no vehicle tracks near it may indicate lack of use). Likewise, it is important to notice what is not said. . . . Another example would be the absence of a reaction when a reaction might be expected.[250]

Duelfer also notes an obsession of U.S. analysts with buildings—both for inspection and bombing purposes—whereas undeclared materials and weapons might more readily be "hidden in plain view" outside buildings or further afield.[251]

It is increasingly being recognized that there is a need to transmit "tacit knowledge," which lies not in handbooks or instructions, but is in large part cultural. Senior inspectors are key to passing on lessons in culture, especially in helping more junior inspectors know how far they can go in being proactive and assertive. The 2005 GAO report said that the Agency did not have a good mentoring system for new inspectors.[252] Many of the most qualified instructors were retiring, while the remainder, who were also full-time inspectors, had increasingly large workloads and less time to devote to teaching. The Agency has undertaken "knowledge management" efforts since 2007 in order to support supervisors in the retention of critical job-related knowledge from staff members retiring or otherwise leaving the department.[253] In 2012, however, the JIU reported that in some cases, the Agency's work had still been negatively affected when key staff left because the outgoing staff did not make available vital information and documentation for new or remaining staff, much less otherwise unattainable tacit knowledge.[254]

As for training to meet new non-proliferation challenges, the Agency reports that it now continuously updates its training program to match the evolution of safeguards implementation.[255] Additional training was provided at short notice in 2016–2017 to support verification in Iran and to address verification challenges at the Fukushima Daiichi site in Japan.[256] Training is also continuing for possible reintroduction of inspections in North Korea.[257]

In addition, the Training Section is cognizant of the need for training to prepare for the arrival of new technologies—both verification technologies to be employed by the Agency and new technologies in the nuclear industry. Facilities that will require the application of safeguards in the future include small and medium-size reactors (SMR); new reactor types including pebble-bed reactors, molten salt reactors, and floating reactors; and new types of spent fuel and waste management facilities.[258]

Despite the significant improvements in the scope and intensity of safeguards training, however, it is still devoted mostly to training in technical skills. Although training seeks to change safeguards culture to accommodate the new espoused values of strengthened safeguards, there is no specific mention of safeguards culture in safeguards training documentation or plans. This sharply contrasts with global practices in nuclear safety and security, where no respectable introductory course would be complete without reference to culture and at least a class on what it is and how to enhance it.

It is not entirely clear, therefore, that despite all the enhanced training efforts, a new safeguards culture has been fully embedded in the inspectorate and the Safeguards Department at large. An article called "A Day in the Life of a Safeguards Inspector," which was published in July 2016 by the IAEA Office of Public Information and Communications, was unintentionally revealing.[259] It still emphasized detecting "diversion" as the aim of safeguards, mentioned correctness and completeness only in relation to facility design information verification, and stressed the need for inspector capabilities that seemed decidedly old school. "Safeguards inspectors need adaptability and good judgement. They need to learn quickly and pay attention to details," according to Hilario Munyaradzi, who worked as an inspector for eight and a half years and spent the past five years training new inspectors.[260] "It's a very physical job," said Abdella Chahid, who had been an inspector for sixteen years. "You really need to be fit and you need to have stamina."[261] There was no mention in the article of inspectors being more investigatory and concerned about undeclared material and activities beyond the inspected facility; no reference to what happens if an anomaly or other problem is discovered, including the need to contact headquarters or seek complementary access; and nothing about involvement in State-Level Approach analysis back at headquarters.

To be fair, the article was meant to be a journalistic account of a typical day in the life of an inspector, but it gave a simplistic impression of a mechanistic process reminiscent of the old safeguards culture. An illustrative inspector schedule at the end of the article was captioned: "walk a mile in the 'overshoes' of a safeguards inspector: the timeline of a routine inspection of an average nuclear power plant goes something like this." It ended with "arrive back at hotel at 5 p.m.," as if inspectors keep office hours and nothing ever requires staying longer on the job.[262] If cultural signals are sent by such characterizations, then this article was unhelpful.

A 2019 report for the Swedish Radiation Safety Authority pointed to examples of inspectors "who are unaware of or non-compliant with the requirements of safety and security in a facility, who are not fully informed of the legal framework (including constraints on the IAEA) or who simply misbehave or engage in combative behaviour with the operator or the State."[263] It concluded that while "luckily, the examples are few," they warrant attention. Obviously, no training program can be perfect, but it is concerning that inspectors may still be sent out into member states' nuclear facilities and territories without proper inculcation in the IAEA's safeguards culture.

The Organization for Economic Cooperation and Development (OECD) recommends that inspector training should result in the trainees' having "beyond technical skills, a real understanding of what inspections and enforcement aim to achieve, how to interact most effectively with regulated subjects (and in particular foster compliance), how to assess and rank risk—as well as all relevant ethical standards."[264] This additional OECD recommendation, although targeted at national regulatory bodies, seems especially relevant to IAEA safeguards culture:

> Effective enforcement cannot take place without discretion—there are too many issues to be decided that require enforcement officials to make a judgement call, to interpret the regulatee's behaviour and intent, to decide on what they think will be the most appropriate course of action in a "responsive regulation" vision, to decide which points to investigate more in depth during an inspection, etc. All this means that discretion needs to be exercised within a clear framework, but early understood to be an inevitable part of the process. This in turn requires inspectors to have a high level of competence and professional standards.[265]

The OECD also suggests that regulatory organizations follow the International Organization for Standardization (ISO) standards for inspections and seek accreditation for this.[266] The IAEA has already done this for its Safeguards

Analytical Laboratory. Given the IAEA's ingrained cultural perception that its safeguards system is unique, it may be unwilling to subject itself to such an accreditation process, although it could learn from the widely accepted standards themselves.

Incentives and Rewards

Another recognized way to change a culture is through incentives and rewards for both individuals and groups. As long ago as the early 1980s, Tom Shea and David Thompson enumerated several elements of the IAEA's personnel policies that could enhance the "human factors" (essentially the culture) that would contribute to improving safeguards.[267] These included recruitment, promotion, and tenure policies; salaries and compensation; pay incentives; and employee evaluation. One of the goals of the Safeguards Department's Strategic Plan is to "work to make the Department more capable of attracting and retaining high-calibre staff."[268]

IAEA personnel are well-compensated, and in most cases their earnings are tax-free (depending on the employee's state of origin). In addition, there are excellent health and pension benefits. It is beyond the scope of this study to consider all of the other rewards and incentives that the IAEA might employ to build a stronger safeguards culture. Many are generic and would therefore apply to all Agency employees, not just the Safeguards Department.

The Agency already has an awards program that, at the highest level, is called the Distinguished Service Award. Certificates from the Director General and monetary awards for performance excellence are also given annually.[269] Defining an optimal IAEA safeguards culture should be a prerequisite for designing specific incentives to foster it, but initiatives could include a safeguards award system, such as Inspector of the Year and Analyst of the Year. More controversial is the idea of rewards for uncovering significant safeguards non-compliance. Reflecting the safeguards culture of the day, in the early 1980s Shea and Thompson suggested a reward of $100,000 tax-free for detecting an "actual diversion of nuclear material."[270] The OECD cautions regulatory bodies that "staff that systematically find the highest possible number of violations and issue the highest possible sanctions" should not be considered "high performers," presumably because of the perverse incentives created by such accolades.[271] On the contrary, it advises that staff that effectively promote compliance and work in line with the principles of "responsive regulation" should be given adequate recognition.[272] Responsive Regulatory Theory, devised in the early 1990s, holds that inspection and compliance outcomes are improved if inspectors are responsive to the degree of compliance demonstrated by those being inspected. In order to do this, they need discretion and professional skills to judge the situation appropriately.[273]

Kenneth Abbott and Duncan Snidal, though confident that some elements of Responsive Regulatory Theory are applicable to international organizations, caution that it was "designed for domestic settings in which a single agency had clear jurisdiction, full regulatory capacity, and extensive information and could (contingently) deploy stringent sanctions against well-defined targets."[274] None of these features pertain in the IAEA's case. Not only do IAEA inspectors lack clear jurisdiction over inspected countries, but they also cannot themselves impose sanctions. If they become too "responsive" to the inspected state, they risk regulatory capture and being lulled into complacency. Any reward system for IAEA inspectors would need to take such considerations into account. As an indication of how complicated (and circular) making changes to organizational culture can become, the OECD opines that "improving the way regulatory enforcement agencies work involves not only changing the statutes, rules and *incentives* [emphasis added], but also in many cases changing cultures and behaviors."[275] This begs the question of what comes first.

Interviews conducted for this book indicate that one compelling incentive is neither monetary nor related to working conditions, but instead involves pride in the significant contribution that IAEA staff believe the Agency makes to international peace and security. Shea says he is amazed that people would become so dedicated to the Agency's mission.[276] Wisely, the Agency features this enthusiasm in its recruitment and training. Profiles of safeguards inspectors chosen for the Agency's recruitment pamphlet emphasize the international character of the work, the multicultural environment, the possibility of travel, the joys of living in Vienna, and the contribution of the work to international peace and security. Below are two examples of the Agency's characterization of the incentives for becoming an IAEA inspector.

> Davide Parise of Italy: "Working for a non-profit organization that benefits people around the globe is his strongest motivation. He also enjoys the good work environment, surrounded by colleagues of different cultures. In addition, the IAEA offers a non-standard experience—constantly learning things that are not written in books. . . . Although being a nuclear safeguards inspector can be stressful at times, he values the possibility of seeing countries from an unusual, non-touristy perspective. Working in an international organization can be very different from working in a domestic field. Every

country has its own way of working, giving Davide the opportunity to develop and gain valuable experience. Davide thinks that Vienna is a wonderful city to live in, with a very high quality of life."[277]

Sahar Shawky of Egypt: "Sahar is extremely committed to her work, and finds it to be interesting as well as challenging. She is proud to be contributing to the international effort to further peace. Nearly four months out of the year, Sahar is travelling around Japan, and other parts of Asia, conducting inspections. This physically demanding role can be considered even more challenging due to language and cultural differences—something that Sahar believes is actually one of the best aspects about working at the IAEA. She believes that the diversity found amongst her colleagues creates an interesting work environment, and can ultimately impact and change the way that people interact with one another. Sahar's work goes above and beyond that of her dedication to her job and to the IAEA."[278]

The website Glassdoor, which hosts reviews of workplaces by employees, has several pertaining to the IAEA.[279] The advantages of working for the Agency are invariably described in terms of the laudable international goals of the Agency, the international work environment, the delights of Vienna, and the pay and benefits, including work-life balance. One interlocutor noted, however, that "about 90 percent of the complaints I hear revolve around human resources management issues such as recruitment and retention; find an easy but fair way to terminate bad employees and offer long-term contracts to the best ones."[280]

The survey that the JIU conducted in 2012, which is the most recent one available, included sixty-two closed questions (with a multiple-choice option) and one open question for qualitative assessments, comments, and suggestions. The report concluded that "the great majority of staff value [the] IAEA as a good place to work and they are motivated and dedicated."[281] Staff overall appreciated the good work-life balance arrangements at the Agency, felt they were treated with respect, and considered that "cultural differences are well respected." Staff obviously valued having their views sought. They felt, and the JIU concurred, that periodic staff surveys should be conducted to "allow staff at large to comment, in a confidential setting, on relevant issues pertaining to their daily work, while at the same time provide administration and management with feedback and input on their work. It may also serve as an additional instrument for consultation and positively affect staff morale, enhance transparency and overall trust in the organization by staff."[282]

Whither Safeguards Culture?

Despite calls for revamping nuclear safeguards culture following the Iraq case in the early 1990s, and unlike in the fields of nuclear safety and nuclear security, cultural change has not been widely recognized as part of the response to sustaining the effectiveness of IAEA safeguards. The Secretariat has not adopted a deliberate cultural change strategy, nor for the most part has it used the language of culture. Yet revolutionary changes in the safeguards system since the Iraq case have inevitably produced changes in safeguards culture in the three key areas identified by organizational theorists: artefacts, espoused values, and underlying assumptions.

The greatest changes in espoused values have concerned correctness and completeness, the need for a more investigatory approach by inspectors, and the value of a more collaborative approach by all staff. The Safeguards Department has begun to make changes that have been culturally sensitive, such as the consultations that produced the Strategic Plan and its updates, improvements in recruitment and training, and reforms resulting from the Quality Management process. Staff turnover and generational change are helping to ensure that a new culture is becoming more widespread over time, but it also means that the culture may change in unexpected ways, especially as the proportion of women and personnel from under-represented countries increases. The department also faces the continuing challenge of integrating or at least reconciling several sub-cultures, especially the bureaucratic and scientific, as well as the inspector and analyst sub-cultures.

Skepticism about the value of the cultural approach still abounds within the Safeguards Department, presumably due to a lack of understanding about the insights it can provide and perhaps fear of what it might reveal. This occurs despite the Agency's routinely urging its member states to attend to culture, not just in the areas of safety and security, but also in undertaking organizational change to strengthen their national nuclear agencies.

The elements of an optimal nuclear safeguards culture should be apparent. Some of these are aspirations that all organizations should have: organizational excellence; a sense of service and loyalty; and a commitment to effectiveness and efficiency. Other values are specific to the Agency as an international organization dedicated to a higher cause than its own well-being, notably international peace and security. An optimal safeguards culture should embody a strong commitment by the entire Agency to the non-proliferation regime.

Despite its best intentions, the Safeguards Department is unable, by itself, to change Agency-wide, much less UN, cultural norms that deeply affect safeguards culture—the most prominent being those related to leadership, management style, recruitment, and promotion. Changes in these areas would require action from the highest levels of the Agency, the Director General, and senior staff, as well as the Board of Governors and the general membership.

As for safeguards culture specifically, the Secretariat should engage the entire safeguards community, including member states, to reach an agreed definition of safeguards culture and identify the elements that constitute an optimal culture, just as the nuclear safety and security communities have done. While such an exercise will not automatically lead to cultural change, it can serve as a guide and inspiration to the Secretariat, member states, and other stakeholders. Furthermore, the Agency should commission a survey and study of its organizational culture by qualified management experts, with a focus on safeguards and related staff. This should include reflections on the impact on safeguards culture of the staff rotation policy, recruitment and training practices, staff assessment and counseling, and the rewards system. The Agency should also seek lessons from other organizations with regulatory functions. When contemplating major organizational change, it should from the outset include consideration of the likely cultural impact and put into place measures to achieve the desired cultural shift.

Six

Conclusions and Recommendations

SAFEGUARDS CULTURE, IT turns out, was likely a part of the explanation for what happened in Iraq. It therefore needs to be part of the solution to the perpetual challenge of sustaining the effectiveness of IAEA safeguards. In the thirty years since the Iraq case broke, the Secretariat has not, for the most part, used the language of culture, nor has it said that it was deliberately seeking cultural change. Yet the revolutionary changes made by the Agency that have been designed to strengthen the safeguards system since the Iraq case, along with the organizational and personnel changes designed to implement them, have inevitably produced changes in safeguards culture.

In recent years, the Safeguards Department has instituted changes that have avowedly been culturally sensitive, notably in strategic planning, quality management, and training. New artefacts and enhanced espoused values have emerged in safeguards culture. The greatest changes in espoused values have concerned correctness and completeness, the need for a more investigatory approach by inspectors, and the value of a more collaborative approach by all personnel and parts of the department.

Overall, safeguards culture today appears to embody values and attitudes that comport well with the Agency's mission to help prevent nuclear weapons proliferation. The Safeguards Department is rightly proud of the way it has handled the most serious non-compliance cases it has confronted since the Iraq case. It is also justifiably pleased with its continuing efforts to strengthen safeguards while enduring mostly zero real growth in its budget.

Cultural change takes time, however, and may not yet be fully reflected in underlying assumptions held by safeguards personnel. Reactions to probing questions about safeguards culture in a group situation reveal a high degree of group solidarity and dedication to the mission. In private, misgivings are aired. This is only natural, as no organization is perfect. Behind the pride in the accomplishments of safeguards, there remains a defensiveness among safeguards employees when questioned about their culture. Attacks on the State-Level Concept and the travails of the Iran non-compliance case have not helped. Only a comprehensive survey and in-depth study, with access to all safeguards personnel, would reveal the true underlying assumptions that currently prevail in safeguards culture.

Staff turnover and generational change help to ensure that a new culture becomes widespread over time. The relatively high turnover of staff at the IAEA, due in part to the rotation policy, means that the culture may be more amenable to deliberate cultural change, but it also means that the culture may change in unexpected ways. The increasing proportion of women and personnel from under-represented and developing countries, albeit slow, will inevitably change the organization. The Safeguards Department also faces the continuing challenge of integrating or at least reconciling several sub-cultures, especially the bureaucratic and scientific, as well as the inspector and analyst sub-cultures. Skepticism about the value of the cultural approach still exists, presumably due to a lack of understanding about the insights it can provide and perhaps fear of what it might reveal. More open attention to both the positives and negatives of safeguards culture, past and present, may alleviate this problem.

Despite its best intentions, however, the Safeguards Department is unable, by itself, to change Agency-wide, much less UN, cultural norms that deeply affect safeguards culture—the most prominent being those related to leadership, management style, recruitment and staff promotion, and the standing of safeguards within the organization and externally. This requires action from the highest levels of the Agency, as well as the Board of Governors and the general membership.

My key recommendations on enhancing safeguards culture at the IAEA are as follows:

- The Agency should engage the entire safeguards community, including member states, in defining nuclear safeguards culture and identifying the elements that constitute an optimal culture, just as the nuclear safety and security communities have done. While such an exercise will not automatically lead to cultural change, it can serve as a guide and an inspiration to the Secretariat, member states, and other stakeholders.
- The Agency should commission a survey and study of its organizational culture by qualified management experts, with a focus on safeguards and related staff.
- The Secretariat and member states should take organizational culture more seriously as a factor that can either hinder or facilitate planned organizational change, depending on how it is handled. When contemplating major organizational change, the Agency should include consideration of the likely cultural impact from the outset and put in place

measures to achieve the desired cultural shift. This includes appointing an officer in charge of cultural change management.

- The Agency should consider how to reinforce the status of its inspectors while taking care to protect the collaborative culture that it is seeking to foster among all safeguards personnel.

- The Agency should commission a study of the impact on safeguards culture of its rotation policy, recruitment practices, and other staff procedures, including assessment and counseling; this would include learning how other international organizations create a loyal, permanent inspectorate with a robust regulatory culture, either within or outside UN employment practices.

- The Agency should consider how its rewards system can be shaped to encourage an optimal safeguards culture: this might involve a Safeguards Award System, including rewards for uncovering significant safeguards non-compliance.

- A new campaign to improve intra-Agency transparency and information-sharing should be launched, accompanied by appropriate incentives and rewards.

The elements of an optimal safeguards culture should now be apparent. Some of these are generic ambitions that all organizations should aspire to: organizational excellence; a sense of service and loyalty; and a commitment to effectiveness and efficiency. The Agency already does reasonably well in promoting organizational loyalty and a sense of belonging, with the exception of the contra-indicated aspects of the rotation and promotion systems. Other values are more specific to the Agency as an international organization dedicated to a higher cause than its own well-being, notably international peace and security. At the level of the individual safeguards employee, the culture already appears to exhibit a strong personal commitment to safeguards as a means of preventing the proliferation of nuclear weapons. This needs to be sustained and promoted, because it is easy in the day-to-day pursuit of technical goals to lose sight of the ultimate aim of the safeguards enterprise.

An optimal safeguards culture should embody a strong commitment to the non-proliferation regime by the entire Agency, not just the Safeguards Department. It is clear that some parts of the IAEA do not see safeguards as the most important role that the Agency performs for the global community. It is up to the Director General to set the cultural tone in this regard. Positioning the Agency as a quasi-development body, while useful in securing the support of developing countries, may not be helpful in the long run. The Agency can certainly contribute to the achievement of development goals, but it is not in charge of them; it is demonstrably in charge of the implementation of nuclear safeguards. Any adjustment of the Agency's self-image should, of course, be accomplished without making developing countries unduly concerned that reduced attention will be paid to their priorities.

In the new safeguards system, unlike in the old, everyone's contribution to safeguards, whether manager, inspector, or analyst, needs to be appropriately acknowledged and rewarded. Given the unique value of the inspectors as the "eyes and ears" of the Agency on the ground, the Secretariat needs to accomplish the tricky feat of reassuring the inspectorate of its continued unique value and importance while also integrating it into more collaborative analytical and planning processes that now involve a much bigger team of players. In addition, ways need to be found to reassure the inspectors of their value in the face of increasing deployment of advanced safeguards technologies that make some aspects of inspectors' jobs redundant.

One cultural value that is vital to an optimal safeguards culture but has not been successfully established to date, despite efforts under the "One House" campaign, is the need for greater internal and external transparency and information-sharing. Opacity helps preserve the Secretariat's autonomy in managing safeguards and avoids member states' second-guessing or seeking to micromanage its decisions, but it also makes it difficult to tell exactly when planned organizational changes have been implemented. Change is characterized as part of "evolving safeguards implementation," but without specific beginning and end points.

In addition, a genuine shift is required from "need to know" to "need to share." Change in this cultural norm can also only come from the top, at both the levels of the Director General and the department heads. It is to be hoped that the latest Director General, Rafael Grossi, with his more open and communicative personal style, will set the cultural tone in this regard. Appropriate incentives and disincentives need to be put into place, but nothing will work as well as leading by example with more transparency from senior management. As for external transparency, overturning the safeguards culture of maximum confidentiality at the expense of keeping stakeholders well informed can only be initiated by the Director General and the Deputy Director General for Safeguards.

The need to ensure an optimal safeguards culture has assumed even greater importance with the advent of the Joint Comprehensive Plan of Action (JCPOA) in 2015 and the continuing non-compliance of Iran. With the rejection of this agreement by the Trump administration in 2017, the case has become even more challenging for the Agency, which faces the invidious situation of verifying compliance with an agreement that is gradually withering away and may collapse. In the meantime, it is left verifying both compliance and non-compliance. The issue of weaponization, which the JCPOA specifically mentions, is another challenge. While the Secretariat has concluded that dealing with weaponization is within its remit, it has limited verification capabilities to do so and walks a fine line between demonstrating its competence and alarming member states. The Agency will have to be nimble and creative in handling all possibilities. The parties to the agreement, the UN Security Council, and the entire international community are watching closely to see how the "nuclear watchdog" performs. Not since the Iraq case have safeguards personnel been required to be as intrusive, inquisitive, investigatory, questioning, and persistent. Their new safeguards culture needs to be on full display.

Other crises will undoubtedly challenge the safeguards system and its culture. Among the "known unknowns" is North Korea. The Agency is standing by to return inspectors to that country if an agreement is struck to freeze or reduce its nuclear weapons capability. Also difficult to predict is the impact of new nuclear technologies, such as advanced reactors, on the credibility of safeguards. The utilization of new verification technologies for safeguards is another challenge, especially if it threatens to replace human inspectors. "Unknown unknowns" can be even more challenging, but can also produce innovation. The global COVID pandemic that began in 2020 has threatened the continuity of safeguards inspections, but the IAEA has risen well to the occasion.

These and future unknown challenges to safeguards and the nuclear nonproliferation regime will test the IAEA and the robustness of its safeguards culture. As this study has shown, culture is woven into virtually all aspects of the safeguards enterprise, from management at headquarters to operations in the field, from the Agency's reputation as a fair and equitable employer to its standing as a contributor to international peace and security. Maximizing the effectiveness of IAEA safeguards therefore requires attention to safeguards culture. The IAEA has made eminently sensible recommendations to its member states about how they should handle culture in the context of organizational change at their national nuclear agencies. It behooves the Agency to apply such admonitions to itself.

Notes

Preface

1. The IAEA Annual Report and Annual Program and Budget are remarkably silent on such matters. Furthermore, organizational matters specific to safeguards go unrecorded in the annual report of the Director General to the IAEA General Conference on "Strengthening the Effectiveness and Improving the Efficiency of the Agency Safeguards System."

2. See Toby McIntosh and William Burr, "IAEA Lacks Transparency, Observers, Researchers Say," FreedomInfo.org, http://www.freedominfo.org/2015/04/iaea-lacks-transparency-observers-researchers-say/. For documents dealing with member states, permission is sought from those concerned, meaning that in practice these are rarely released. In 1996, the Agency's Board of Governors approved the derestriction of its documents after two years, except for the Safeguards Implementation Report (SIR), closed-session records, and documents subject to legal impediments. Paradoxically, the decision was classified "Restricted." In any event, such a reform has not been implemented. See IAEA, Board of Governors, Proposal for the Derestriction of Board Documents, GOV/2843, February 15, 1996; and "Leaked Document Describes IAEA Transparency Proposal," FreedomInfo.org, April 24, 2015, http://www.freedominfo.org/2015/04/leaked-document-describes-iaea-transparency-proposal/.

Introduction

1. Gudrun Harrer, *Dismantling the Iraqi Nuclear Program: The Inspections of the International Atomic Energy Agency, 1991–1998* (Abingdon, Oxon: Routledge, 2014), p. 91.

2. For a study of all safeguards non-compliance cases to date, see Trevor Findlay, "Proliferation Alert! The IAEA and Non-Compliance Reporting," Report No. 2015–04, Project on Managing the Atom, Belfer Center for Science and International Affairs, Harvard University, Cambridge, Mass., October 2015.

3. Hans Blix, *Disarming Iraq: The Search for Weapons of Mass Destruction* (London: Bloomsbury, 2005), p. 24.

4. Michael Wilson, "Safeguards and the IAEA Board of Governors: 1991–1993: Iraq, A Necessary Stimulus for Handling the DPRK," in David Fischer, ed., *International Atomic Energy Agency: Personal Reflections* (Vienna: IAEA, 1997), p. 130.

5. David Fischer, *History of the International Atomic Energy Agency: The First Forty Years* (Vienna: IAEA, 1997), pp. 285–286.

6. Ibid., p. 103.

7. Paul C. Szasz, "The Adequacy of International Nuclear Safeguards," *Journal of International Law and Economics*, Vol. 10 (1975), p. 435.

8. Although ElBaradei says that "these questions have good answers," he does not, at least in his autobiography, provide them in any detail. Mohamed ElBaradei, *The Age of Deception: Nuclear Diplomacy in Treacherous Times* (New York: Metropolitan Books, 2011), p. 10.

9. For further discussion of this definition, see Chapter 3 in this volume.

10. The literature on organizational culture is vast. The founding guru is Edgar Schein. See E.H. Schein, "The Three Cultures of Management: Implications for Organizational Learning," *Sloan Management Review*, No. 38, pp. 9–20; and E.H. Schein, *Organizational Culture and Leadership*, 4th ed. (San Francisco: Jossey-Bass, 2016). For a variety of perspectives, see Deborah G. Ancona, Thomas A. Kochan, Maureen Scully, John Van Maanen, and Eleanor Westney, *Managing for the Future: Organizational Behavior and Processes*, 3rd ed. (Boston: South-Western College Publishing, 2004).

11. John S. Carroll, "Understanding Safety Management Through Strategic Design, Political, and Cultural Approaches," in J.U. Hagen, ed., *How Could This Happen? Managing Errors in Organizations* (Cham, Switzerland: Springer Nature Switzerland AG, 2018), p. 132.

12. Ibid., p. 133.

13. Ibid.

14. John Child, *Organization: A Guide to Problems and Practice*, 2nd ed. (London: Harper and Row, 1984), p. 9. An apt example is Barbara Koremenos, Charles Lipson, and Duncan Snidal, eds., *The Rational Design of International Institutions* (Cambridge: Cambridge University Press, 2004).

15. John S. Carroll, "Introduction to Organizational Analysis: The Three Lenses," unpublished class paper, Sloan School of Management, MIT, revised June 2006, p. 3; and Jeffrey Pfeffer, *Power: Why Some People Have It—And Others Don't* (New York: HarperCollins, 2010).

16. Approaches range from realists such as Hans Morgenthau (Hans. J. Morgenthau, *Politics Among Nations: The Struggle for Power and Peace* [New York: Alfred A. Knopf, 1962]) to soft power advocate Joseph Nye (Joseph S. Nye, Jr., *The Future of Power* [New York: Public Affairs, 2011]).

17. Oran R. Young, "The United Nations and the International System," in Robert S. Wood, ed., *The Process of International Organization* (New York: Random House, 1971), p. 10.

18. Ian Hurd, *International Organizations: Politics, Law, and Practice*, 2nd ed. (Cambridge: Cambridge University Press, 2014), p. 7.

19. Diane Vaughan, *The Challenger Launch Decision: Risky Technology, Culture, and Deviance at NASA* (Chicago: University of Chicago Press, 1991), p. 64.

20. Carroll, "Understanding Safety Management Through Strategic Design, Political, and Cultural Approaches," p. 135.

21. Quote Investigator, "Culture Eats Strategy for Breakfast," https://quoteinvestigator.com/2017/05/23/culture-eats/.

22. Vaughan, pp. 62–64.

23. Charles Perrow, *Normal Accidents: Living with High-Risk Technologies* (Princeton, N.J.: Princeton University Press, 1999).

24. See, for example, Institute of Nuclear Power Operations (INPO), "Principles for a Strong Nuclear Safety Culture," INPO, November 2004; and World Institute for Nuclear Security (WINS), "Nuclear Security Culture," a WINS International Best Practice Guide, WINS, Vienna, September 2011.

25. See especially Stephen V. Mladineo, Karyn R. Durbin, and Andrew Van Duzer, "Changing the Safeguards Culture: Broader Perspectives and Challenges," INMM/ ESARDA Workshop on Changing the Safeguards Culture: Broader Perspectives and Challenges, October 30–November 2, 2005, Santa Fe, New Mexico; and unpublished presentations from Safeguards Culture Workshop, Texas A&M University, College Station, Texas, April 26–27, 2016, http://www.inmm.org/Content/NavigationMenu /Events/UpcomingEvents/SafeguardsCultureWorkshop_Agenda_as_of_04-14-16.pdf.

Chapter One

1. Statute of the International Atomic Energy Agency, Article II.

2. IAEA, The Annual Report for 2019, GC(64)/3, Vienna, 2020, p. vi, https://www .iaea.org/sites/default/files/publications/reports/2019/gc64-3.pdf.

3. IAEA, Safeguards Statement for 2020, para. 60, https://www.iaea.org/sites/default /files/21/06/statement-sir-2020.pdf.

4. IAEA, The Agency's Safeguards System (1965, as provisionally extended in 1966 and 1968), INFCIRC/66/Rev. 2, September 16, 1968. INFCIRC is an abbreviation of Information Circular.

5. IAEA, The Agency's Inspectorate: Memorandum by the Director General; and Annex, The Agency's Inspectors, GC(V) INF/39, Vienna, August 28, 1961 (hereafter the "Inspectors' Document"). For background, see David Fischer, *History of the International Atomic Energy Agency: The First Forty Years*, (Vienna: IAEA, 1997), p. 247.

6. Ibid., p. 1.

7. Ibid.

8. Ibid., p. 2.

9. Statute of the International Atomic Energy Agency, Article XII.A.6.

10. David Fischer, *History of the International Atomic Energy Agency: The First Forty Years*, p. 82.

11. Ibid., p. 248.

12. Ibid., p. 82; and Carlos L. Büchler, "Safeguards: The Beginnings," in IAEA, *International Atomic Energy Agency: Personal Reflections* (Vienna: IAEA, 1997), p. 48.

13. Ibid., p. 49.

14. Ibid.

15. Treaty on the Non-Proliferation of Nuclear Weapons, also known as the Nuclear Non-Proliferation Treaty (NPT), http://disarmament.un.org/treaties/t/npt.

16. "The Structure and Content of Agreements between the Agency and States Required in Connection with the Treaty on the Non-Proliferation of Nuclear Weapons [NPT]," INFCIRC/153 (Corrected), IAEA, Vienna, June 1972 (hereafter INFCIRC/153), Part II, para. 28, https://www.iaea.org/sites/default/files/publications/documents/infcircs/1972/infcirc153.pdf.

17. See Allan McKnight, *Atomic Safeguards: A Study in International Verification* (New York: United Nations Institute for Training and Research, 1971), p. 117.

18. IAEA, The Standard Text of Safeguards Agreements in Connection with the Treaty on the Non-Proliferation of Nuclear Weapons [NPT], Note by the Director General, GOV/INF/276, August 22, 1974; IAEA Department of Safeguards, Subsidiary Arrangement to the Agreement by the Government of [. . . . .] and the International Atomic Energy Agency for the Application of Safeguards in Connection with the Treaty on the Non-Proliferation of Nuclear Weapons [NPT], Fifth Revision, SG-FM-1170, November 2, 2011.

19. INFCIRC/153, Part II, para. 29.

20. IAEA Safeguards Glossary, 2001 ed., International Nuclear Verification Series, No. 3, IAEA, Vienna, 2001, p. 46.

21. Joseph Pilat, "The International Atomic Energy Agency (IAEA) and its Safeguards: A Fact Sheet," U.S. Congressional Research Service, Washington, D.C., May 6, 1982, p. 22.

22. Robert Pendley, Lawrence Scheinman, and Richard W. Butler, "International Safeguarding as Institutionalized Collective Behavior," *International Organization*, Vol. 29, No. 3 (June 1975), p. 614.

23. Lawrence Scheinman, *The International Atomic Energy Agency and World Nuclear Order* (Washington, D.C.: Resources for the Future, 1987), p. 156.

24. This was embodied in the IAEA Statute, Article VII.F.

25. IAEA, Code 3.1, Subsidiary Arrangement to the Agreement between the Government of [. . .] and the International Atomic Energy Agency for the Application of Safeguards in Connection with the Treaty on the Non-Proliferation of Nuclear Weapons, Fifth Revision, March 12, 1974, https://www.iaea.org/sites/default/files/sg-fm-1170-subsidiary-arrangement-code-1-9.pdf.

26. INFCIRC/153, Article 91.

27. IAEA Safeguards Glossary, pp. 23–24, para. 3.15–3.17.

28. IAEA Safeguards Glossary, Table 1, p. 23, para. 3.14.

29. David Fischer and Paul Szasz, *Safeguarding the Atom: A Critical Appraisal* (London and Philadelphia: Taylor and Francis for the Stockholm International Peace Research Institute [SIPRI], 1985), p. 32.

30. IAEA, IAEA Safeguards, Aims, Limitations, and Achievements, IAEA/SG/INF.4, Vienna, 1983, p. 26. These values have remained unchanged despite the urging of some critics to revise them in the light of reassessments of the amount of material required for a crude nuclear explosive device. See Henry Sokolski, "Assessing the IAEA's Ability to Verify the NPT," and Thomas B. Cochran, "Adequacy of IAEA's Safeguards for

Achieving Timely Detection," in Henry Sokolski, ed., *Falling Behind: International Scrutiny of the Peaceful Atom* (Carlisle, Penn.: Nonproliferation Policy Education Center, Strategic Studies Institute, U.S. Army War College, 2008).

31. See John P. Boright, "Historical Development and Current Trends in Nuclear Safeguards," Lecture at the International Training Course on Implementation of State Systems of Accounting for and Control of Nuclear Materials, March 15–30, 1982, reproduced in U.S. Congressional Research Service, *Nuclear Safeguards: A Reader*, Report Prepared by the U.S. Congressional Research Service for the Subcommittee on Energy Research and Production transmitted to the Committee on Science and Technology, U.S. House of Representatives, 98th Cong., 1st sess., December 1983, U.S. Government Printing Office, Washington, D.C., December 1983, p. 45. This point has been widely misunderstood by critics and supporters alike.

32. It superseded the "Inspectors' Document," which remained current for INCIRC/66/Rev. 2 agreements.

33. Scheinman, p. 153.

34. INFCIRC/153, Article 84.

35. Olli Heinonen, "International Atomic Energy Agency Inspections in Perspective," in Henry Sokolski, ed., *Moving Beyond Pretense: Nuclear Power and Nonproliferation* (Arlington, Va.: Strategic Studies Institute, U.S. Army War College, 2014), p. 305.

36. INFCIRC/153, Article 71.

37. Fischer, *History of the International Atomic Energy Agency: The First Forty Years*, p. 256.

38. IAEA Safeguards Glossary, p. 87.

39. Fischer, *History of the International Atomic Energy Agency: The First Forty Years*, p. 257.

40. INFCIRC/153, Article 37. Material that that Agency deemed irrecoverable could also be excluded at the state's request.

41. IAEA Safeguards Glossary, p. 16, para. 2.11.

42. INFCIRC/153, Article 42.

43. Fischer and Szasz, p. 80.

44. IAEA Statute, Article XII.

45. Hans Blix, *Disarming Iraq: The Search for Weapons of Mass Destruction* (London: Bloomsbury, 2005), p. 18.

46. Ibid.

47. Scheinman, pp. 16–17.

48. Observers in the mid-1970s predicted that satellites would one day be able to verify that all significant nuclear plants had been declared to the IAEA. France even proposed an International Satellite Monitoring Agency (ISMA) partly for this purpose in 1973.

49. Blix, p. 18.

50. The Agency did occasionally advance changes in this direction. In its 1981 Introduction to Safeguards, it declared that it was "necessary for *all* nuclear material in a

country's fuel cycle to be safeguarded if the IAEA is to be in a position to give assurances of non-diversion for the State as a whole." As usual, however, the emphasis was on diversion. IAEA, IAEA Safeguards: An Introduction, IAEA/SG/INF/3, October 1981, p. 16.

51. U.S. Arms Control and Disarmament Agency (ACDA), "Review of the Negotiating History of the IAEA Safeguards Document INFCIRC/153," AC2NC103, Washington, D.C., July 30, 1984, p. vii.

52. McKnight, p. 111.

53. IAEA Statute, Article II.

54. Fischer and Szasz, p. 6.

55. Ibid., pp. 5–6.

56. IAEA, The Annual Report for 1990, GC(XXXV)/953, Vienna, October 1991, p. 107, https://www.iaea.org/About/Policy/GC/GC35/GC35Documents/English/gc35-953 _en.pdf.

57. Ibid., p. 119.

58. Ibid., p. 120.

59. The Agency's Accounts for 1990, GC(XXXV)/954, Vienna, August 1991, Table 1, p. 5, https://www.iaea.org/About/Policy/GC/GC35/GC35Documents/English/gc35 -954_en.pdf.

60. See the efforts of successive Deputy Director Generals for Safeguards in "Paying Tribute to 25 Years of Safeguards Leadership," *IAEA Bulletin*, No. 3 (1994), pp. 14–15.

61. Myron B. Kratzer, "Prospective Trends in International Safeguards," *Nuclear News*, October 1980, p. 57.

62. Although Committee II agreed on this language for inclusion in the conference's final document, the plenary failed to agree on the document as a whole and the conference ended in failure. See Fourth Review Conference of the Parties to the Treaty on the Non-Proliferation of Nuclear Weapons (NPT), Report of Main Committee II, Geneva, August 20–September 14, 1990, NPT/CONF.IV/MC.II/l, September 10, 1990, pp. 5–6.

63. This statement was said to be "based on all the information available to the Agency, including information derived from safeguards activities conducted in the field and at Headquarters and information provided in reports submitted by States." See IAEA, The Annual Report for 1989, GC(XXXIV)/915, Vienna, July 1990, p. 102, https://www.iaea.org/About/Policy/GC/GC34/GC34Documents/English/gc34-915 _en.pdf.

64. Ibid., p. 113.

65. Målfrid Braut-Hegghammer, *Unclear Physics: Why Iraq and Libya Failed to Build Nuclear Weapons* (Ithaca, N.Y.: Cornell University Press, 2016), p. 44.

66. Ibid.

67. Ibid.

68. See U.S. Senate Committee on Foreign Relations, Hearings, The Israeli Airstrike, 97th Cong., 1st sess., June 18, 19, and 25, 1981, U.S. Government Printing Office, Washington, D.C., 1981.

69. For an account of Iraq's nuclear program from its inception, see Comprehensive Report of the Special Advisor to the Director of Central Intelligence (DCI) on Iraq's WMD, Vol. II, "Nuclear," Washington, D.C., September 30, 2004 (hereafter Duelfer Report, Vol. II, "Nuclear"), pp. 1–130; and Addendums to the Comprehensive Report of the Special Advisor to the DCI on Iraq's WMD, Washington, D.C., March 2005 (hereafter Duelfer Report, Addendums).

70. Ibid., p. 28.

71. Fischer, *History of the International Atomic Energy Agency: The First Forty Years*, p. 316.

72. Braut-Hegghammer, p. 45.

73. International Institute for Strategic Studies (IISS), "Iraq's Weapons of Mass Destruction: A Net Assessment," Strategic Dossier, September 9, 2002.

74. The reactor was named after the month in which the Ba'ath Party had come to power. An alternative English spelling is Tammuz. The French and other Westerners called it Osiraq (or Osirak), as it was based on the French Osiris design.

75. IAEA, The Annual Report for 1990, p. 126; David Albright, Corey Gay, and Khidhir Hamza, "Development of the Al-Tuwaitha Site: What if the Public or the IAEA had Overhead Imagery?" Institute for Science and International Security (ISIS), April 26, 1999, p. 3, http://isis-online.org/isis-reports/detail/development-of-the-al -tuwaitha-site-what-if-the-public-or-the-iaea-had-over/9.

76. Khidhir Hamza with Jeff Stein, *Saddam's Bombmaker: The Terrifying Inside Story of the Iraqi Nuclear and Biological Weapons Agenda* (New York: Scribner, 2000), pp. 22, 80–83. See also "From Aflaq to Tammuz: Iraq's Nuclear Weapons Program," p. 3, http://nuclearweaponarchive.org/Iraq/IraqAtoZ.html, accessed January 4, 2018. Hamza's claims should be treated skeptically. He has been largely discredited over his accounts of a continuing Iraqi nuclear program after 1991 and for his connections with the Iraqi National Council, which lobbied for the ill-fated U.S. invasion of the country in 2003. See Gudrun Harrer, *Dismantling the Iraqi Nuclear Program: The Inspections of the International Atomic Energy Agency, 1991–1998* (Abingdon, Oxon: Routledge, 2014), pp. 188–193. Saddam Hussein's brother-in-law, Hussein Kamil, told UN inspectors after his 1995 defection to Jordan that Hamza was a "professional liar." Braut-Hegghammer, p. 101.

77. Albright, Gay, and Hamza, pp. 3–4. Hamza claims that he was involved in planning clandestine plutonium production in this reactor and seeking to "defeat" IAEA safeguards (p. 70), but at other times he scoffed at the idea that Iraq could get away with violating them. He claims that the real aim in obtaining the reactor was to replicate it outside the safeguards regime (p. 120), although there is no evidence that this was attempted.

78. See J.P. Schwartz, "Uranium dioxide Caramel fuel for an alternative fuel cycle for research and test reactors," Commissariat a l'Energie Atomique, Saclay, France, 1978, http://www.iaea.org/inis/collection/NCLCollectionStore/_Public/26/016/26016740 .pdf.

79. Hamza, p. 74.

80. John Hall, "The International Atomic Energy Agency: Origins and Early Years," *IAEA Bulletin*, No. 2 (1987), p. 52.

81. See Robert L. Brown and Jeffrey M. Kaplow, "Talking Peace, Making Weapons: IAEA Technical Cooperation and Nuclear Proliferation," *Journal of Conflict Resolution*, April 14, 2014, p. 412, http://jcr.sagepub.com/content/58/3/402. See also Hamza, pp. 225, 227, 229, 231, 233, and 234.

82. See "The 1970 Safeguards Committee," *IAEA Bulletin*, No. 1, 1990, p. 4.

83. Hamza, p. 74.

84. An Iraqi delegation comprising Hamza, Dr. Moyasser al-Mallah, and Minister of Higher Education Dr. Hisham al-Shawi visited the IAEA in September 1973 to lobby for an Iraqi seat; al-Shawi was elected. See Hamza, pp. 74–77; and "From Aflaq to Tammuz," pp. 3–4.

85. Iraq has been elected to the Board in 1960–1962, 1974–1976, 1980–1982, 1983–1985, 1986–1988, 1989–1991, and 2007–2008. IAEA, Office of Legal Affairs, Country Fact Sheets, https://ola.iaea.org/ola/FactSheets/CountryDetails.asp?country =IQ, accessed December 13, 2021.

86. Hamza, p. 74.

87. Ibid.

88. Interview with Dieter Goethel by Elisabeth Roehrlich, IAEA Oral History Series, Vienna, August 18, 2016.

89. Interview with Dieter Goethel by Elisabeth Roehrlich.

90. Treaty on the Non-Proliferation of Nuclear Weapons (NPT), http://disarmament .un.org/treaties/t/npt, accessed August 22, 2017.

91. Ibid., Article II.

92. Ibid.

93. Source material is defined as "uranium containing the mixture of isotopes occurring in nature; uranium depleted in the isotope 235; thorium; any of the foregoing in the form of metal, alloy, chemical compound, or concentrate," or any other material that the Board of Governors determines. Special fissionable material is defined as "plutonium-239; uranium-233; uranium enriched in the isotopes 235 or 233; any material containing one or more of the foregoing," and any material that the Board determines. IAEA Safeguards Glossary, pp. 30–31.

94. IAEA Office of Legal Affairs, Country Fact Sheets, https://ola.iaea.org/ola/Fact Sheets/CountryDetails.asp?country=IQ, accessed August 22, 2017. See Text of the Agreement between Iraq and the Agency for the Application of Safeguards in Connection with the Treaty on the Non-Proliferation of Nuclear Weapons [NPT], INFCIRC/172, February 22, 1973, https://www.iaea.org/sites/default/files/publica tions/documents/infcircs/1973/infcirc172a1.pdf.

95. Safeguards would continue if the IAEA-Iraq safeguards agreement ended in order to ensure "the peaceful and non-explosive utilization of the materials, nuclear materials, installations, equipment and technical know-how provided by France to

Iraq under the Franco-Iraqi agreement." See Exchange of Letters between the Government of the French Republic and the Government of the Republic of Iraq supplementary to the Franco-Iraqi co-operation agreement for the peaceful utilization of nuclear energy, November 18, 1975, INFCIRC/172/Add 1, September 1979, p. 2, https://www .iaea.org/sites/default/files/publications/documents/infcircs/1973/infcirc172a1.pdf.

96. It contains a typing error, however; in Article 8(a), Iraq is misspelled as "raq."

97. Iraq failed, however, to disclose that Saddam Hussein himself became the IAEC's chair in 1973. "From Aflaq to Tammuz," p. 4.

98. See http://www-ns.iaea.org/projects/iraq/tuwaitha/irt5000.asp?l=66, accessed August 22, 2017. The IRT-5000 was destroyed in the 1991 Gulf War. See IAEA, Consolidated Report on the First Two IAEA Inspections Under Security Council Resolution 687 (1991) of Iraqi Nuclear Capabilities, July 11, 1991, Security Council, Note by the Secretary-General, S/22788, July 15, 1991.

99. Fischer and Szasz, p. 7.

100. U.S. Senate Committee on Foreign Relations, Hearings, The Israeli Airstrike, 97th Cong., 1st sess., June 18, 19, and 25, 1981, Testimony of Roger Richter, Washington, D.C., 1981 (hereafter Richter testimony), p. 135.

101. The Iran-Iraq War lasted from 1980 to 1988. See Richter testimony, p. 135.

102. Senator Alan Cranston, U.S. Senate Committee on Foreign Relations, Hearings, The Israeli Airstrike, 97th Cong., 1st sess., June 18, 19, and 25, 1981, p. 26; Frank Barnaby, *The Invisible Bomb: The Nuclear Arms Race in the Middle East* (London: I.B. Tauris, 1989), p. 94; and Fischer and Szasz, p. 64. See also U.S. Senate Committee on Foreign Relations, Hearings, The Israeli Airstrike, 97th Cong., 1st sess., June 18, 19, and 25, 1981, Appendix, IAEA Comments on the Testimony of Roger Richter on June 19, 1981, before the Senate Foreign Relations Committee, U.S. Government Printing Office, Washington, D.C., 1981 (hereafter IAEA comments on Richter testimony), pp. 125–128. Inspectors G. Rabot of France and V. Seleznev of the Soviet Union inspected Tuwaitha in November 1981, after the Israeli bombing. "An IAEA Inspection of the Iraqi Nuclear Complex 'Revealed No Non-Compliance'," *Nucleonics Week*, November 26, 1981, p. 3. Richter told the Senate Foreign Relations Committee that a French inspector had been designated for Iraq at the time of the Israeli attack but had not yet carried out an inspection. Richter testimony, p. 215.

103. Fischer and Szasz, p. 21, n. 8.

104. Hans Grümm, "Safeguards and Tamuz: Setting the Record Straight," *IAEA Bulletin*, No. 4, 1992, pp. 10–11.

105. Approximately 25 kilograms of 20-percent-enriched HEU is the minimum required for a single nuclear weapon.

106. Fischer and Szasz, p. 18.

107. Grümm, pp. 10–11.

108. Braut-Hegghammer, pp. 57–58.

109. Grümm, p. 13.

110. Duelfer Report, Vol. II, "Nuclear," p. 13.

111. Hamza claims that the damage amounted to "only a few holes in a couple of barrels of radioactive waste." Hamza, p. 128.

112. "Damage to Iraqi Reactor Still Unknown as Controversy Continues," *Nucleonics Week,* June 18, 1981, p. 2, cited in www.lexis-nexis.com. Israel also alleged that the hot cell facility provided by Italy was a proliferation risk. (Hot cells are shielded nuclear radiation containment chambers used in both the nuclear energy and the nuclear medicine industries. They provide a safe containment box in which personnel can manipulate radioactive materials.) See Fischer and Szasz, p. 80.

113. Ibid., p. 18.

114. Sigvard Eklund, "Peaceful Nuclear Development Must Continue," *IAEA Bulletin,* No. 3 (1981), p. 3; and Fischer, *History of the International Atomic Energy Agency: The First Forty Years,* p. 104.

115. For significant sources on this, see Scheinman, pp. 262–263, n. 57–67.

116. Ibid., pp. 234–241.

117. Sandy Cannon, "U.S. at Odds over Need for Osirak Strike," *Nuclear Fuel,* June 22, 1981, cited in www.lexisnexis.com.

118. Marsha McGraw, "No More Safeguards," *Christian Science Monitor,* July 14, 1981, cited in www.lexis-nexis.com. Soon afterward, Richter, the only U.S. inspector in the South and South East Asia section, which covered Iraq, was fired for breaching safeguards confidentiality by leaking documents to the United States. He had been with the Agency for three years. He had not inspected facilities in Iraq. Statement by Dr. Sigvard Eklund, Director General of the International Atomic Energy Agency to the IAEA Board of Governors, IAEA Press Release, PR 81/6, Vienna, July 6, 1981.

119. Fischer and Szasz, p. 18.

120. "France Sees Obligation to Supply Iraq with New Reactor," *Nucleonics Week,* June 25, 1981, p. 1, cited in www.lexis-nexis.com.

121. This was to be included in the facility attachment to Iraq's Comprehensive Safeguards Agreement (CSA).

122. Richter testimony, pp. 297–298.

123. Report prepared by Warren Donnelly, Foreign Affairs and National Defense Division, U.S. Congressional Research Service, Library of Congress, June 16, 1981, U.S. Senate Committee on Foreign Relations, Hearings, The Israeli Airstrike, 97th Cong., 1st sess., June 18, 19, and 25, 1981, U.S. Government Printing Office, Washington, D.C., 1981, pp. 72–84.

124. Grümm, p. 12.

125. Ibid., p. 10.

126. "Al Tuwaitha-Tamuz 2 Reactor," http://www-ns.iaea.org/projects/iraq/tuwaitha/tamuz-2.asp?s=8&l=66, accessed August 22, 2017.

127. IAEA, The Annual Report for 1989, pp. 121, 124, and 126.

128. Memo by Laura Rockwood, Office of Legal Affairs, IAEA, "What the Agency Has Missed in Iraq," March 17, 2003 (archive of Hans Blix), cited in Harrer, p. 32.

129. Ibid., p. 31. Blix later conceded that Jennekens had said some "naive things" about Iraq. Quoted by Harrer, p. 45, note 39.

130. Transcript of Panel Discussion with Myron Kratzer, Rich Hooper, and Ambassador Norman Wolf, "Retrospective of INFCIRCs 153 and 540," 4.0, Foundations of International Safeguards video series, Pacific Northwest Nuclear Laboratories (PNNL), Santa Fe, New Mexico, October 2005, p. 1.

131. Interview with John Carlson by Elisabeth Roehrlich, IAEA Oral History Series, Vienna, November 13, 2015.

132. Matthew Fuhrmann, *Atomic Assistance: How "Atoms for Peace" Programs Cause Nuclear Insecurity* (Ithaca, N.Y.: Cornell University Press, 2012), p. 231.

133. Albright, Gay, and Hamza, p. 4.

134. "From Aflaq to Tammuz," p. 3. See also Braut-Hegghammer, p. 48.

135. "From Aflaq to Tammuz," p. 3.

136. See Duelfer Report, Vol. II, "Nuclear," pp. 1–130; and Duelfer Report, Addendums, pp. 28–30.

137. Al-Sa'idi alleged that prior to the attack, Israel had sabotaged key equipment on its way to Iraq and assassinated Iraqi nuclear scientists. See Duelfer Report, Addendums, p. 28.

138. For a detailed chronology, see Nuclear Threat Initiative (NTI), Iraq Nuclear Chronology, http://www.nti.org/media/pdfs/iraq_nuclear.pdf?_=1316466791.

139. See Duelfer Report, Vol. II, "Nuclear"; and Duelfer Report, Addendums.

140. Joseph Cirincione, Jon B. Wolfsthal, and Miriam Rajkumar, *Deadly Arsenals: Nuclear, Biological, and Chemical Threats* (Washington, D.C.: Carnegie Endowment for International Peace, 2005), p. 388.

141. Duelfer Report, Addendums, p. 29.

142. Fischer, *History of the International Atomic Energy Agency: The First Forty Years*, p. 284.

143. Jacques C. Hymans, "Spinning in Place: Iraq's Fruitless Search for Nuclear Weapons," in Jacques C. Hymans, *Achieving Nuclear Ambitions: Scientists, Politicians, and Proliferation* (Cambridge: Cambridge University Press, 2012), pp. 79–123.

144. Braut-Hegghammer, p. 21.

145. Eighth Report by the Director General of the International Atomic Energy Agency on the Implementation of the Agency's Plan for Future Ongoing Monitoring and Verification of Iraq's Compliance with Paragraph 12 of Resolution 687 (1991), Appendix, Note by Secretary-General, UN Security Council, S/995/844, October 6, 1995.

146. Had the process been successful, there would have been 25 kilograms of HEU, with an average enrichment of 86 percent. Report of the Twenty-Eighth IAEA On-Site Inspection in Iraq under Security Council Resolution 867 (1991), Attachment, Note by the Secretary-General, UN Security Council S/1995/1003, December 1, 1995, pp. 4–6, and S/995/844, p. 3.

147. Ibid., p. 5.

148. Ibid., p. 3.

149. Hymans, p. 112. Indicating how incoherent the plans were, no decision seems to have been made on whether a demonstration test would be conducted, or alternatively, a weapon lobbed on Israel or invading U.S. forces. The demonstration test would obviously have expended all the available HEU, leaving Iraq without a weapon.

150. See Hymans' scathing assessment in ibid., pp. 111–114.

151. Contracts were signed with European firms to acquire key component manufacturing technology and critical equipment. As late as August 1988, German engineers visited Baghdad with centrifuge design data from the URENCO consortium, which was immediately photocopied by the Iraqis to advance their "otherwise slow progress in developing centrifuge equipment." Before 1991, several more German engineers reportedly provided additional URENCO documents. See Duelfer Report, Vol. II, "Nuclear," p. 4.

152. See Harrer, pp. 55–57.

153. Albright, Gay, and Hamza, p. 1.

154. Fischer, *History of the International Atomic Energy Agency: The First Forty Years*, p. 274. See also Leonard S. Spector, "Iraq," in Leonard S. Spector, *The Undeclared Bomb* (Cambridge, Mass.: Ballinger, 1988), pp. 207–218. Although he notes that Iraq's possession of 12.5 kilograms of HEU, albeit under safeguards, was "a source of concern" because it was enough for one "carefully constructed bomb," Spector concluded that "there is little risk that Iraq's now dormant nuclear program could lead to the production of nuclear arms" (p. 213). Frank Barnaby, writing in 1989, was less sanguine, predicting correctly that Iraq would "revitalize" its nuclear program once the Iran-Iraq War (1980–1988) had ended. Barnaby, pp. 995–996.

155. Harrer, p. 210; and Braut-Heghammer, p. 110.

156. Harrer, p. 55.

157. British journalist Farzad Bazoft, who traveled to Iraq to investigate, was apprehended and executed. Braut-Hegghammer, p. 115.

158. Ibid.

159. "From Aflaq to Tammuz," p. 4, citing William Burrows and Robert Windrem, *Critical Mass: The Dangerous Race for Superpowers in a Fragmenting World* (New York: Simon and Schuster, 1994), p. 37.

160. Duelfer Report, Vol. II, "Nuclear," p. 5.

161. As Braut-Hegghammer points out, however, these experiments had little relevance to the Iraqis' main nuclear weapons program, which was focused on enriching uranium (p. 116).

162. Interview with Pier Roberto Danesi by Elisabeth Roehrlich, IAEA Oral History Series, Vienna, April 23, 2015.

163. Interview with Shirley Johnson by Elisabeth Roehrlich, IAEA Oral History Series, Vienna, August 20, 2015.

164. Braut-Hegghammer, p. 55.

165. Ibid., p. 110.

166. Blix, p. 19.

167. "National Technical Means" is an arms control euphemism for any methods that states use to acquire intelligence information.

168. Judith Miller, "Was Iraq Planning to Make the Bomb? Debate by Experts Seems Inconclusive," *New York Times*, June 19, 1981, http://www.nytimes.com/1981/06/19 /world/was-iraq-planning-to-make-the-bomb-debate-by-experts-seems-inconclusive .html?pagewanted=all.

169. Albright, Gay, and Hamza, p. 5.

170. Braut-Hegghammer, p. 14.

171. Interview with Shirley Johnson by Elisabeth Roehrlich.

172. Interview with John Carlson by Elisabeth Roehrlich.

173. Interview with Shirley Johnson by Elisabeth Roehrlich.

174. Former Deputy Director for Finance and Administration David Waller says that "we were always dealing with silos." Interview with David Waller by Elisabeth Roehrlich, IAEA Oral History Series, Washington, D.C., November 21, 2015. Even the Seibersdorf laboratories were "siloed" off from Vienna headquarters. Interview with Pier Roberto Danesi by Elisabeth Roehrlich.

175. Braut-Hegghammer, p. 48.

176. Moyasser al-Mallah, Secretary-General of the IAEC, to A.E. Cairo, IAEA, Vienna, September 12, 1973 (TA/IRQ/3, IAEA Archives), cited in Braut-Hegghammer, p. 49.

177. Braut-Hegghammer, p. 87.

178. "From Aflaq to Tammuz," p. 7.

179. Blix, p. 20.

180. Comprehensive Report of the Special Advisor to the Director of Central Intelligence (DCI) on Iraq's WMD, Washington, D.C., Vol. I, September 30, 2004 (hereafter Duelfer Report, Vol. I), p. 3.

181. Hymans, p. 112.

182. Eighth Report by the Director General of the International Atomic Energy Agency on the Implementation of the Agency's Plan for Future Ongoing Monitoring and Verification of Iraq's Compliance with Paragraph 12 of Resolution 687 (1991), Appendix, Note by Secretary-General, UN Security Council, S/995/844, October 6, 1995, p. 4, para. 8.

183. Safeguards do not require the Secretariat to determine that diverted or undeclared nuclear materials are intended for a nuclear weapons program, but only that they are for "purposes unknown."

184. Fischer and Szasz, p. 18.

185. Grümm, pp. 10–11.

186. Ibid., p. 11.

187. Ibid.

188. During training overseas, Iraqi nuclear personnel were approached by what they believed were foreign intelligence agents asking questions about the direction of the Iraqi nuclear program. Braut-Hegghammer, p. 56.

189. Duelfer Report, Vol. I, p. 2.

Chapter Two

1.   See Michael Wilson, "Safeguards and the IAEA Board of Governors: 1991–1993: Iraq, A Necessary Stimulus for Handling the DPRK," in IAEA, *International Atomic Energy Agency: Personal Reflections* (Vienna: IAEA, 1997), p. 130; David Fischer, *History of the International Atomic Energy Agency: The First Forty Years* (Vienna: IAEA, 1997), pp. 285–286; Tim Trevan, *Saddam's Secrets: The Hunt for Iraq's Hidden Weapons* (London: HarperCollins, 1999), pp. 47–49; and Hans Blix, *Disarming Iraq: The Search for Weapons of Mass Destruction* (London: Bloomsbury, 2005), p. 20.

2.   See Treaty on the Non-Proliferation of Nuclear Weapons, also known as the Nuclear Non-Proliferation Treaty (NPT), http://disarmament.un.org/treaties/t/npt; Lawrence Scheinman, *The International Atomic Energy Agency and World Nuclear Order* (Washington, D.C.: Resources for the Future, 1987), especially Chapter 7, "Problems Facing the IAEA," pp. 207–266; David B. Waller, "Atoms for Peace and the International Atomic Energy Agency," in Joseph F. Pilat, ed., *Atoms for Peace: A Future after Fifty Years?* (Baltimore: Johns Hopkins University Press, 2007), pp. 21–38; Allan McKnight, *Atomic Safeguards: A Study in International Verification* (New York: United Nations Institute for Training and Research, 1971), especially Part One, "Political Development of the International Atomic Energy Safeguards," pp. 1–90; and David Fischer, *History of the International Atomic Energy Agency: The First Forty Years* (Vienna: IAEA, 1997), especially Chapter 5, "A Changing Political and Technical Environment," pp. 71–141. For political machinations among the United States and other countries and within the U.S. administration, see Glenn T. Seaborg, with Benjamin T. Loeb, *Stemming The Tide: Arms Control in the Johnson Years* (Lexington, Mass: Lexington Books, 1971).

3.   U.S. Arms Control and Disarmament Agency (ACDA), "Review of Negotiating History of International Atomic Energy Agency [IAEA] Document INFCIRC/66/ REV.2, 'The Agency's Safeguards System,'" AC2NC109, Washington, D.C., September 5, 1984, p. 1.

4.   Ibid.

5.   Blix, pp. 17–18.

6.   For the argument about sovereignty, see McKnight, pp. 24–25; the IAEA Statute requires the Agency to carry out its activities "with due observance of the sovereign rights of States" (Art. 3.D). For the argument about discrimination, see McKnight, pp. 25–26. Confidentiality of safeguards-derived information is embodied in the IAEA Statute and in safeguards agreements.

7.   Fischer and Szasz, *Safeguarding the Atom: A Critical Appraisal* (London: Taylor and Francis for the Stockholm International Peace Research Institute [SIPRI], 1985), pp. 79–80.

8. Scheinman, p. 152; and Fischer and Szasz, p. 80. On the strange case of Sweden, see Seaborg, pp. 270–273. For the policies of various states toward the NPT in the early years of the treaty, see Jozef Goldblat, ed., *Nonproliferation: The Why and the Wherefore* (London and Philadelphia: Taylor and Francis for SIPRI, 1985).

9. Fischer and Szasz, p. 80.

10. ACDA, "Review of the Negotiating History of the IAEA Safeguards Document INFCIRC/153," AC2NC103, Washington, D.C., July 30, 1984, p. 6; and Fischer and Szasz, p. 80.

11. ACDA, "Review of the Negotiating History of the IAEA Safeguards Document INFCIRC/153," p. xi; and Fischer and Szasz, pp. 80–81.

12. Craig Everton, "In Defence of the Evolution of IAEA Safeguards," in *Verification and Implementation* (London: Verification Research, Training and Information Center [VERTIC], 2015).

13. Ibid.

14. ACDA, "Review of the Negotiating History of the IAEA Safeguards Document INFCIRC/153," p. 7.

15. Treaty on the Non-Proliferation of Nuclear Weapons, also known as the Nuclear Non-Proliferation Treaty (NPT), Article IV, http://disarmament.un.org/treaties/t/npt.

16. See ACDA, "Review of the Negotiating History of the IAEA Safeguards Document INFCIRC/153," pp. 124–125. Veteran non-proliferation expert Josef Goldblat noted in 1985 that to improve safeguards, it would be necessary to "free the designation of inspectors from the political constraints imposed by a number of states." Goldblat, p. 36.

17. Blix, p. 18.

18. Report of the Advisory Committee on U.S. Policy Toward the International Atomic Energy Agency (the Smyth Report), U.S. Government Printing Office, Washington, D.C., 1962, p. 12.

19. ACDA, "Review of the Negotiating History of the IAEA Safeguards Document INFCIRC/153," p. 11.

20. Interview with safeguards official by the author, IAEA Safeguards Symposium, Vienna, October 20, 2014.

21. Slobodan Nakićenović, "Comments on IAEA Safeguards," U.S. Congressional Research Service, Washington, D.C., August 1981, p. 28, reproduced in U.S. Congressional Research Service, *Nuclear Safeguards: A Reader*, Report Prepared by the U.S. Congressional Research Service for the Subcommittee on Energy Research and Production transmitted to the Committee on Science and Technology, U.S. House of Representatives, 98th Cong., 1st sess., U.S. Government Printing Office, Washington, D.C., December 1983 (hereafter *Nuclear Safeguards: A Reader*), pp. 661–697.

22. Nakićenović clearly had some axes to grind when he left the Agency in 1977. He was excluded from communicating with IAEA personnel after his departure and seemed aggrieved that his former division had been split in two without his being appointed to head either one. Over the years, he claims to have produced over two

dozen papers on "safeguards limitations" and, on the instruction of Director General Eklund, provided them to the U.S., Soviet, and other missions in Vienna, but to no apparent effect. Nakićenović, pp. 6–7.

23. Ibid., p. 17.

24. Ibid., p. 28.

25. Ibid., p. 23.

26. McKnight, p. 45.

27. Testimony of Dr. Thomas E. Shea, IAEA Programs of Safeguards, Hearing before the U.S. Senate Committee on Foreign Relations, 97th Cong., 1st sess., December 2, 1981, reproduced in *Nuclear Safeguards: A Reader*, p. 755.

28. Blix, p. 20.

29. "Paying Tribute to 25 Years of Safeguards Leadership," *IAEA Bulletin*, No. 3 (1994), p. 14.

30. Jon Jennekens, "IAEA Safeguards: A Look at 1970–1990 and Future Prospects," *IAEA Bulletin*, No. 1, 1990, p. 1.

31. "Paying Tribute to 25 Years of Safeguards Leadership," p. 19.

32. Blix, pp. 17–18.

33. "Paying Tribute to 25 Years of Safeguards Leadership," p. 14.

34. Peter Pringle and James Spigelman, *The Nuclear Barons* (New York: Avon Books, 1981), p. 204.

35. The Soviets refused to provide HEU, but instead offered fuel rods with plutonium that had to be returned after use to prevent their misuse. Ibid.

36. Nakićenović confirms that "the Soviets, despite their somewhat misdirected policy towards [the] IAEA, are strong supporters of safeguards, although their goals, strategy and tactics differ from those of the United States." Nakićenović, p. 4. Coincidentally, the Agency held two introductory courses for new inspectors in the German Democratic Republic and the Soviet Union in 1989. See IAEA, The Annual Report for 1989, GC(XXXIV)/915, July 1990, p. 132, https://www.iaea.org/About/Policy/GC/GC34/GC34Documents/English/gc34-915_en.pdf.

37. "IAEA Comments on the Testimony of Roger Richter on June 19, 1981, before the Senate Foreign Relations Committee," Appendix, U.S. Congress, Senate Committee on Foreign Relations, Hearings, The Israeli Airstrike, 97th Cong., 1st sess., June 18, 19, and 25, 1981 (hereafter IAEA comments on Richter testimony), pp. 297–298.

38. Richter says he was the "token American," replacing a previous U.S. inspector, as he announced in introducing himself in his Senate testimony. "Testimony of Roger Richter," U.S. Senate Committee on Foreign Relations, Hearings, The Israeli Airstrike, 97th Cong., 1st sess., June 18, 19, and 25, 1981 (hereafter Richter testimony), p. 134. Moreover, it was an Eastern European inspector, Vladimir Ruhklo, who reportedly "got nosy" in North Korea and helped detect its non-compliance in 1993. Interview with Tom Shea by the author, Vienna, June 26, 2014.

39. A declassified U.S. government report on the negotiations on INFCIRC/153 safeguards affirms that "negotiation of the [NPT] Treaty made it clear that several key

potential parties, including the FRG [Federal Republic of Germany] and Japan, would be unlikely to adhere to the Treaty without assurances that Agency safeguards would be applied in a manner consistent with the fact that all peaceful nuclear activities in the state would be under safeguards . . . and espoused, at the FRG initiative, [the] 'principle of safeguarding effectively, the flow of nuclear material . . . at certain strategic points." ACDA, "Review of the Negotiating History of the IAEA Safeguards Document INFCIRC/153," p. 7. For analysis of special inspections and strategic points, see p. 233.

40. At least one member state, the United States, had developed nuclear weapon proliferation indicators and observables, some of which could have been used by the IAEA if it had been permitted to do so. See Richard R. Paternoster, "Nuclear Weapons Proliferation Indicators and Observables," Los Alamos National Laboratory, New Mexico, LA-12430-MS, UC-700, December 1992.

41. There is a vast contemporary and historic literature on this phenomenon. Select examples are: Ian Hurd, *International Organizations: Politics, Law, and Practice*, 2nd ed. (Cambridge: Cambridge University Press, 2014); Darren G. Hawkins, David A. Lake, Daniel L. Nielson, and Michael J. Tierney, eds., *Delegation and Agency in International Organizations* (Cambridge: Cambridge University Press, 2006); Michael Barnett and Martha Finnemore, *Rules for the World: International Organizations in Global Politics* (Ithaca, N.Y.: Cornell University Press, 2004); Philip E. Jacob and Alexine L. Atherton, *The Dynamics of International Organization: The Making of World Order* (Homewood, Ill.: Dorsey Press, 1965); Edward C. Luck and Michael W. Doyle, eds., *International Law and Organization: Closing the Compliance Gap* (Lanham, Md.: Rowman and Littlefield, 2002); and Volker Rittenberger, Bernhard Zangl, and Andreas Kruck, *International Organization*, 2nd ed. (Houndsmills, UK: Palgrave Macmillan, 2012).

42. McKnight records that "it has been argued that any individual inspector can insist on transmittal to the Board of his opinion that there has been a non-compliance [incident]." But "the most likely event in practice will be an inability on the part of the inspectors to verify the accounting for nuclear material. The inspector is most unlikely to have any evidence as to the use of the deficient material." McKnight, p. 126.

43. For a comprehensive study of the IAEA's non-compliance reporting, see Trevor Findlay, "Proliferation Alert! The IAEA and Non-Compliance Reporting," Report No. 2015–04, Project on Managing the Atom, Belfer Center for Science and International Affairs, Harvard University, Cambridge, Mass., October 2015.

44. Scheinman, p. 235.

45. Ibid.

46. Ibid., p. 237.

47. Nakićenović, p. 20.

48. Ibid.

49. Hans Blix became Director General on December 1, 1981.

50. See Findlay, "Proliferation Alert! The IAEA and Non-Compliance Reporting," p. 22.

51. Testimony of Dr. Thomas E. Shea, reproduced in *Nuclear Safeguards: A Reader*, p. 755; and Transcript of Panel Discussion with Myron Kratzer, Rich Hooper, and Ambassador Norman Wolf, "Retrospective of INFCIRCs 153 and 540," 4.0, Foundations of International Safeguards video series, Pacific Northwest Nuclear Laboratories (PNNL), Santa Fe, New Mexico, October 2005.

52. Unfortunately, Eklund did not write about his time at the IAEA, and there is apparently no biography of him in English that might explain his safeguards philosophy or his decisions while he was Director General.

53. Eklund's efforts in this regard are allegations by Nakićenović.

54. Blix, p. 18. Shea says he had a conversation with Blix about the possibilities of undeclared production in 1980. Interview with Tom Shea by the author.

55. Gudrun Harrer, *Dismantling the Iraqi Nuclear Program: The Inspections of the International Atomic Energy Agency, 1991–1998* (Abingdon, Oxon: Routledge, 2014), p. 33. Harrer uses the word "reserve," but presumably "reservations" is more appropriate.

56. Scheinman, p. 237.

57. Ibid.

58. Blix, p. 20.

59. See "Bashing Blix and ElBaradei," Chapter 10, in Blix, p. 215.

60. Interview with David Waller by Elisabeth Roehrlich, IAEA Oral History Series, Washington, D.C., November 21, 2015.

61. See Findlay, "Proliferation Alert! The IAEA and Non-Compliance Reporting."

62. See ibid., Table 1, p. 24.

63. Stockholm International Peace Research Institute (SIPRI), *Safeguards Against Nuclear Proliferation* (Stockholm: Almqvist and Wiksell for SIPRI, 1975), p. 52.

64. ACDA, "Review of the Negotiating History of the IAEA Safeguards Document INFCIRC/153," p. 11.

65. For an extreme version of the animus against the UN system as a whole, see Shirley Hazzard, *Defeat of an Ideal: The Self-Destruction of the United Nations* (London: Macmillan, 1937). For an illustration of the UN-IAEA connection, see the description of the relationship between UNSCOM and the UN and UNSCOM and the IAEA in Jean E. Krasno and James S. Sutterlin, *The United Nations and Iraq: Defanging the Viper* (Westport, Conn.: Praeger, 2003), pp. 18–25. For disparaging comments about the IAEA being a "typical" UN organization, see Richard Armitage's "the IAEA isn't worth a bucket of warm spit," and John Bolton's remarks about "a typical UN battle" and the IAEA Board breaking for two hours for lunch, "this being a UN agency, after all," in John Bolton, *Surrender is Not an Option: Defending America at the United Nations and Abroad* (New York: Simon and Schuster, 2007), p. 142. To be fair, some U.S. government bodies have lauded the IAEA as an exception to the general rule about ineffective and inefficient UN bodies; in 2006, the U.S. Office of Management and Budget gave the IAEA a virtually unprecedented rating of 100 in terms of value for money. "Contributions to the IAEA," Office of Management and Budget, The White House, Washington, D.C., 2006, http://www.whitehouse.gov/omb/expectmore/summary/10004639.2006

.html, cited in IAEA, "20/20 Vision for the Future: Background Report by the Director General for the Commission of Eminent Persons," Annex, "Report of the Commission of Eminent Persons on the Future of the Agency," GOV/2008/22-GC(52)/INF/4, IAEA, Vienna, 2008, p. 24.

66. The United States and Singapore have been major critics of UNESCO on these grounds, and both have left the organization.

67. As well as reporting to the UN Security Council "when appropriate," the IAEA Statute provides that "if in connection with the activities of the Agency there should arise questions that are within the competence of the Security Council, the Agency shall notify the Security Council, as the organ bearing the main responsibility for the maintenance of international peace and security." IAEA Statute, Art. III.B.4.

68. The IAEA is not exactly like a national regulator because it cannot issue regulations unilaterally, without the agreement of the regulated—that is, its member states—and has no enforcement powers itself, but must rely on the UN Security Council. Even national regulators, however, struggle to impose unpopular regulations, even in autocratic systems, and enforcement is never total.

69. Email from John Tilemann to the author, May 30, 2019.

70. Transcript of Discussion with Laura Rockwood, "Legal Foundations of Safeguards," 3.0, Foundations of International Safeguards video series, PNNL, Richland, Wash., June 2006, p. 13.

71. See, for example, United Nations, Report of the Secretary-General's High-Level Panel on Threats, Challenges and Change, New York, 2004; and International Commission on Nuclear Non-Proliferation and Disarmament, *Eliminating Nuclear Threats: A Practical Agenda for Global Policymakers* (Canberra: Paragon, 2009). These documents, however, were produced well after the Iraq case.

72. "The International Atomic Energy Agency: An Appraisal with Recommendations for United States Policy, a Report to the Secretary of State of the Panel to Review the Activities of the International Atomic Energy Agency," U.S. State Department, Washington, D.C., 1972, p. 5.

73. The 1972 report for the State Department recommended that "in the formulation of United States policy for the Agency, particular care should be exercised to prevent the application of indiscriminate across-the-board cuts to the Agency." Ibid., p. 4. In 1972, the U.S. Congress imposed a 25-percent limit on assessed contributions to international organizations, but exempted the IAEA due to the importance of safeguards. Comptroller General of the United States, Assessment of U.S. and International Controls Over the Peaceful Uses of Nuclear Energy, Report to the Congress by the Comptroller General of the United States, ID-76–60, Appendix II, September 14, 1976, p. 88.

74. For the year before the Iraq revelations, see IAEA, The Agency's Accounts for 1980, GC(XXV)/645, Vienna, July 1981, http://www.iaea.org/inis/collection/NCLCol lectionStore/_Public/40/086/40086035.pdf.

75. Thomas E. Shea and David A. Thompson, "Human Factors Affecting IAEA Inspector Performance," Prepared for the U.S. Arms Control and Disarmament Agency (ACDA), AC2NC106 (unpublished), Washington, D.C., 1982/1983, p. 5.

76. Fischer, *History of the International Atomic Energy Agency: The First Forty Years*, p. 83.

77. Ibid., p. 301.

78. Berhanykun Andemicael and John Mathiason, *Eliminating Weapons of Mass Destruction: Prospects for Effective International Verification* (London: Palgrave, 2005), p. 214.

79. See Richard G. Hewlett and Jack M. Holl, *Atoms for Peace and War, 1953–1961* (Berkeley: University of California Press, 1989), pp. 442–444.

80. Fischer, *History of the International Atomic Energy Agency: The First Forty Years*, p. 75.

81. Interview with James Goodby by Elisabeth Roehrlich, IAEA Oral History Series, Washington, D.C., November 17, 2015.

82. Interview with Dieter Goethel by Elisabeth Roehrlich, IAEA Oral History Series, Washington, D.C., August 18, 2016.

83. This was reported by an unnamed former IAEA legal affairs director, cited in Mark Hibbs and Andreas Persbo, "The ElBaradei Legacy," *Bulletin of the Atomic Scientists*, September–October 2009, p. 12.

84. Interview with Hans Blix by Elisabeth Roehrlich, IAEA Oral History Series, Stockholm, April 13, 2015.

85. M. Campbell, et al., *At What Cost Success? Final Report of the External Review on the Management Processes of the International Atomic Energy Agency*, MANNET, Geneva, 2002 (hereafter MANNET Report), p. 6.

86. Interview with Dieter Goethel by Elisabeth Roehrlich, IAEA Oral History Series.

87. Interview with John Carlson by the author, June 5, 2014.

88. Interview with James Goodby by Elisabeth Roehrlich, IAEA Oral History Series.

89. See Findlay, "Proliferation Alert! The IAEA and Non-Compliance Reporting," pp. 39–49.

90. Hans Blix, IAEA Director General Statement, "The Current and Future Safeguards Role of the IAEA: Challenges and Opportunities," Vienna, January 27, 1994, pp. 6–7, https://www.iaea.org/newscenter/statements/seminar-current-and-future-safeguards-role-iaea-challenges-and-opportunities.

91. Blix himself says that after Iraq "we were wide awake." Interview with Hans Blix by Elisabeth Roehrlich, IAEA Oral History Series.

92. See Findlay, "Proliferation Alert! The IAEA and Non-Compliance Reporting," pp. 39–49.

93. The Safeguards Department was originally called the Inspector General's Department or the Department of Safeguards and Inspection. McKnight, p. 151. See also IAEA, IAEA Safeguards: An Introduction, IAEA/SG/INF/3, October 1981, p. 27.

94. One recommendation was to hire a management expert to implement the other recommendations. Interview with William Lichliter by Elisabeth Roehrlich, IAEA Oral History Series, Vienna, March 13, 2015. It is not clear whether these changes drew on the consultant's recommendations, because their report was kept confidential. The author has not been able to access it. See also IAEA, The Annual Report for 1983, GC(XXVIII)/713, July 1984, Vienna, p. 66, https://www.iaea.org/About/Policy/GC/GC28/GC28Documents/English/gc28-713_en.pdf.

95. Other states included in Operations B at the time were Australia, Greece, India, Indonesia, Iran, Israel, North Korea, Pakistan, the Philippines, Thailand, Turkey, and Yugoslavia. IAEA, IAEA Safeguards: An Introduction, p. 30.

96. Email from John Tilemann to the author, May 30, 2019.

97. The other departments deal with safety and security, the promotion of the peaceful uses of nuclear energy, and technical assistance to developing countries.

98. Shea and Thompson, p. 5.

99. Jeffrey Lewis, "Replacing Olli," Arms Control Wonk, July 3, 2010, and comments by Mark Hibbs, http://lewis.armscontrolwonk.com/archive/2792/replacing-olli.

100. Benjamin N. Schiff, *International Nuclear Technology Transfer: Dilemmas of Dissemination and Control* (Lanham, Md.: Rowman and Littlefield, 1983), cited by Jeffrey Lewis in "Replacing Olli," Arms Control Wonk, July 3, 2010, http://lewis.armscontrolwonk.com/archive/2792/replacing-olli.

101. Nakićenović accused Eklund of ignoring and keeping confidential the recommendations of a 1974 Panel of Experts on the structure of the Safeguards Department. In 1977, Eklund replaced all members of the panel except those from the Soviet Union and the chair, "who was his best friend," and used them to "accomplish a peculiar redistribution of supervisory posts in safeguards." Nakićenović, p. 16. These allegations may have been prompted by the fact that the newly created post of Director of Safeguards Operations, which Nakićenović was offered, was a level lower than the position that his Yugoslav compatriot, Dragoslav Popovích, had occupied as Director of Safeguards. Nakićenović had previously been a senior administrator in the Yugoslav nuclear program, rising to become director of the Vinča atomic research institute and later director of the Yugoslav Federal Nuclear Energy Commission. Jacques C. Hymans, *Achieving Nuclear Ambitions: Scientists, Politicians, and Proliferation* (Cambridge: Cambridge University Press, 2012), pp. 194–195. It is not clear whether Nakićenović chose to retire or was dismissed.

102. Email from John Tilemann to the author, May 30, 2019.

103. Prior to the Iraq case, the lone exception was Dragoslav Popovích of Yugoslavia, who was allegedly "socialized into the international nuclear community" by his work in Norway. Hymans, pp. 188–189.

104. Barnett and Finnemore, p. 30.

105. Ibid.

106. See "Paying Tribute to 25 Years of Safeguards Leadership," pp. 13–15.

107. John Mathiason, *Invisible Governance: International Secretariats in Global Politics* (Bloomfield, Conn.: Kumarian Press, 2007), p. 71.

108. Shea and Thompson, p. 32.

109. "Paying Tribute to 25 Years of Safeguards Leadership," p. 14.

110. Andemicael and Mathiason, p. 200.

111. Ibid.

112. Nakićenović, p. 5.

113. Ibid., p. 2 and pp. 10–11.

114. Ibid., pp. 5, 7–9, and 21.

115. Ibid., p. 11.

116. Shea and Thompson, p. 34.

117. Interview with Dieter Goethel by Elisabeth Roehrlich, IAEA Oral History Series.

118. Ibid.

119. IAEA Safeguards: An Introduction, p. 25.

120. Ibid.

121. ACDA, "Review of the Negotiating History of the IAEA Safeguards Document INFCIRC/153," p. 58.

122. Nakićenović, p. 21.

123. John P. Boright, "Historical Development and Current Trends in Nuclear Safeguards," Lecture at the International Training Course on Implementation of State Systems of Accounting for and Control of Nuclear Materials, March 15–30, 1982, reproduced in *Nuclear Safeguards: A Reader*, p. 45.

124. See John Carlson, "SAGSI: Its Role and Contribution to Safeguards Development," Australian Safeguards and Non-Proliferation Office, Canberra, 2006, p. 2.

125. Interview with John Carlson by Elisabeth Roehrlich, IAEA Oral History Series, Vienna, November 13, 2015.

126. Fischer and Szasz, p. 68.

127. Unfortunately for researchers, the SIR is confidential, although in some years it has been leaked.

128. Testimony of Dr. Thomas E. Shea, reproduced in *Nuclear Safeguards: A Reader*, p. 755.

129. Nakićenović, pp. 5, 7–9, and 21.

130. See Letter from U.S. Nuclear Regulatory Commission to Hon. Larry Pressler, November 27, 1981, reproduced in *Nuclear Safeguards: A Reader*, p. 727.

131. Testimony of Dr. Thomas E. Shea, reproduced in *Nuclear Safeguards: A Reader*, p. 754.

132. Ibid., p. 755.

133. IAEA, INFCE Summary Volume, Vienna, 1980 (Excerpt), reproduced in *Nuclear Safeguards: A Reader*, p. 603.

134. David Fischer, "Safeguards under the Non-Proliferation Treaty," *Disarmament*, Vol. 3, No. 2 (July 1980), p. 39.

135. Ibid.

136. IAEA, "The Evolution of IAEA Safeguards," Vienna, 1998, p. 54, www.pub.iaea .org/MTCD/publications/PDF/NVS2_web.pdf.

137. Interview with Hans Blix by Gudrun Harrer, quoted in Harrer, p. 42.

138. "Paying Tribute to 25 Years of Safeguards Leadership," p. 14.

139. Ibid., p. 15.

140. Jennekens, p. 9.

141. Jennekens seemed frustrated by the process of reaching agreement, complaining that "the two terms 'streamlining' and 'strengthening' of safeguards have been mentioned repeatedly and incessantly during the last few years, in many instances by people who don't understand the issue and are not interested in understanding the issue." "Paying Tribute to 25 Years of Safeguards Leadership," p. 15.

142. McKnight, p. 60.

143. U.S. national nuclear laboratories, including Lawrence Livermore National Laboratory, Pacific Northwest Nuclear Laboratory, and Sandia National Laboratory, have been significant innovators in safeguards techniques and technologies. The American firm Canberra Industries has traditionally been a major supplier of safeguards equipment, as well as engaging in research and development; it was absorbed by Mirion Technologies in 2016. See http://www.canberra.com/about/, accessed January 5, 2018.

144. L. Manning Muntzing, "Safeguards and Nuclear Safety: A Personal Perspective," *IAEA Bulletin*, Vol. 24, No. 4 (1982), pp. 7–9.

145. Myron B. Kratzer, "Prospective Trends in International Safeguards," *Nuclear News*, October 1980, p. 57.

146. Interview with Shirley Johnson by Elisabeth Roehrlich, IAEA Oral History Series. Johnson, with twenty-five years of experience as a safeguards inspector, was a Senior Nuclear Safeguards Officer at the IAEA from March 1982 to April 2007.

147. Interview with John Carlson by the author, June 5, 2014.

148. Interview with Shirley Johnson by the author, INMM Conference, Atlanta, July 22, 2014.

149. Interview with William Lichliter by Elisabeth Roehrlich, IAEA Oral History Series.

150. Nakićenović, pp. 13–14.

151. IAEA, The Annual Report for 1983, p. 61.

152. "Paying Tribute to 25 Years of Safeguards Leadership," pp. 14–15.

153. Shea and Thompson, p. 17. They further note that within the spectrum of normal behavior, one aspect of human personality is particularly important to inspector performance: for safeguards to be effective, "the inspector must be sufficiently assertive to accomplish his appointed tasks and to resolve any anomalies identified. At the opposite extreme, the inspector must not be so arrogant that cooperation is inhibited."

154. Interview with William Lichliter by Elisabeth Roehrlich, IAEA Oral History Series.

155. Jennekens, p. 9.

156. Interview with Shirley Johnson by the author.

157. IAEA, The Annual Report for 1990, GC(XXXV)/953, October 1991, p. 118, https://www.iaea.org/About/Policy/GC/GC35/GC35Documents/English/gc35-953_en.pdf.

158. Ibid., p. 135.

159. Ibid., p. 116.

160. Ibid., p. 136.

161. Ibid., p. 137.

162. Ibid., p. 116.

163. Even the U.S. Congressional Research Service, which has consistently been a source of expert external assessment, seemed to go quiet.

164. Blix, p. 19.

165. Ibid.

166. A Report on the International Control of Atomic Energy, Prepared for the U.S. Secretary of State's Committee on Atomic Energy, Washington, D.C., March 16, 1946 (Excerpt), reproduced in *Nuclear Safeguards: A Reader*, pp. 46–56.

167. Ibid.

168. McKnight, p. 153.

169. SIPRI, *Safeguards Against Nuclear Proliferation*, p. 44. Reflecting the values of the era, the report assumed that they would all be "young men."

170. Emanuel R. Morgan, an IAEA inspector from May 1977 to August 1980, told the U.S. Senate Foreign Relations Committee that during the three years he was employed there, the Agency did not have a single professional accountant on the staff of the then–Division of Safeguards. Prepared Statement of Emanuel R. Morgan, IAEA Programs of Safeguards, Hearing before the U.S. Senate Committee on Foreign Relations, 97th Cong., 1st sess., December 2, 1981, reproduced in *Nuclear Safeguards: A Reader*, p. 751.

171. Jennekens, p. 8.

172. Interview with Shirley Johnson by the author.

173. Interview with Shirley Johnson by Elisabeth Roehrlich, IAEA Oral History Series.

174. Transcript of Panel Discussion with Current and Former SAGSI Representatives, "International Perspectives on Safeguards," 5.2, Foundations of International Safeguards video series, Tucson, Arizona, July 2007, p. 12.

175. Shea and Thompson, p. 4. Their report was intended for ACDA's use in assessing the effectiveness of safeguards. It is not known whether it was eventually submitted to ACDA or shared with the Agency. The only copy available has several pages missing.

176. Ibid.

177. Ibid., p. 35.

178. Interview with Shirley Johnson by the author.

179. Shea and Thompson, p. 73.

180. Email from Shirley Johnson to the author, October 21, 2020.

181. Shea and Thompson say, diplomatically, that the result was that the quality of new staff was "less than it could be." Shea and Thompson, p. 37.

182. Ibid., p. 38.

183. See IAEA, "Personnel: Staffing the Agency's Secretariat," GOV/2013/34/GD(57)/14, July 23, 2013.

184. IAEA Statute, Art. VII.

185. Interview with William Lichliter by Elisabeth Roehrlich, IAEA Oral History Series.

186. Ibid.

187. This consisted of two two-year contracts plus one five-year extension, since reduced to seven years in total.

188. Interview with William Lichliter by Elisabeth Roehrlich, IAEA Oral History Series.

189. Interview with Hans Blix by Elisabeth Roehrlich, IAEA Oral History Series.

190. Interview with Dimitri Perricos by Elisabeth Roehrlich, IAEA Oral History Series, Vienna, June 27, 2015.

191. McKnight, p. 153. Fischer says the first training course was in 1969. Fischer, *History of the International Atomic Energy Agency: The First Forty Years*, p. 252.

192. Interview with Dimitri Perricos by Elisabeth Roehrlich, IAEA Oral History Series.

193. SIPRI, *Safeguards Against Nuclear Proliferation*, p. 44.

194. One recommendation was to hire a management expert to implement the other recommendations. Interview with William Lichliter by Elisabeth Roehrlich, IAEA Oral History Series. The author has not been able to access this document.

195. Scheinman, p. 239.

196. The idea of an International School of Safeguards run by the IAEA, proposed in 1967 by the U.S. Atomic Energy Commission's *Ad Hoc* Advisory Panel on Safeguarding Special Nuclear Material, was never taken up. Report to the Atomic Energy Commission by the *Ad Hoc* Advisory Panel on Safeguarding Special Nuclear Material, March 10, 1967, reproduced in *Nuclear Safeguards: A Reader*, p. 252.

197. Shea and Thompson, p. 18.

198. Email from Tariq Rauf to the author, June 1, 2015.

199. Fischer and Szasz, p. 19.

200. Shea and Thompson, p. 3.

201. Lichliter notes that the Agency, recognizing the potential problem that inspectors might begin to identify with their "clients," proposed a rotation policy between operational divisions and between operational and support divisions. Interview with William Lichliter by Elisabeth Roehrlich, IAEA Oral History Series.

202. Interview with Shirley Johnson by the author.

203. Jennekens, p. 10.

204. Herman Nackaerts, "The Future of Safeguards: Adapting to Change," June 7, 2011, published in *Journal of Nuclear Materials Management*, Fall 2011; Herman Nackaerts, "IAEA Safeguards: Cooperation as the Key to Change," Presented at INMM 52nd

Annual Meeting, Palm Desert, Calif., July 18, 2011, published in *Journal of Nuclear Materials Management*, Fall 2011.

205. Diane Vaughan, *The Challenger Launch Decision: Risky Technology, Culture, and Deviance at NASA* (Chicago: University of Chicago Press, 1991), Chapters 4 and 5.

206. While it might be thought that the non-use of special inspections and the focus on declared materials and facilities to the detriment of concern about undeclared materials and facilities were examples of the normalization of deviance, these were never embodied in safeguards practice to begin with, so they can hardly be considered examples of normalization. Rather, this author argues that they were a form of cultural drift, which is examined in Chapter 3 of this volume.

207. IAEA Safeguards: An Introduction, pp. 13–14.

208. Testimony of Dr. Thomas E. Shea, reproduced in *Nuclear Safeguards: A Reader*, p. 755.

209. Ibid., p. 754.

210. Ibid.

211. Fischer, *History of the International Atomic Energy Agency: The First Forty Years*, pp. 269–270.

212. Michael D. Rosenthal, et al., *Deterring Nuclear Proliferation: The Importance of IAEA Safeguards: A Textbook* (Upton, N.Y.: Brookhaven National Laboratory, 2019), p. 142.

213. Ibid., p. 144.

214. Blix, p. 20.

Chapter Three

1.   Michael Wilson, "Safeguards and the IAEA Board of Governors: 1991–1993: Iraq, A Necessary Stimulus for Handling the DPRK," in IAEA, *International Atomic Energy Agency: Personal Reflections* (Vienna: IAEA, 1997), p. 130.

2.   Christine Wing and Fiona Simpson, *Detect, Dismantle, and Disarm: IAEA Verification, 1992–2005* (Washington, D.C.: U.S. Institute of Peace, 2013), p. 27.

3.   U.S. Office of Technical Assessment, "Nuclear Safeguards and the International Atomic Energy Agency," OTA-ISS-615, Washington, D.C., April 1995, p. 40.

4.   This report is cited in Slobodan Nakićenović, "Comments on IAEA Safeguards," U.S. Congressional Research Service, Washington, D.C., August 1981, p. 18, reproduced in U.S. Congressional Research Service, *Nuclear Safeguards: A Reader*, Report Prepared by the U.S. Congressional Research Service for the Subcommittee on Energy Research and Production transmitted to the Committee on Science and Technology, U.S. House of Representatives, 98th Cong., 1st sess., December 1983, U.S. Government Printing Office, Washington, D.C., December 1983 (hereafter *Nuclear Safeguards: A Reader*).

5.   See Clifford Geertz, *The Interpretation of Cultures* (New York: Basic Books, 1973).

6. See Edgar H. Schein, *The Corporate Culture Survival Guide* (San Francisco: Jossey-Bass, 1999).

7. Adam Lowther, "A Year Later: Responding to Problems in the ICBM Force," *Bulletin of the Atomic Scientists*, February 20, 2015, http://thebulletin.org/year-later-responding -problems-icbm-force7984; "Secret Service Chief Says He's Working on Culture," *Washington Times*, March 17, 2015, http://www.washingtontimes.com/news/2015/mar/17 /secret-service-director-says-hes-working-on-agency/?page=all; and "Uber Report: Eric Holder's Recommendations for Change," *New York Times*, June 14, 2017.

8. Schein, *The Corporate Culture Survival Guide*, p. 14.

9. Edgar H. Schein, with Peter Schein, *Organizational Culture and Leadership* (Hoboken, N.J.: John Wiley and Sons, 2017), p. 6.

10. Schein, with Peter Schein, *Organizational Culture and Leadership*, p. 20.

11. Others identify these as "attitudes and behaviors." See IAEA, "Managing Organizational Change in Nuclear Organizations," *IAEA Nuclear Energy Series*, No. NG-T-1.1, Vienna, 2014, p. 21. Schein, however, sees behavior as an outcome of culture.

12. Carroll, John S., "Introduction to Organizational Analysis: The Three Lenses," unpublished class paper, Sloan School of Management, MIT, revised June 2006, p. 10.

13. IAEA Safety Glossary: Terminology Used in Nuclear Safety and Radiation Protection, 2007 ed., p. 175, http://www-pub.iaea.org/MTCD/publications/PDF/Pub1290 _web.pdf.

14. IAEA, "Nuclear Security Culture: Implementing Guide," IAEA Nuclear Security Series No. 7, IAEA, Vienna, 2008, p. 3, https://www-pub.iaea.org/MTCD/Publications /PDF/Pub1347_web.pdf.

15. Frazar and Mladineo have proposed that safeguards culture broadly defined is: "A shared belief among individuals, organizations, and institutions that strict attention to international safeguards requirements and affirmative cooperation with safeguards authorities will enhance their nonproliferation stature and benefit their missions." S.L. Frazar and S.V. Mladineo, "Safeguards Culture: Lessons Learned," *ESARDA Bulletin*, No. 44 (June 2010). This does not quite fit the IAEA specifically, as it is one of the "authorities." As a definition of IAEA safeguards culture, they have proposed: "A shared belief by IAEA safeguards authorities, inspectors, and analysts that the people, processes, and procedures used by the IAEA to draw safeguards conclusions are effective and efficient." Ibid., p. 52. This, however, seems less a definition of desirable safeguards culture than a description of an existing assumption about that culture— which may or may not be true.

16. Some of those interviewed for this project preferred the term "non-proliferation culture" rather than "safeguards culture," as the former is broader.

17. John Carlson, "Changing the Safeguards Culture: Views on and Expectations of the Workshop," INMM/ESARDA Workshop on Changing the Safeguards Culture: Broader Perspectives and Challenges, Santa Fe, New Mexico, October 30–November 2, 2005.

18. Daryl R. Conner, *Managing At the Speed of Change: How Resilient Managers Succeed and Prosper Where Others Fail* (New York: Villart Books, 1993), p. 163.

19. See John Mathiason, "Evolution of the International Public Service (1919–2006)," Chapter 2, in John Mathiason, *Invisible Governance: International Secretariats in Global Politics* (Bloomfield, Conn.: Kumarian Press, 2007).

20. Ibid., p. 41.

21. Alisa Clarke, "Organizational Culture, System Evolution, and the United Nations of the 21st Century," *Journal of Organizational Studies*, Vol. 4, No. 1 (2013), p. 129.

22. Mac Darrow and Louise Arbour, "The Pillar of Glass: Human Rights in the Development Operations of the United Nations," *American Journal of International Law*, Vol. 103, No. 3, p. 452.

23. Clarke, p. 129.

24. For a somewhat hysterical account of this phenomenon, see Shirley Hazzard, *Defeat of an Ideal: The Self-Destruction of the United Nations* (London: Macmillan, 1973).

25. The situation was made worse at the IAEA's inception by the Austrian government's insistence that in return for hosting the Agency in Vienna, the organization should employ large numbers of unskilled Austrians in its support divisions.

26. John Hall, "The International Atomic Energy Agency: Origins and Early Years," *IAEA Bulletin*, No. 2 (1987), pp. 47–54. See also interview with Dieter Goethel by Elisabeth Roehrlich, IAEA History Research Project, August 18, 2016.

27. Ibid.

28. Ibid.

29. Ibid.

30. Ibid.

31. Ibid.

32. David Fischer, *History of the International Atomic Energy Agency: The First Forty Years* (Vienna: IAEA, 1997), p. 89.

33. According to dictionary definitions, "autonomous" means "self-governing," while "independent" means "not bound or subject to another." See G. N. Garmonsway, *The Penguin English Dictionary* (Harmondsworth, UK: Penguin Books, 1965). This seems like a distinction without a difference.

34. This dates back to negotiations on the IAEA Statute, in which the United States and its allies pressed for the Agency to be the master of its own destiny (or, more cynically, one controlled by a Western-dominated board). They clearly wished to avoid the fate of the short-lived UN Atomic Energy Commission, which was established by and beholden to the majority-ruled UN General Assembly. The Soviet Union, neutrals like Sweden, and India and other developing states wanted closer ties to the General Assembly, but were overruled. The Western powers also wished to avoid the IAEA becoming a UN specialized agency reporting to the UN Economic and Social Council (ECOSOC).

35. Specialized agencies, so designated under Article 57, Chapter IX of the UN Charter, are "autonomous organizations working with the UN and each other through the coordinating machinery of ECOSOC at the intergovernmental level and through the Chief Executives Board for coordinating (CEB) machinery at the intersecretariat level." "The United Nations System," www.un.org. Examples are the UN Economic, Social and Cultural Organization (UNESCO) and the UN Food and Agricultural Organization (FAO). Historically, however, the specialized agencies have acted as independently as the IAEA, their directors perceiving themselves as "operating their own fiefdoms." Margaret P. Karns and Karen A. Mingst, *International Organizations: The Politics and Processes of Global Governance*, 2nd ed. (Boulder, Colo.: Lynne Rienner, 2010), p. 116.

36. See IAEA Fact Sheet on DPRK Safeguards, https://www.iaea.org/newscenter /focus/dprk/fact-sheet-on-dprk-nuclear-safeguards, accessed November 1, 2016.

37. See Tariq Rauf and Jan Lodding, "UNSCR 1540 and the Role of the IAEA," in Olivia Bosch and Peter van Ham, eds., *Global Non-Proliferation and Counter-Terrorism: The Impact of UNSCR 1540* (Baltimore, Md.: Brookings Institution Press, 2007).

38. For details of the IAEA/UNSCOM relationship, see Gudrun Harrer, "Tasking the IAEA," Chapter 2, in Gudrun Harrer, *Dismantling the Iraqi Nuclear Program: The Inspections of the International Atomic Energy Agency, 1991–1998* (London: Routledge, 2014).

39. Ibid., p. 31.

40. For the Agency's public presentation of its view of the relationship (but reading between the lines), see Rauf and Lodding, "UNSCR 1540 and the IAEA." Rauf reports that under Director General Yukiya Amano, the IAEA increased its cooperation with the Security Council's 1540 Committee. Email from Tariq Rauf to the author, June 1, 2015.

41. See Jack Boureston and Yana Feldman, "Verifying Libya's Nuclear Disarmament," in Trevor Findlay, ed., *Verification Yearbook 2004* (London: Verification Research, Training and Information Center [VERTIC], 2004), pp. 90–91.

42. See U.S. Arms Control and Disarmament Agency (ACDA), "Review of Negotiating History of International Atomic Energy Agency [IAEA] Document INFCIRC/66/ REV.2, "The Agency's Safeguards System," Washington, D.C., September 5, 1984; and ACDA, "Review of the Negotiating History of the IAEA Safeguards Document INFCIRC/153," AC2NC103, Washington, D.C., July 30, 1984.

43. These states were Australia, Belgium, Canada, France, Portugal, South Africa, and the United Kingdom.

44. See Fischer, *History of the International Atomic Energy Agency: The First Forty Years*, pp. 57–70.

45. Treaty on the Non-Proliferation of Nuclear Weapons, also known as the Nuclear Non-Proliferation Treaty (NPT), http://disarmament.un.org/treaties/t/npt; and Robert Pendley, Lawrence Scheinman, and Richard W. Butler, "International Safeguarding as

Institutionalized Collective Behavior," *International Organization*, Vol. 29, No. 3 (June 1975), p. 614.

46. Ibid.

47. Ibid.

48. The first head of the Safeguards Department was Canadian physicist Roger Smith; the chair of the experts groups that first advised on safeguards was Dr. Gunnar Randers, Director of the Norwegian Atomic Energy Institute. The first "dry run" of an INFCIRC/153 agreement was done in Finland. Much of the work on NPT safeguards techniques was carried out at the Nuclear Research Center in Karlsruhe in the Federal Republic of Germany by Professors Wolf Haefele and Karl Wirtz. Fischer, *History of the International Atomic Energy Agency: The First Forty Years*, pp. 246, 253–254. Carlos Büchler has declared that the U.S. contribution was profound and lasting: "in my view the highly qualified and devoted assistance which we received from our U.S. colleagues . . . was essential in the process of developing the Agency's safeguards into a professional and technically respectable activity. We learned from them the theory and practice of nuclear materials accountability, as well as audit and inventory taking procedures. The methods of work which we developed on that basis, though improved through the years, are still in use today." Carlos L. Büchler, "Safeguards: The Beginnings," in IAEA, *International Atomic Energy Agency: Personal Reflections* (Vienna: IAEA, 1997), p. 51.

49. Interviews with key negotiators have confirmed this. See Pendley, Scheinman, and Butler, p. 613, n. 63.

50. Email from Tariq Rauf to the author, June 1, 2015.

51. Another international organization, the Nuclear Energy Agency (NEA) of the Organization for Economic Cooperation and Development (OECD), has nuclear research and development projects, including collaboratively with the IAEA, that affect safeguards and perhaps culture, although it is beyond the scope of this book to investigate this.

52. It appears that their joint workshop on "Changing Safeguards Culture: Broader Perspectives and Challenges," in Santa Fe, New Mexico, from October 30 to November 2, 2005, was the only such event ever held on the subject prior to the May 2014 Harvard workshop held as part of the research for this book.

53. Email from Tariq Rauf to the author, June 1, 2015.

54. The former is clearly inaccurate because the Agency does not inspect nuclear weapons. The latter neglects, *inter alia*, the role of the UN Security Council, the Conference on Disarmament, groups of states, and individual members.

55. See Harrer, p. 36.

56. John Carlson says that ElBaradei took a "legalistic" approach to the Iranian weaponization issue. Interview with John Carlson by the author, June 5, 2014. In an interview with Gudrun Harrer, Hans Blix reportedly denied that the IAEA has a "legal" culture, although probably what is meant is "legalistic." See Harrer, p. 33.

57. The 2002 MANNET Report on the IAEA said that under Blix, "the Agency [had] consistently shown its determination to enhance its effectiveness and efficiency." M. Campbell, et al., *At What Cost Success? Final Report of the External Review on the Management Processes of the International Atomic Energy Agency*, MANNET, Geneva, 2002 (hereafter MANNET Report), p. 6.

58. Mathiason, p. 41.

59. Berhanykun Andemicael and John Mathiason, *Eliminating Weapons of Mass Destruction: Prospects for Effective International Verification* (London: Palgrave, 2005), p. 214.

60. Ibid.

61. See MANNET Report, pp. 17 and 22.

62. IAEA, Annual Report for 1990, p. 143, https://www.iaea.org/About/Policy/GC /GC35/GC35Documents/English/gc35-953_en.pdf; and information provided informally by the IAEA Secretariat. The Agency does not regularly publicize the number of inspectors, preferring "person days of inspection" (PDI), partly because the numbers are in constant flux, but also because in theory anyone can be designated an inspector if approved by the Board of Governors. It is also due to a less fathomable reluctance, which is partly cultural, to identify an "inspectorate." (See Chapter 5 in this volume for further analysis of this phenomenon.)

63. More staff are authorized to conduct inspections if needed but they usually work on safeguards at headquarters, so they are not included in the official numbers. In addition to the regular staff, at the end of 2020 there were approximately 110 (in total) extrabudgetary consultants, staff with temporary contracts, cost-free experts, and junior professional officers working in the Safeguards Department. In these categories, there are no designated inspectors. Information provided informally by the IAEA Secretariat, June 23 2021.

64. IAEA, Annual Report for 1990, p. 143; information provided informally by the IAEA Secretariat; and "IAEA at a Glance," February 2021, https://www.iaea.org/sites /default/files/21/02/iaea-at-a-glance_2021.pdf.

65. Jeffrey Lewis, "Replacing Olli," Arms Control Wonk, July 3, 2010, http://lewis.arms controlwonk.com/archive/2792/replacing-olli.

66. Goethel notes this of the Agency as a whole. Interview with Dieter Goethel by Elisabeth Roehrlich, IAEA Oral History Series.

67. Dieter Goethel, IAEA Director of Personnel from 1989 to 2002, claimed without elaborating that "women bring a different approach to management." Interview with Dieter Goethel by Elisabeth Roehrlich, IAEA Oral History Series.

68. As of December 31, 1990, in posts subject to geographical distribution. IAEA, Annual Report for 1990, p. 139.

69. As of December 31, 1990, in posts subject to geographical distribution. IAEA, Annual Report for 1990, p. 139.

70. Although not specifically related to safeguards, he confessed that he himself had to get used to the "urgent" approach of the Germans. Interview with Dieter Goethel by Elisabeth Roehrlich, IAEA Oral History Series.

71. There are currently five offices reporting to the IAEA Director General: the Director General's Office, the Secretariat of the Policy-Making Organs, the Office of Internal Oversight Services, the Office of Legal Affairs, and the Office of Public Information and Communication; see https://www.iaea.org/about/organizational-structure/offices-reporting-to-the-director-general, accessed March 16, 2021.

72. IAEA, The Agency's Inspectorate: Memorandum by the Director General; and Annex, The Agency's Inspectors, GC(V) INF/39, Vienna, August 28, 1961 (hereafter the "Inspectors' Document"). Although it was superseded by INFCIRC/153, this early document greatly influenced the content of the latter.

73. See Trevor Findlay, "Proliferation Alert! The IAEA and Non-Compliance Reporting," Report No. 2015–04, Project on Managing the Atom, Belfer Center for Science and International Affairs, Harvard University, Cambridge, Mass., October 2015.

74. Tariq Rauf, Head of the Verification and Security Policy Coordination Office from 2002 to 2011, claims that public affairs had a "poor understanding of safeguards and even worse way of explaining" them. Email from Tariq Rauf to the author, June 1, 2015.

75. The IAEA's fiftieth anniversary publication described the Agency as "the world's nuclear watchdog: the custodian of the Nuclear Nonproliferation Treaty [NPT]." IAEA, *Atoms for Peace: A Pictorial History of the International Atomic Energy Agency* (Vienna: IAEA, 2007), p. 8.

76. Büchler, p. 48.

77. IAEA Statute, Article II.

78. In his address to the General Conference in 1961, Glenn Seaborg, head of the U.S. delegation, exclaimed that "to my Government it is clear that the Agency, being charged in its Statute with world-wide peaceful expansion of the greatest scientific-technological developments of our time, is in a position to be a powerful catalyst of peace and international co-operation." Quoted in Hall, p. 53.

79. As well as reporting to the Security Council "when appropriate," the IAEA Statute provides that "if in connection with the activities of the Agency there should arise questions that are within the competence of the Security Council, the Agency shall notify the Security Council, as the organ bearing the main responsibility for the maintenance of international peace and security." IAEA Statute, Article III.B.4.

80. "Preface by the Director General," in IAEA, *International Atomic Energy Agency: Personal Reflections* (Vienna: IAEA, 1997).

81. Thomas E. Shea and David A. Thompson, "Human Factors Affecting IAEA Inspector Performance," Prepared for the U.S. Arms Control and Disarmament Agency (ACDA), AC2NC106 (unpublished), Washington, D.C., 1982/1983, p. 59.

82. Richard G. Hewlett and Jack M. Holl, *Atoms for Peace and War, 1953–1961: Eisenhower and the Atomic Energy Commission* (Berkeley: University of California Press, 1989), p. 314; the conference was called the Meeting of Six Governments (ibid., p. 628, n. 24).

83. Allan McKnight, *Atomic Safeguards: A Study in International Verification* (New York: United Nations Institute for Training and Research, 1971), p. 66. The short-lived United Nations Atomic Energy Commission, by contrast, had identified three distinct types of proliferation risk (although they did not use the word, which only came into vogue in the 1970s): "(1) diversion of materials from known and declared activities; (2) clandestine operations; and (3) unauthorized seizure of material or plants, thus removing them from the purview of the control agency." McKnight, p. 14. From the outset, IAEA safeguards only dealt with the first scenario.

84. Büchler, p. 48.

85. Hans Grümm called it the principle of "equality of misery." "Paying Tribute to 25 Years of Safeguards Leadership," *IAEA Bulletin*, No. 3 (1994), p. 14.

86. McKnight, p. 37.

87. The "Inspectors' Document," Article II.7.

88. "The Structure and Content of Agreements between the Agency and States Required in Connection with the Treaty on the Non-Proliferation of Nuclear Weapons [NPT]," INFCIRC/153 (Corrected), June 1972, IAEA, Vienna (hereafter INFCIRC/153), Article 4(c), https://www.iaea.org/sites/default/files/publications/documents/infcircs/1972/infcirc153.pdf.

89. Stockholm International Peace Research Institute (SIPRI), *Safeguards Against Nuclear Proliferation* (Stockholm: Almqvist and Wiksell for SIPRI, 1975), p. 33.

90. McKnight, p. 45.

91. Ibid., p. 37.

92. Shea and Thompson, p. 4.

93. Ibid., p. 43.

94. "Testimony of Roger Richter," U.S. Senate Committee on Foreign Relations, Hearings, The Israeli Airstrike, 97th Cong., 1st sess., June 18, 19, and 25, 1981 (hereafter Richter testimony), pp. 112–157.

95. Prepared Statement of Roger Richter, Richter testimony, p. 127.

96. There is a troubling circularity here: even if there were suspicions about a state's compliance, the Safeguards Department could not act without firm evidence; however, because it was not permitted to "discriminate" against a state by paying special attention to it in order to try to confirm its suspicions, it was not able to uncover firm evidence if it existed.

97. IAEA Mission Statement, http://www.iaea.org/About/mission.html. The IAEA was especially encouraged in this quest by the Kennedy administration (1961–1963), which had its own soaring scientific ambitions, not least the moon mission, and in its turn by the Johnson administration. Interview with James Goodby by Elisabeth Roehrlich, IAEA Oral History Series, Washington, D.C., November 17, 2015.

98. Reinhard Loosch, "The Emergence of the Group of 77 as a Major Player in the Board of Governors," in IAEA, *International Atomic Energy Agency: Personal Reflections* (Vienna: IAEA, 1997), p. 65.

99. Ibid., p. 66.

100. Paul Jolles, "From the East River to the Danube: Preparatory Commission and First Years in Vienna," in IAEA, *International Atomic Energy Agency: Personal Reflections* (Vienna: IAEA, 1997), p. 34.

101. Fischer, *History of the International Atomic Energy Agency: The First Forty Years*, pp. 87–88.

102. Loosch, p. 65.

103. Raja Ramanna, "Positive Aspects of the Work of the International Atomic Energy Agency," in IAEA, *International Atomic Energy Agency: Personal Reflections* (Vienna: IAEA, 1997), p. 291. The SAC was wound up in 1988 in favor of more focused scientific and technical advisory groups. Fischer, *History of the International Atomic Energy Agency: The First Forty Years*, p. 79.

104. The marine research facility became the International Laboratory of Marine Radioactivity (ILMR), later the IAEA Marine Environment Laboratory. Today it is part of the IAEA Environment Laboratories.

105. Fischer, *History of the International Atomic Energy Agency: The First Forty Years*, p. 89.

106. Emil Keblúšek, "The Spirit of Vienna," in IAEA, *International Atomic Energy Agency: Personal Reflections* (Vienna: IAEA, 1997), pp. 107–114.

107. Interview with Hans Blix with Elisabeth Roehrlich, IAEA Oral History Project, Stockholm, April 13, 2015.

108. Today, one can see this scientific legacy in the way the Agency runs its international conferences. They feature multiple technical sessions and scientific posters. Even the listing of participants and paper presenters follows scientific practice.

109. James Goodby recalls urging the first U.S. ambassador to the IAEA to stress the scientific capability of the IAEA "in order to gain respect in the world." The "idea fell flat" at the time but was taken up later. Interview with James Goodby with Elisabeth Roehrlich, IAEA Oral History Series.

110. Michael Barnett and Martha Finnemore, *Rules for the World: International Organizations in Global Politics* (Ithaca, N.Y.: Cornell University Press, 2004), pp. 20–29.

111. Ibid.

112. Michael Barnett and Liv Coleman, "Designing Police: Interpol and the Study of Change in International Organizations," *International Studies Quarterly*, Vol. 49 (2005), p. 595.

113. Oscar A. Quihillalt and Carlos L. Büchler, "The Fifth General Conference of the IAEA (October 1961)," in IAEA, *International Atomic Energy Agency: Personal Reflections* (Vienna: IAEA, 1997), p. 61.

114. Keblúšek, "The Spirit of Vienna," p. 109.

115. IAEA Statute, Articles 12.A.6 and B.

116. See Barnett and Finnemore, pp. 20–29. For a detailed discussion of the theory as applied to the IAEA, see Robert L. Brown, *Nuclear Authority: The IAEA and the Absolute Weapon* (Washington, D.C.: Georgetown University Press, 2015).

117. In the case of a much younger organization, the Organization for the Prohibition of Chemical Weapons (OPCW), in which a cultural assumption of autonomy had not yet embedded itself, the assertion of the Technical Secretariat's prerogative to manage its own operations contributed to the dismissal of Director General José Bustani in 2002. The experience, Maurizio Barbeschi says, "illustrates how an element of the technical dimension (budget authority, a part of control systems) is inseparable from the political/cultural dimension." Bustani's actions in denying the Executive Council "became a powerful symbol to the Secretariat of having a leader that was willing to stand up to states parties to defend the integrity of the Secretariat staff and independence of the organization." Maurizio Barbeschi, "Organizational Culture of the OPCW Secretariat," *Disarmament Forum*, No. 4 (2002), p. 51.

118. Mohamed ElBaradei, *The Age of Deception: Nuclear Diplomacy in Treacherous Times* (New York: Metropolitan Books, 2011), p. 215.

119. "Challenges in Verification," presentation by IAEA Director General Yukiya Amano to the Center for Strategic and International Studies (CSIS), Washington, D.C., April 5, 2019, https://www.iaea.org/newscenter/statements/challenges-in-nuclear-verification.

120. IAEA Statute, Article VII.F.

121. By 1958, EURATOM had already developed and was applying a comprehensive safeguards system to the nuclear activities of its member states, but this was a solo effort—not in collaboration with the IAEA. Fischer, *History of the International Atomic Energy Agency: The First Forty Years*, p. 245.

122. Even the United States, despite President Eisenhower's Atoms for Peace proposal, had little idea of how multilateral safeguards would work. At a special meeting called after the first Geneva Conference on the Peaceful Uses of Atomic Energy in 1955, the U.S. delegation was, embarrassingly, at a loss to explain the system to the Soviets, who pressed them on the details. Hewlett and Holl, p. 314.

123. McKnight, p. 47.

124. David Fischer and Paul Szasz, *Safeguarding the Atom: A Critical Appraisal* (London: Taylor and Francis for the Stockholm International Peace Research Institute [SIPRI], 1985), p. 97.

125. Interview with Mohamed Shaker by Elisabeth Roehrlich, IAEA Oral History Series, Berlin, September 1, 2017. His study, *The Nuclear Non-Proliferation Treaty: Origins and Implementation, 1959–1979*, is out of print but available online at https://nonproliferation.org/the-nuclear-non-proliferation-treaty-origins-and-implementation-1959-1979/.

126. INFCIRC discussions, including correspondence with Australia. See Box 02571_S.Comm.5.2, IAEA Archives, Vienna.

127. Fischer and Szasz, p. 97.

128. Ibid.

129. See Lawrence Scheinman, *The International Atomic Energy Agency and World Nuclear Order* (Washington, D.C.: Resources for the Future, 1987), pp. 244–245. The Secretariat included such language in draft agreements that it was negotiating with Argentina and Spain. In 1979, it also took the initiative to exclude assistance for "peaceful" nuclear explosions from new technical assistance guidelines.

130. IAEA, The Annual Report for 1990, p. 106.

131. For a detailed analysis of the Safeguards Implementation Report (SIR) see Trevor Findlay, "Proliferation Alert! The IAEA and Non-Compliance Reporting," pp. 111–119. Tariq Rauf notes that "some years ago," the SIR originally carried more information, listing each state by name along with its record of cooperation in implementing safeguards, but this was rejected mainly by Western states and the SIR reverted to its traditional format. Email from Tariq Rauf to the author, June 1, 2015.

132. Heinonen concurs, saying "it is worth considering making the entire Safeguards Implementation Report available to the public" provided that it does not contain sensitive proprietary information. Olli Heinonen, "IAEA Safeguards: Evolving its 40-year old Obligations to Meet Today's Verification Undertakings," Belfer Center for Science and International Affairs, July 12, 2013, p. 6, https://www.belfercenter.org/publication/iaea-safeguards-evolving-meet-todays-verification-undertakings.

133. Transcript of Panel Discussion with Jacques Baute, Rich Hooper, and Dimitri Perricos, "The Path Toward Strengthened Safeguards: Verification Challenges of the 1990s," 7.0, Foundations of International Safeguards video series, Pacific Northwest Nuclear Laboratories (PNNL), Vienna, December 2009, p. 3.

134. Hooper is quoted in Harrer, p. 42.

135. Heinonen, "IAEA Safeguards—Evolving its 40-Year old Obligations to Meet Today's Verification Undertakings," pp. 6–7.

136. Ibid.

137. Email from Tariq Rauf to the author, March 25, 2021.

138. "National Technical Means" is an arms control euphemism for any methods that states use to acquire intelligence information.

139. Hans Blix, *Disarming Iraq: The Search for Weapons of Mass Destruction* (London: Bloomsbury, 2005), p. 20.

140. Schein, *The Corporate Culture Survival Guide*, p. 107.

141. Peter M. Haas, "Introduction: Epistemic Communities and International Policy Coordination," *International Organization*, Vol. 46, No. 1 (Winter 1992), pp. 1–35.

142. Errol Meidinger, "Regulatory Culture: A Theoretical Outline," *Law and Policy*, October 1987, p. 374.

143. For glimpses of the pride of inspectors in their work, see "Life of the Inspector in the Field" and "Life of an Inspector and Analyst in Headquarters," Symposium on International Safeguards, "Preparing for Future Verification Challenges," IAEA, Vienna, November 1–5, 2010.

144. Shea and Thompson, p. 43.

145. Interview with Shirley Johnson by Elisabeth Roehrlich, IAEA Oral History Series, Vienna, August 20, 2015.

146. Ibid.

147. Sometimes more than twenty people attended. All of the divisions, including consultants, were welcome. Sometimes even DDG for Safeguards Heinonen attended. Shirley "went around the room and asked everyone how their week had been." The fact that both professional and support staff attended was "the really good thing . . . we were all on the same level in that room with a glass of wine or juice and we not only talked about work but also personal events." Email from Shirley Johnson to the author, October 21, 2020.

148. Interview with Shirley Johnson by the author, Atlanta, July 22, 2014.

149. Olli Heinonen, "Five Decades of IAEA Safeguards," Paper Presented to INMM Meeting, Session C, Atlanta, Georgia, July 2014, www.belfercenter.org/publication /five-challenging-decades-iaea-safeguards.

150. Workshop on Nuclear Safeguards Culture, Project on Managing the Atom, Belfer Center for Science and International Affairs, Harvard University, Cambridge, Mass., May 9, 2014. This claim is somewhat overstated. Weapons-related issues in all of these cases were dealt with by newly recruited inspectors or cost-free experts from the nuclear weapon states, not the regular IAEA inspectorate. Regular Agency inspectors only did normal safeguards activities on the civilian side. But a cultural myth has been perpetuated, partly due the media, that IAEA inspectors did "weapons inspections." Email from Tariq Rauf to the author, June 1, 2015.

151. Interview with Shirley Johnson by the author.

152. Email from Shirley Johnson to the author, October 21, 2020.

153. Ibid.

154. Interview with Shirley Johnson by the author.

155. Shea and Thompson, p. 60.

156. U.S. Congressional Research Service, *Nuclear Safeguards: A Reader*, Report Prepared by the U.S. Congressional Research Service for the Subcommittee on Energy Research and Production transmitted to the Committee on Science and Technology, U.S. House of Representatives, 98th Cong., 1st sess., December 1983, U.S. Government Printing Office, Washington, D.C., December 1983, p. 1.

157. Shea and Thompson, p. 60.

158. Current statistics on the professional backgrounds of IAEA inspectors are not publicly available, but some idea may be gleaned from 2009 statistics on U.S. international safeguards specialists: 22 percent were nuclear engineers; 16 percent were physicists, and 14 percent were chemists. Social scientists constituted 9 percent, and no other individual engineering or scientific discipline contributed more than 6 percent. D. Lockwood, M. Scholtz, L. Blair, and E. Wonder, "Next Generation Safeguards Initiative: Human Capital Development Programs," IAEA, IAEA-CN-184/108, November 2010.

159. See the section on recruitment below for details of gender balance at the Agency.

160. MANNET Report, pp. 17–18.

161. Ibid.

162. Nakićenović, p. 19.

163. Shea and Thompson, p. 34.

164. Ibid., pp. 32–33.

165. See Scott Ritter, *Iraq Confidential: The Untold Story of the Intelligence Conspiracy to Undermine the UN and Overthrow Saddam Hussein* (New York: Nation Books, 2005).

166. See C.P. Snow, *The Two Cultures: A Second Look* (New York: New American Library, 1959).

167. Ibid., p. 14.

168. Ibid., p. 12.

169. Organization for Economic Cooperation and Development, *OECD Best Practice Principles for Regulatory Policy: Regulatory Enforcement and Inspections* (Paris: OECD, 2014), p. 48.

170. One exception is the OECD report cited above.

171. IAEA Safeguards Glossary, 2001 ed., International Nuclear Verification Series, No. 3 (2001), p. 89.

172. The Secretariat annually provides member states with a list of the safeguards staff who are designated as inspectors; there are usually about 250.

173. This has been attenuated in recent years by IAEA safety and security personnel also donning IAEA helmets and white uniforms.

174. Blix, p. 52.

175. Ibid.

176. Interview with Olli Heinonen by the author, June 5, 2015. At the Fourth NPT Review Conference in 1990, several countries expressed serious doubts about Iraq's non-proliferation credentials. Canada, Hungary, the Netherlands, and the United Kingdom proposed that the IAEA conduct a special inspection, but this idea was rejected. See Harrer, p. 32.

177. Robert Kelley recalls a visit to the KANUPP reactor in Pakistan when he accompanied an IAEA inspector who did not pay much attention to the facility but simply changed the batteries in the equipment. Interview with Robert Kelley by the author, Vienna, March 19, 2014.

178. See IAEA, "IAEA Safeguards: Staying Ahead of the Game," Vienna, 2007, pp. 10–11. Even then, assessing design often involves subjective judgment, so for some inspectors it was easier to ignore, pay lip service to, or even prevaricate about it.

179. U.S. Senate Committee on Foreign Relations, Hearings, The Israeli Airstrike, 97th Cong., 1st sess., June 18, 19, and 25, 1981, Appendix, IAEA Comments on the Testimony of Roger Richter on June 19, 1981, before the Senate Foreign Relations Committee, U.S. Government Printing Office, Washington, D.C., 1981, pp. 297–298.

180. John Carlson, "Changing the Safeguards Culture: Views on and Expectations of the Workshop," p. 1.

181. Interview with Shirley Johnson by Elisabeth Roehrlich, IAEA Oral History Series.

182. Interviews with U.S. safeguards officials by the author, U.S. State Department, Washington, D.C., May 19, 2014.

183. Interview with Shirley Johnson by the author.

184. Interview with Shirley Johnson by the author. Johnson confirms that they still do, although they now recognize that specialized inspectors should not be rotated so often, in order to maintain continuity of knowledge of both the project and of sensitive facilities. She says that "the rotation policy is basically good, even though it is somewhat disruptive." Email from Shirley Johnson to the author, October 21, 2020.

185. Office of the Comptroller and Auditor, Report of the External Auditor on the Audit of the Financial Statements of the International Atomic Energy Agency for the Year Ended 31 December 2013, Part V, IAEA, The Agency's Financial Statements for 2013, GC(58)/5, p. 145, https://www.iaea.org/About/Policy/GC/GC58/GC58Docu ments/English/gc58-5_en.pdf.

186. Interview with Pier Roberto Danesi by Elisabeth Roehrlich, IAEA Oral History Series, Vienna, April 23, 2015. The Safeguards Analytical Laboratory (SAL) at Seibersdorf has since been brought under the authority of the Safeguards Department, strengthening the link between them but presumably involving cultural change at the SAL, which was previously managed as a technical service operation.

187. Transcript of Panel Discussion with Current and Former SAGSI Representatives, "International Perspectives on Safeguards," 5.2, Foundations of International Safeguards video series, PNNL, Tucson, Arizona, July 2007, p. 10. IAEA Statute, Article XII.

188. Treaty on the Non-Proliferation of Nuclear Weapons (NPT), Article III.

189. Transcript of Panel Discussion with Myron Kratzer, Rich Hooper, and Ambassador Norman Wolf, "Retrospective of INFCIRCs 153 and 540," 4.0, Foundations of International Safeguards video series, PNNL, Santa Fe, New Mexico, October 2005, p. 7.

190. Transcript of Panel Discussion with Myron Kratzer, Rich Hooper, and Ambassador Norman Wolf, "Retrospective of INFCIRCs 153 and 540," p. 12.

191. Transcript of Panel Discussion with Jacques Baute, Rich Hooper, and Dmitri Perricos, "The Path Toward Strengthened Safeguards: Verification Challenges of the 90s," p. 8.

192. Report on International Control of Atomic Energy, Prepared for the Secretary of State's Committee on Atomic Energy (The Acheson-Lilienthal Report), Department of State Publication 2498, U.S. Government Printing Office, Washington, D.C., March 16, 1946, p. 7, http://www.learnworld.com/ZNW/LWText.Acheson-Lilienthal.html.

193. Report of the Advisory Committee on U.S. Policy Toward the International Atomic Energy Agency, U.S. Government Printing Office, Washington, D.C., 1962. In summarizing his report for the Joint Committee on Atomic Energy, chairman Henry D. Smyth noted that one of the problems arising from the expected increase in the use of atomic energy as a source of electric power was the "question of 'safeguards' against the diversion of nuclear materials to military uses." United States Policy Toward the International Atomic Energy Agency, Hearing before the Joint Committee on Atomic

Energy, Congress of the United States, Eighty-Seventh Congress, August 2, 1962, U.S. Government Printing Office, Washington, 1962, p. 3.

194. There have been persistent rumors that this material ended up in Israel.

195. Myron B. Kratzer, "Prospective Trends in International Safeguards," *Nuclear News*, October 1980, p. 56.

196. U.S. bilateral agreements were also concerned with physical protection of the materials supplied, but the IAEA did not replicate this feature because it regarded this as the responsibility of states. Member states agreed.

197. McKnight, p. 15. The 1967 Treaty of Tlatelolco, which established the Latin American Nuclear Weapon Free Zone and which was being negotiated contemporaneously with the NPT, does so. It provides for special inspections both of the IAEA type and when any party "suspects that some activity prohibited by this Treaty has been carried out or is about to be carried out" by an accused party, presumably to prove its innocence. There is no mention of diversion from declared activities (Article 16).

198. Treaty on the Non-Proliferation of Nuclear Weapons (NPT), Article III.1.

199. The solution was that such material would be withdrawn from safeguards for the period of its non-explosive military use and returned to safeguards thereafter. See McKnight, pp. 129–130.

200. Treaty on the Non-Proliferation of Nuclear Weapons (NPT), Article III.1.

201. ACDA, "Review of the Negotiating History of the IAEA Safeguards Document INFCIRC/153," p. 33.

202. "Paying Tribute to 25 Years of Safeguards Leadership," p. 14.

203. See Laura Rockwood, "The IAEA's State-Level Concept and the Law of Unintended Consequences," *Arms Control Today*, Vol. 44, No. 7 (September 2014), p. 30. She cites International Energy Associates Ltd., "Review of the Negotiating History of the IAEA Safeguards Document INFCIRC/153," July 30, 1984, pp. 33–44, http://cgs.pnnl.gov/fois/doclib/INFCIRC153Ch1-3.pdf. See also interview with Laura Rockwood by Elisabeth Roehrlich, IAEA Oral History Series, Vienna, August 3, 2016.

204. ACDA, "Review of the Negotiating History of the IAEA Safeguards Document INFCIRC/153," p. 35.

205. INFCIRC/153, para. 2.

206. ACDA, "Review of the Negotiating History of the IAEA Safeguards Document INFCIRC/153," p. 35.

207. Ibid.

208. Ibid., pp. 40–41.

209. Ibid., p. 43.

210. Ibid., p. ii.

211. Fischer and Szasz, p. 36.

212. "Paying Tribute to 25 Years of Safeguards Leadership," p. 14.

213. Ibid.

214. Interview with Laura Rockwood by the author, Vienna, July 3, 2013.

215. Tariq Rauf and Robert Kelley, "Nuclear Verification in Iraq," *Arms Control Today*, September 2014, p. 14.

216. Kirkpatrick is quoted in Blix, p. 19.

217. "Safeguards against Nuclear Proliferation," Remarks by Myron B. Kratzer, Director, Division of International Affairs, U.S. Atomic Energy Commission for presentation at George Washington University—American Assembly, Airlie House, Warrenton, Va., May 11, 1967, p. 6, reproduced in *Nuclear Safeguards: A Reader*, pp. 253–266.

218. "The International Atomic Energy Agency (IAEA) and Its Safeguards: A Fact Sheet," Joseph F. Pilat, Office of Senior Specialists, U.S. Congressional Research Service, May 6, 1982, reproduced in *Nuclear Safeguards: A Reader*, p. 22.

219. Garmonsway, *The Penguin English Dictionary*, p. 215.

220. See, for instance, the discussion of undeclared material. ACDA, "Review of the Negotiating History of the IAEA Safeguards Document INFCIRC/153," pp. 33–43.

221. Ibid., p. 41.

222. *Nuclear Safeguards: A Reader.*

223. Harrer, p. 145.

224. Nic von Wielligh and Lydia von Wielligh-Steyn, *The Bomb: South Africa's Nuclear Weapons Programme* (Pretoria: Litera for the Institute for Security Studies, 2015), p. 237.

225. Ibid.

226. Email from Robert Kelley to the author, April 24, 2015.

227. Interview with Olli Heinonen by the author, June 5, 2015.

228. John Carlson, "Safeguards in a Broader Policy Perspective: Verifying Treaty Compliance," INMM/ESARDA Workshop on Changing the Safeguards Culture, Santa Fe, New Mexico, October 30–November 2, 2005, p. 2. One result of this mindset, Carlson says, was that provisions in INFCIRC/153 that allowed for differentiation in safeguards application to reflect state-specific factors were substantially underutilized.

229. Interview with Olli Heinonen by the author.

230. Shea and Thompson, p. 35.

231. Ibid.

232. John Carlson, "Safeguards in a Broader Policy Perspective: Verifying Treaty Compliance."

233. Charles Duelfer, *Hide and Seek: The Search for Truth in Iraq* (New York: PublicAffairs, 2009), p. 97.

234. Christopher Chabris and Daniel Simons, *The Invisible Gorilla: How Our Intuitions Deceive Us* (New York: Broadway Paperbacks, 2009), pp. 5–8.

235. IAEA, Strengthening the Effectiveness and Improving the Efficiency of the Safeguards System, Report of the Director General, GC(39)/17, August 22, 1995, Annex 3, p. 54.

236. The cases of Pakistan's KANUPP reactor and India's equivalent facility in the early 1970s should have rung more insistent alarm bells. For details, see Trevor Findlay, "Proliferation Alert! The IAEA and Non-Compliance Reporting," p. 22.

237. Interview with Mohamed Elbaradei by Elisabeth Roehrlich, IAEA Oral History Series, Vienna, May 31, 2016.

238. See, for instance, the work of Daniel Kahneman and Amos Tversky, et al.: Daniel Kahneman, Paul Slovie, and Amos Tversky, eds., *Judgement Under Uncertainty: Heuristics and Bias* (Cambridge: Cambridge University Press, 1982); Thomas Gilovich, Dale W. Griffin, and Daniel Kahneman, *Heuristics and Biases: The Psychology of Intuitive Judgement* (Cambridge: Cambridge University Press, 2002); Daniel Kahneman and Amos Tversky, eds., *Choices, Values, and Frames* (Cambridge: Cambridge University Press, 2000); and Daniel Kahneman, *Thinking Fast . . . Thinking Slow* (New York: Farrar, Strauss, and Giroux, 2011).

239. Nakićenović, p. 23. He further reports an attempt to collect information about an "unpublished" [presumably meaning unpublicized] nuclear program and individual unsafeguarded facilities in 1973. Again, he says he was "bypassed" by this instruction. Director General Eklund refused to halt this activity or inform the Board. It only stopped when Nakićenović informed the U.S. and Soviet representatives to the IAEA.

240. Special inspections were possible under both INFCIRC/66-type agreements dating from 1965–1968 and INFICIRC/153 agreements after 1971.

241. Laura Rockwood, "Safeguards 101," Presentation at the Belfer Center for Science and International Affairs, Harvard University, Cambridge, Mass., May 22, 2015.

242. Email from Laura Rockwood to the author, October 19, 2020.

243. John Carlson, "Special Inspections Revisited," Paper Presented to Annual Meeting of the Institute of Nuclear Materials Management (INMM), Phoenix, Arizona, July 10–14, 2005, p. 1.

244. Interview with Hans Blix by Elisabeth Roehrlich, IAEA Oral History Series.

245. "Viewpoints," *IAEA Bulletin*, No. 3 (1994), p. 19.

246. Romania voluntarily requested a special inspection in 1992. Laura Rockwood explains that she was informed that prior to 1991, the IAEA had carried out three inspections formally referred to as "special inspections," but they were all at declared locations. Email from Laura Rockwood to the author, October 19, 2020.

247. Transcript of Panel Discussion with Myron Kratzer, Rich Hooper, and Ambassador Norman Wolf, "Retrospective of INFCIRCs 153 and 540," p. 7.

248. Blix is quoted in Fischer, *History of the International Atomic Energy Agency: The First Forty Years*, p. 295.

249. Blix, p. 18.

250. Ibid.

251. ElBaradei, p. 10.

252. Blix, p. 20.

253. Harrer, p. 45, note 39.

254. Ibid., pp. 17–18.

255. Transcript of Panel Discussion with Jacques Baute, Rich Hooper, and Dimitri Perricos, "The Path Toward Strengthened Safeguards: Verification Challenges of the 90s," p. 8.

Chapter Four

1. Model Protocol Additional to the Agreement(s) between State(s) and the International Atomic Energy Agency for the Application of Safeguards, INFCIRC/540 (Corrected), September 1997 (hereafter Additional Protocol).

2. For brief definitions of "artefacts," "espoused values," and "underlying assumptions," see Chapter 3 in this volume. For further elaboration of these concepts, see Edgar H. Schein, with Peter Schein, *Organizational Culture and Leadership* (Hoboken, N.J.: John Wiley and Sons, 2017),

3. Hans Blix was apparently unaware of the machinations until after the resolution was adopted. Tim Trevan, *Saddam's Secrets: The Hunt for Iraq's Hidden Weapons* (London: HarperCollins, 1999), pp. 47–49. See also Hans Blix, *Disarming Iraq: The Search for Weapons of Mass Destruction* (London: Bloomsbury, 2005), p. 20.

4. Gudrun Harrer, *Dismantling the Iraqi Nuclear Program: The Inspections of the International Atomic Energy Agency, 1991–1998* (Abingdon, Oxon: Routledge, 2014), p. 29.

5. Ibid., p. 33.

6. This assessment ignores the fact that U.S. intelligence had also failed to detect the program and had not even shared its suspicions with the IAEA. Trevan, p. 47.

7. Ibid. Some U.S. officials did not share this critical view of the IAEA. Ambassador Michael Newlin, UNSCOM Deputy Executive Chairman, said in a luncheon briefing at the Washington Council on Non-Proliferation that in describing Iraq's "very expensive and ambitious secret nuclear weapons program, I do not imply any criticism, failure, or dereliction on the part of the IAEA." "Disarming Iraq: Preparing for the Long-Term Monitoring of Iraq's Nuclear Weapons Capability," Washington Council on Non-Proliferation, Washington, D.C., October 26, 1992.

8. Interview with Robert Gallucci by Gudrun Harrer, Washington, D.C., June 3, 2005, quoted in Harrer, p. 33.

9. It was, however, argued that to sideline the Agency might jeopardize the crucial 1995 Nuclear Non-Proliferation Treaty (NPT) Review Conference, at which an indefinite extension of the treaty would be considered.

10. Even after the issue was settled, the administration, at Gallucci's urging, continued to threaten to withdraw the nuclear disarmament file from the IAEA. Harrer, p. 90.

11. Blix, p. 22.

12. Over the twelve years or so of the Iraq Action Team's existence, the leaders were: David Kay, IAEA Department of Technical Cooperation; Robert Kelley, Los Alamos National Laboratory; Paul Stokes, Sandia National Laboratory; Jay Hyland, Pacific Northwest National Laboratory; and Robert Kelley again.

13. These outside experts were not certified by the Board as IAEA inspectors until approximately 2002, eleven years later. Email from Robert Kelley to the author, April 24, 2015.

14. See Trevor Findlay, "The Lessons of UNSCOM and UNMOVIC," in Trevor Findlay, ed., *Verification Yearbook 2004* (London: Verification Research, Training, and Information Center [VERTIC], 2004).

15. Transcript of Panel Discussion with Jacques Baute, Rich Hooper, and Dmitri Perricos, "The Path Toward Strengthened Safeguards: Verification Challenges of the 90s," 7.0, Foundations of International Safeguards video series, Pacific Northwest Nuclear Laboratories (PNNL), Vienna, December 2009, p. 4.

16. Blix, p. 22.

17. There is evidence that personal rivalries and differences between the two Swedes, Blix and Ekéus, played some role in the IAEA-UNSCOM clash, so not everything can be attributed to organizational cultural differences.

18. Jean E. Krasno and James S. Sutterlin, *The United Nations and Iraq: Defanging the Viper* (New York: Praeger, 2003), p. 23.

19. Trevan, pp. 99–100. IAEA Inspection 4 was conducted from July 27 to August 10, 1991.

20. Interview with Rolf Ekéus by Gudrun Harrer, Vienna, January 12, 2005, in Harrer, p. 90.

21. Ibid., p. 87.

22. Krasno and Sutterlin, p. 22.

23. Yale-UN Oral History interview with Scott Ritter, Larchmont, N.Y., October 27, 1998, quoted in Harrer, p. 24.

24. Yale-UN Oral History interview with Scott Ritter.

25. Ibid.

26. For example, French nuclear weapons expert Jacques Baute was recruited to UNSCOM as a nuclear expert in 1992 and participated in IAEA inspections. He joined the IAEA in 1994.

27. See Richard Butler, *Saddam Defiant: The Threat of Weapons of Mass Destruction, and the Crisis of Global Security* (London: Weidenfeld and Nicholson, 2000), especially pp. 107 and 201.

28. For a straightforward account of IAEA inspections during this period, see Eric Chauvistré, "The Implications of IAEA Inspections under Security Council Resolution 687," Research Paper No. 11, United Nations Institute for Disarmament Research (UNIDIR), Geneva, 1992.

29. Yale-UN Oral History interview with Rolf Ekéus, February 3, 1998, quoted in Krasno and Sutterlin, p. 23.

30. Stockholm International Peace Research Institute (SIPRI), *Safeguards Against Nuclear Proliferation* (Stockholm: Almqvist and Wiksell for SIPRI, 1975), p. 9.

31. Ibid., p. 40.

32. Blix, p. 18.

33. IAEA, "The Evolution of IAEA Safeguards," IAEA, Vienna, 1998, p. 14. "National Technical Means" is an arms control euphemism for any methods that states use to

acquire intelligence information. High-resolution satellite reconnaissance was not declassified until the mid-1980s.

34. Treaty on the Non-Proliferation of Nuclear Weapons, also known as the Nuclear Non-Proliferation Treaty (NPT), http://disarmament.un.org/treaties/t/npt.

35. Lawrence Scheinman, *The International Atomic Energy Agency and World Nuclear Order* (Washington, D.C.: Resources for the Future, 1987), p. 123. This is a reference to the 1946 Acheson-Lilienthal Report, the 1946 Baruch Plan, and the deliberations of the short-lived UN Atomic Energy Commission (1946–1952), all of which contemplated a vastly more intrusive form of "safeguards" for verifying complete nuclear disarmament.

36. David Fischer and Paul Szasz, *Safeguarding the Atom: A Critical Appraisal* (London: Taylor and Francis for SIPRI, 1985), p. 18.

37. Warren Donnelly, "Nuclear Weapons Proliferation and the International Atomic Energy Agency: An Analytic Report," prepared for the U.S. Senate Committee on Governmental Operations, Washington, D.C., 1976, p. 92.

38. Slobodan Nakićenović, "Comments on IAEA Safeguards," U.S. Congressional Research Service, Washington, D.C., August 1981, reproduced in U.S. Congressional Research Service, *Nuclear Safeguards: A Reader*, Report Prepared by the U.S. Congressional Research Service for the Subcommittee on Energy Research and Production transmitted to the Committee on Science and Technology, U.S. House of Representatives, 98th Cong., 1st sess., December 1983, U.S. Government Printing Office, Washington, D.C., December 1983.

39. Ibid.

40. UN Security Council Resolution 687 of April 3, 1991, which had been adopted under Chapter VII, the enforcement chapter of the UN Charter, was considered legally binding on Iraq.

41. See First semi-annual report (covering the period 17 June–17 December 1991) on the implementation by the IAEA of the plan for the destruction, removal or rendering harmless of items listed in paragraph 12 of UN Security Council Resolution 687 (1991), S/23295, December 17, 1991, p. 4.

42. Ibid., p. 3.

43. IAEA, Official Records, GOV/OR.748, May 6, 1991, cited in Harrer, p. 47, n. 76.

44. These CSAs were of the INFCIRC/153 variety, such as the agreement that the IAEA had with Iraq. Interview with Hans Blix by Gudrun Harrer, Stockholm, February 18, 2005; see Harrer, p. 33.

45. For details of all IAEA inspections in Iraq from 1991–1997, see Fourth Consolidated Report by the Director General of the International Atomic Energy Agency under Paragraph 16 of Security Council Resolution 1051 (1996), Appendix to Letter dated 6 October 1997 from the Director General of the International Atomic Energy Agency, Annex to UN Security Council, Note by Secretary General, S/1997/779, October 8 1997, https://nsarchive2.gwu.edu/NSAEBB/NSAEBB80/wmd07.pdf.

46. Harrer, p. 52.

47. Ibid., p. 42.

48. Ibid. Harrer's account implies that Kay had not seen the report himself, although this is difficult to believe.

49. Contrary to Harrer's account, these three were not UNSCOM inspectors but U.S. Department of Energy laboratory experts—George Anzelon, Bill Nelson, and Jeff Bedell—seconded to the Iraq Action Team, on its payroll, and under the leadership of Zifferero. Email from Robert Kelley to the author, April 24, 2015.

50. Ibid.

51. Harrer, pp. 54–55.

52. Ibid., p. 91. Ultimately the IAEA was right to discount the importance of the plutonium in the Iraq nuclear weapons program, but this was not known at the time.

53. Interview with Olli Heinonen by the author, June 5, 2015.

54. Transcript of Panel Discussion with Jacques Baute, Rich Hooper, and Dimitri Perricos, "The Path Toward Strengthened Safeguards: Verification Challenges of the 1990s," p. 5.

55. Transcript of Panel Discussion with Myron Kratzer, Rich Hooper, and Ambassador Norman Wolf, "Retrospective of INFCIRCs 153 and 540," 4.0, Foundations of International Safeguards video series, PNNL, Santa Fe, New Mexico, October 2005.

56. Harrer, p. 54.

57. Ibid., pp. 42–43.

58. Ibid., p. 34.

59. Transcript of Panel Discussion with Jacques Baute, Rich Hooper, and Dimitri Perricos, "The Path Toward Strengthened Safeguards: Verification Challenges of the 90s," p. 4.

60. Trevan, p. 132.

61. Ibid.

62. Harrer, p. 107.

63. Email from Robert Kelley to the author, April 24, 2015.

64. The IAEA and UNSCOM, tellingly, adopted different numbering systems for their inspections. These additional experts were paid from New York by non-Agency funds and were on contracts that limited them to working for the Iraq Action Team and not for the Agency itself, although their employment conditions were the same as extra-budgetary Agency employees.

65. Email from Robert Kelley to the author, April 24, 2015.

66. Ibid.

67. Krasno and Sutterlin, p. 25.

68. Harrer, p. 261.

69. Interview with Dimitri Perricos by the author, Baden, Austria, June 20, 2014.

70. Harrer, p. 67. Harrer notes the inconsistencies in various accounts of the incident, but these do not detract from its unprecedented character.

71. Note by the President of the Security Council, S/22746, June 28, 1991, p. 1.

72. Ibid.

73. Harrer, pp. 60–67. Harrer quotes Gallucci as reporting that, during IAEA 2, when the Iraqi driver of the bus carrying UNSCOM and IAEA inspectors missed a turn, only the UNSCOM people protested. The IAEA inspectors, including team leader Maurizio Zifferero, preferred to stay silent. Ibid., p. 69.

74. Scott Ritter, *Iraq Confidential: The Untold Story of the Intelligence Conspiracy to Undermine the UN and Overthrow Saddam Hussein* (New York: Nation Books, 2005), p. 13. Kay later fell out with the IAEA, despite being given a distinguished service award by the Agency for his services in IAEA 6. Blix, p. 25.

75. Ibid., p. 22.

76. Harrer, p. 67.

77. Yale-UN Oral History interview with David Kay, McLean, Va., April 10, 2000, quoted in Krasno and Sutterlin, p. 23.

78. Harrer, p. 67.

79. Blix, p. 23. Blix told Harrer that during the high-level mission sent to Iraq after the Fallujah incident during IAEA 2, a U.S. inspector "mistreated" an Iraqi: "this big American sort of shook the Iraqi scientist and I said to David Kay, can't you cool this fellow, and Kay did cool him." Interview with Hans Blix by Gudrun Harrer, Stockholm, February 18, 2005, quoted in Harrer, p. 69.

80. Mohamed ElBaradei, *The Age of Deception: Nuclear Diplomacy in Treacherous Times* (New York: Metropolitan Books, 2011), pp. 22–23. He recalls that he and Blix took David Kay aside after he interrogated a senior Iraqi scientist on the spot, telling him that was "not the way we performed inspections. Our aim, in this case, was to work towards full cooperation on the part of the Iraqis. Intimidation and humiliation were not, in our view, useful tactics."

81. The term "cowboys" is attributed to Gallucci, quoted in ibid., p. 23. UN Secretary-General Kofi Annan also used the term in a brief to the UN Security Council. See Charles Duelfer, *Hide and Seek: The Search for Truth in Iraq* (New York: PublicAffairs, 2009), p. 142. For the "bunny huggers" reference, see Harrer, p. 33. Rolf Ekéus has also alluded to IAEA jokes at the expense of UNSCOM's intrusive inspection practices and its attitude that "we are the mature guys" in comparison to UNSCOM; he is quoted in Harrer, p. 37.

82. Dimitri Perricos interview with Gudrun Harrer, quoted in ibid., p. 33.

83. Per reports in ibid., p. 67. Gallucci was reportedly angry when Zifferero seemed to acquiesce to an inspection-free Islamic holiday.

84. Ekéus is quoted in Krasno and Sutterlin, p. 23.

85. Harrer, p. 84. Although self-serving, this observation was subsequently supported by Jafar Dhia Jafar, director of Petrochemical Project 3 (PC-3), the code name for the Iraqi nuclear program. He claimed that what had been discovered by the aggressive inspection methods would have been uncovered anyway, as one question led to the next in a logical train of questions and answers.

86. For the official IAEA report, see First Report on the Sixth IAEA On-Site Inspection in Iraq under Security Council Resolution 687 (1991), S/23122, October 8, 1991.

See also Ritter, pp. 13–15; and Trevan, pp. 105–111. Harrer also provides a detailed account; see Harrer, pp. 106–130.

87. See Ritter, pp. 10–13, for his account of intelligence agency involvement.

88. Interview with David Kay by Gudrun Harrer, Washington, D.C., June 3, 2005, cited in Harrer, p. 107.

89. Ibid.

90. Ibid., p. 106.

91. Iraq Action Team archive, cited by Harrer, p. 106. Uniquely, the official report, First Report on the Sixth IAEA On-Site Inspection in Iraq under Security Council Resolution 687 (1991), S/23122, October 8, 1991, does not give the number or nationalities of the personnel involved.

92. This list is adapted from Chauvistré, "The Implications of IAEA Inspections under Security Council Resolution 687," p. 15.

93. UN Security Council, Report of the First Panel Established pursuant to the Note by the President of the Security Council on January 30, 1999 (S/1999/100), concerning disarmament and current and future ongoing monitoring and verification issues (hereafter Amorim Report), Annex 1, Letter dated March 27 and 30, 1999, respectively, from the Chairman of the Panels established Pursuant to the Note by the President of the Security Council of January 30, 1999 (S/1999/100) addressed to the President of the Security Council, S/1999/356, March 30, 1999, p. 13.

94. Ibid., p. 26.

95. Krasno and Sutterlin, p. 25.

96. Email from John Tilemann to the author, May 30, 2019. See also Butler, *Saddam Defiant*, pp. 89, 107, and 201.

97. When UNSCOM was succeeded by UNMOVIC, the IAEA team was renamed the Iraq Nuclear Verification Office (INVO).

98. See Trevor Findlay, "A Standing United Nations Verification Body: Necessary and Feasible," *Compliance Chronicles*, No. 1, Canadian Center for Treaty Compliance, Ottawa, December 2005; and Trevor Findlay, "Lessons of UNSCOM and UNMOVIC for WMD Non-proliferation, Arms Control and Disarmament," in Waheguru Pal Singh Sidhu and Ramesh Thakur, eds., *Arms Control After Iraq: Normative and Operational Challenges* (Tokyo: United Nations University Press, 2006).

99. Harrer, p. 261.

100. IAEA, "IAEA Safeguards: Staying Ahead of the Game," 2007, p. 15.

101. Interview with Dimitri Perricos by the author, June 20, 2014.

102. Michael Wilson, "Safeguards and the IAEA Board of Governors: 1991–1993: Iraq, A Necessary Stimulus for Handling the DPRK," in IAEA, *International Atomic Energy Agency: Personal Reflections* (Vienna: IAEA, 1997), p. 133.

103. ElBaradei, p. 48.

104. Thomas E. Shea, "Financing IAEA Verification of the Nuclear Nonproliferation Treaty," in Henry Sokolski, ed., *Falling Behind: International Scrutiny of the Peaceful*

Atom (Carlisle, Penn.: Nonproliferation Policy Education Center, Strategic Studies Institute, U.S. Army War College, 2008), p. 323.

105. James Tape and Joseph Pilat, "Nuclear Safeguards and the Security of Nuclear Materials," in James Doyle, ed., *Nuclear Safeguards, Security, and Nonproliferation: Achieving Security with Technology and Policy* (Amsterdam: Elsevier for Los Alamos National Laboratory, 2008), p. 29.

106. Ibid.

107. IAEA Press Release, PR/91–24, July 18, 1991.

108. Michael D. Rosenthal, et al., *Deterring Nuclear Proliferation: The Importance of IAEA Safeguards* (Upton, N.Y.: Brookhaven National Laboratory, 2019), p. 142.

109. The Council did not, however, "recast the President's statement into a more formal and binding commitment." Fischer, *History of the International Atomic Energy Agency: The First Forty Years* (Vienna: IAEA, 1997), p. 295.

110. This was the so-called Amended Code 3.1.

111. Interview with Jacques Baute by Gudrun Harrer, July 21, 2005, quoted in Harrer, p. 122.

112. Ibid.

113. Statement of Hans Blix at informal consultations in the UN Security Council, October 8, 1991, in Archive of Hans Blix, quoted and cited in Harrer, p. 122.

114. Interview with Jacques Baute by Gudrun Harrer, quoted in Harrer, p. 122.

115. See Trevor Findlay, "Proliferation Alert! The IAEA and Non-Compliance Reporting," Report No. 2015–04, Project on Managing the Atom, Belfer Center for Science and International Affairs, Harvard University, Cambridge, Mass., October 2015, pp. 39–50.

116. For details, see Nic Von Wielligh and Lydia von Wielligh-Steyn, *The Bomb: South Africa's Nuclear Weapons Programme* (Pretoria: Litera for the Institute for Security Studies, 2015).

117. Further strengthening occurred after revelations about A.Q. Khan's illicit nuclear trading network in December 2003 and the Libyan case from 2003–2008.

118. Fischer, *History of the International Atomic Energy Agency: The First Forty Years*, p. 297.

119. 1995 Review and Extension Conference of the Parties to the Treaty on the Non-Proliferation of Nuclear Weapons [NPT] Final Document, Decision 2: Principles and Objectives for Nuclear Non-Proliferation and Disarmament, para. 11, http://www.un.org/Depts/ddar/nptconf/2142.htm.

120. Email from John Tilemann to the author, May 30, 2019.

121. Ibid.

122. IAEA, "IAEA Safeguards: Staying Ahead of the Game," p. 12.

123. Transcript: The Politics of Safeguards, Carnegie International Nuclear Policy Conference, Washington, D.C., March 24, 2015, p. 4.

124. In addition, the clearly inadequate Small Quantities Protocol was revised, giving the Agency more powers in states with only small quantities of nuclear material.

125. Rosenthal, et al., p. 150.

126. Ibid.

127. Ibid.

128. Ibid. Cuba acceded to the NPT in 2002.

129. Often used interchangeably, the State-Level Concept and State-Level Approach were only gradually defined by the Safeguards Department and continue to be confusing to many, including some member states. State-Level Concept refers to "the general notion of implementing safeguards in a manner that considers a State's nuclear and nuclear-related activities and capabilities as a whole." Supplementary Document to the Report on The Conceptualization and Development of Safeguards Implementation at the State Level (GOV/2013/38): Report by the Director General, GOV/2014/41, August 13, 2014, para. 126. The State-Level Approach refers to a "customized approach to implementing safeguards for an individual State. An SLA is detailed in an internal document developed by the Secretariat." Ibid., para. 217. The SLA is not shared with the state, although the state is consulted in its preparation. For an excellent analysis of evolving terminology, see Noah Mayhew, "A Lexical History of the State-Level Concept and Issues for Today," Vienna Center for Disarmament and Non-Proliferation (VCDNP), Vienna, December 2020, https://vcdnp.org/wp-content/uploads/2021/01/Lexical-History-of-the-State-level-Concept_Final.pdf.

130. South Africa was next.

131. See IAEA, "The Conceptualization and Development of Safeguards Implementation at the State Level," GOV/2013/38, August 12, 2013, p. 3.

132. These safeguards were known as "integrated" because the framework integrated the requirements of INFCIRC/153 with those of the Additional Protocol to produce a single approach. Rosenthal, et al., p. 166, n. 334.

133. See M. Hosoya, E. Franklin Saburido, H. Nackaerts, G. Bernasconi, V. Wong, and K. Warthan, "Implementing an Integrated Safeguards Approach at Multi-Unit CANDU Stations: Potential Savings," IAEA-CN-148/60, "Addressing Verification Challenges," Proceedings of an International Safeguards Symposium, STI/PUB/1289, IAEA, Vienna, October 16–20, 2009.

134. Rosenthal, et al., p. 171. In states with nuclear fuel cycles consisting largely of power reactors and no sensitive facilities, the reduction can be considerably larger. In Canada, for example, the reduction in inspection effort was 70 percent.

135. Transcript of Interview with Rich Hooper, "Concluding Remarks," 7.26, Foundations of International Safeguards video series, PNNL, Vienna, December 2009.

136. ElBaradei, pp. 28–29; and Interview with Hans Blix by Elisabeth Roehrlich, IAEA Oral History Research Project, Stockholm, April 13, 2015, http://iaea-history.univic.at/oral-history-videos/.

137. Transcript: The Politics of Safeguards, Carnegie International Nuclear Policy Conference, p. 17.

138. *INMM Chronicle*, January 2006, citing Roland Schenkel, Presentation at INMM/ESARDA Workshop on Changing the Safeguards Culture: Broader Perspectives and Challenges, Santa Fe, New Mexico, October 30–November 2, 2005.

139. Jill Cooley, "Addressing Verification Challenges," Proceedings of an International Safeguards Symposium, IAEA, Vienna, October 16–20, 2006, p. 142, http://www-pub.iaea.org/MTCD/publications/PDF/P1298/P1298_Book.pdf.

140. Herman Nackaerts, "Evolution of Safeguards Implementation," Presentation to Nuclear Safeguards Culture Workshop, Harvard University, Cambridge, Mass., May 9, 2014.

141. As DDG for Safeguards from 2010 to 2013, Herman Nackaerts brought a different approach drawn from his experience at EURATOM. Its safeguards culture is, however, quite different, partly because its safeguards have a different legal basis. EURATOM can order operators to comply, so it does not need to convince them of the value of safeguards. Its inspectors are more inclined to question the operators and are much more open to a dialogue with them. The attempt to graft EURATOM practices onto the IAEA did not, some inspectors recall, go well. Interview with IAEA safeguards official, 2014.

142. *INMM Chronicle*, January 2006, p. 3.

143. Rosenthal, et al., p. 159.

144. Additional Protocol, Article 18.b.

145. Ibid., Article 2.a.iii.

146. Ibid., Article 4.b.

147. Rosenthal, et al., p. 155.

148. John Carlson, "Safeguards in a Broader Policy Perspective: Verifying Treaty Compliance," INMM/ESARDA Workshop on Changing the Safeguards Culture: Broader Perspectives and Challenges, Santa Fe, New Mexico, October 30–November 2, 2005, p. 3.

149. Ibid.

150. Panel Discussion, "Retrospective of INFCIRCs 153 and 540," Foundations of International Safeguards video series, 4.0, Pacific Northwest Nuclear Laboratories, October 30, 2005, http://cgs.pnnl.gov/fois/videos/foundations/.

151. Ibid.

152. Interview with U.S. safeguards expert by the author, U.S. State Department, Washington, D.C., May 19, 2014.

153. Rosenthal, et al., p. 166.

154. IAEA, *IAEA Safeguards Glossary*, 2001 ed., International Nuclear Verification Series, No. 3 (2002), p. 91; and Additional Protocol, Article 2.

155. INFCIRC/540, Article 4.

156. Grigory Berdennikov, John Carlson, Thomas Countryman, and Anton Khlopkov, "Principles and Recommendations for Implementation of the IAEA Safeguards System," in Nuclear Threat Initiative (NTI) and Center for Energy and Security Studies

(CNESS), "The Future of IAEA Safeguards: Rebuilding the Vienna Spirit through Russian-U.S. Expert Dialogue," Washington, D.C., November 2020, p. 11.

157. "Games" include sponsorship games, alliance-building games, empire-building games, budgeting games, expertise games, management versus staff games, whistle-blowing games, gender games, regional group games, and nationality games. See Maurizio Barbeschi, "Organizational Culture of the OPCW Secretariat," *Disarmament Forum*, No. 4 (2002).

158. Edgar H. Schein, *The Corporate Culture Survival Guide* (San Francisco: Jossey-Bass, 1999), pp. 122–123.

159. James A. Casterton, "Using State-Level Approaches in International Safeguards," INMM/ESARDA Joint Workshop on Changing Safeguards Culture: Broader Perspectives and Challenges, Santa Fe, New Mexico, October 30–November 1, 2005, p. 2.

160. Interview with Shirley Johnson by the author, Atlanta, July 22, 2014.

161. Annette Berriman, Russel Leslie, and John Carlson, "The Role of 'Safeguards Criteria' in an Evolving Safeguards Environment," Australian Safeguards and Non-Proliferation Office, Canberra, 2016, p. 1, https://www.dfat.gov.au/sites/default/files/inmm2006_safeguards_criteria.pdf.

162. Ibid., p. 2.

163. Ibid.

164. IAEA, *IAEA Safeguards Glossary*, 2001 ed., p. 21.

165. Ibid., p. 5.

166. Interview with John Carlson by the author, June 5, 2014.

167. David Kay was an evaluator for the Agency's TC program at the time, so resentment at his making recommendations about safeguards from outside the safeguards "community" may also have been a factor.

168. Interview with Shirley Johnson by the author, Atlanta, July 22, 2014.

169. Email from Shirley Johnson to the author, October 21, 2020.

170. In 2015, Olli Heinonen was still questioning whether the pendulum had swung too far away from the criteria-based approach. Interview with Olli Heinonen by the author, June 5, 2015.

171. Interview with IAEA official by the author, Safeguards Symposium, Vienna, October 20–24, 2014.

172. As of 2019, the Safeguards Document Manager System contained about 3,000 active documents such as departmental policies, procedures, instructions, guidelines, and other safeguards documents. In 2019, the external auditor described it as a "legacy system" that had not been updated for many years and recommended an evaluation by the department to upgrade it. See The Audit Board of the Republic of Indonesia, Audit Report on the International Atomic Energy Agency, 2019, Part V, IAEA, The Agency's Financial Statements for 2019, GC(64)/4, Vienna, July 2020, pp. 203–204. Personnel: Staffing of the Agency's Secretariat, Report by the Director General, August 2, 2017, GOV/2017/38/GC(61)/18, pp. 17–175.

173. Interview with Olli Heinonen by the author, May 22, 2014.

174. Informal discussion with IAEA official by the author, IAEA Safeguards Symposium, Vienna, October 20–24, 2014.

175. *INMM Communicator*, January 2006, p. 3.

176. Interview with IAEA safeguards official, July 2014.

177. Ibid.

178. Interview with Olli Heinonen by the author, May 22, 2014.

179. Scott Shane, "Agents Enjoy Status, But Intelligence Analysts Gain Attention," *New York Times*, March 27, 2015, p. A15.

180. IAEA Statute, Article XII.c.

181. Interview with IAEA safeguards official, July 2014.

182. Comments by Jon Jennekens in "Paying Tribute to 25 Years of Safeguards Leadership," *IAEA Bulletin*, No. 3 (1994), p. 15.

183. At the end of 2020, the number of regular staff in the Safeguards Department was about 770, of whom 285 were inspectors (information provided informally by the IAEA Secretariat, June 23, 2021).

184. By 2019, the number of surveillance cameras installed at facilities where nuclear material is present had grown by a third to nearly 1,600 since 2010. The number of unattended monitoring systems had risen by 16 percent to 171, while the number of remotely readable, tamper-proof seals placed on nuclear material had jumped by nearly 280 percent to 560. "Challenges in Verification," presentation by IAEA Director General Yukiya Amano to the Center for Strategic and International Studies (CSIS), Washington, D.C., April 5, 2019, https://www.iaea.org/newscenter/statements/challenges-in-nuclear-verification.

185. Ibid.

186. See Sections N to S, Joint Comprehensive Plan of Action (JCPOA), Vienna, 14 July 2015, Communication dated 24 July 2015 received from China, France, Germany, the Russian Federation, the United Kingdom, the United States of America (the E3/EU+3) and the Islamic Republic of Iran concerning the text of the Joint Comprehensive Plan of Action (JCPOA), INFCIRC/887, July 31, 2015.

187. One of the latter was Jacques Baute, who was an UNSCOM inspector before joining the IAEA.

188. IAEA, "20/20 Vision for the Future: Background Report by the Director General for the Commission of Eminent Persons," Annex, "Report of the Commission of Eminent Persons on the Future of the Agency," GOV/2008/22-GC(52)/INF/4, IAEA, Vienna, 2008, pp. 20–21.

189. IAEA, "The Agency's Program and Budget 2016–2017," July 2015, Vienna, GC(59)/2, p. 152.

190. Email from former IAEA official to the author, April 24, 2015.

191. Olli Heinonen, "Five Decades of IAEA Safeguards," Paper Presented to INMM Meeting, Atlanta, July 2014, www.belfercenter.org/publication/five-challenging-decades-iaea-safeguards, p. 6.

192. These practices included Results-Based Management (RBM), Quality Management Systems (QMS), strategic planning, and Business Impact Management (BIM), all of which the Agency eventually adopted.

193. Kelley recommends a "matrix management" approach, whereby technical specialists work on a multitude of programs across the Agency, in the same way that the Department of Management provides common services to all parts of the Agency. This system, he argues, would encourage managers to restructure their programs to be competitive and innovative to attract the best technical staff. Robert Kelley, "Creating a 'One House' Culture at the IAEA through Matrix Management," *SIPRI Policy Brief*, SIPRI, Stockholm, January 2014, pp. 2–3.

194. Email from former IAEA official to the author, April 24, 2015.

195. Ibid.

196. Email from Tariq Rauf to the author, March 25, 2021. This occurred even though Division B had the largest number of states, followed by Divisions A and C. Rauf claims that "at one level there was a tendency to gloss over certain open questions on the assurance of a country officer/director that these would be resolved. [It was] only a matter of time. And on the other hand, there also was questioning mainly regarding open questions that had been carried over from previous years and in some cases a conclusion was not agreed and the matter kicked over to the next year. In other cases, the responses and promised corrective actions were accepted and a conclusion was granted. This seemingly questioning and confrontational tendency seemed to weaken over time."

197. Discussion with Tariq Rauf, Vienna, June 24, 2014.

198. John Patten, "The Role of Quality Management in Changing Safeguards Culture," INMM/ESARDA Joint Workshop on Changing Safeguards Culture: Broader Perspectives and Challenges, Santa Fe, New Mexico, October 30–November 1, 2005.

199. The International Organization for Standardization (ISO) is an independent, non-governmental international organization with a membership of 165 national standards bodies. With a central secretariat in Geneva, it "brings together experts to share knowledge and develop voluntary, consensus-based, market relevant International Standards that support innovation and provide solutions to global challenges." See "About Us," International Organization for Standardization, https://www.iso.org /about-us.html, accessed April 22, 2021.

200. The IAEA's Safeguards Analytical Laboratory had already gone through such a process and been certified against the ISO 9001:2000 standards. Patten, p. 2.

201. This statement is cited and quoted in ibid.

202. Ibid., pp. 3, 4.

203. Ibid., p. 3.

204. Ibid.

205. Ibid., p. 1.

206. Ibid., p. 3.

207. Ibid., p. 4.

208. IAEA, "Managing Organizational Change in Nuclear Organizations," *IAEA Nuclear Energy Series*, No. NG-T-1.1, Vienna, 2014, p. 21.

209. Ibid.

210. Ibid.

211. Participant at the Nuclear Safeguards Culture Workshop, Harvard University, Cambridge, Mass., May 9, 2014.

212. Panel Discussion, "Retrospective of INFCIRCs 153 and 540."

213. Christine Wing and Fiona Simpson, *Detect, Dismantle, and Disarm: IAEA Verification, 1992–2005* (Washington, D.C.: U.S. Institute of Peace, 2013), p. 27.

214. U.S. Office of Technical Assessment, "Nuclear Safeguards and the International Atomic Energy Agency," OTA-ISS-615, Washington, D.C., April 1995, p. 40.

215. IAEA, "20/20 Vision for the Future: Background Report by the Director General for the Commission of Eminent Persons," GOV/2008/22-GC (52)/INF/4, May 23, 2008, p. 18. Commission advisor Matthew Bunn of the Project on Managing the Atom (MTA) at the Belfer Center for Science and International Affairs played a key role in researching and drafting the report; it is not clear to what extent Commission members understood or paid attention to this seemingly harmless recommendation.

216. "Eliminating Nuclear Threats: A Practical Agenda for Global Policymakers," *Report of the International Commission on Nuclear Non-Proliferation and Disarmament* [ICNND] (Canberra and Tokyo: Paragon Press, 2009), pp. 85–86 and 91–92.

217. One exception is Kelley, "Creating a 'One House' Culture at the IAEA through Matrix Management."

218. Discussion with Tariq Rauf, Vienna, June 24, 2014.

219. Personal communication with Dimitri Perricos by the author.

220. Harrer, pp. 33–34.

221. Interview with Shirley Johnson by Elisabeth Roehrlich, IAEA Oral History Series, Vienna, August 20, 2015. When ElBaradei became Director General, he declared that "we are One House and should start acting like that."

222. IAEA, "20/20 Vision for the Future: Background Report by the Director General for the Commission of Eminent Persons," Annex, Report of the Commission of Eminent Persons on the Future of the Agency, GOV/2008/22-GC(52)/INF/4, IAEA, Vienna, 2008, p. 21.

223. Blix, p. 26.

224. Blix, p. 52.

225. ElBaradei, pp. 17–18.

226. Ibid., pp. 22–23.

227. See Proceedings, INMM/ESARDA Workshop on Changing the Safeguards Culture: Broader Perspectives and Challenges, Santa Fe, New Mexico, October 30–November 2, 2005.

228. The main exceptions were John Carlson, "Views on and Expectations from the Workshop," INMM/ESARDA Workshop on Changing the Safeguards Culture: Broader Perspectives and Challenges, Santa Fe, New Mexico, October 30–November 2,

2005, and Stephen V. Mladineo, Karyn R. Durbin, and Andrew Van Duzer, "Changing the Safeguards Culture: Broader Perspectives and Challenges," INMM/ESARDA Workshop on Changing the Safeguards Culture: Broader Perspectives and Challenges, Santa Fe, New Mexico, October 30–November 2, 2005.

229. Heinonen's remarks were delivered by then–Director of Safeguards Concepts and Planning Division Jill Cooley. See Jill N. Cooley, "Views on and Expectations from the Workshop," INMM/ESARDA Workshop on Changing the Safeguards Culture: Broader Perspectives and Challenges, Santa Fe, New Mexico, October 30–November 2, 2005.

230. Doris Ellis and Sara Scott, "Views on and Expectations from the Workshop," INMM/ESARDA Workshop on Changing the Safeguards Culture: Broader Perspectives and Challenges, Santa Fe, New Mexico, October 30–November 2, 2005.

231. Carlson, "Views on and Expectations from the Workshop."

232. Report of Working Group 1, The Further Evolution of Safeguards, Working Group Reports from INMM/ESARDA Workshop on Changing the Safeguards Culture: Broader Perspectives and Challenges, Santa Fe, New Mexico, October 30–November 2, 2005.

233. The *INMM Communicator* repeated the same message: it recognized that "changing safeguards culture requires commitment at all levels: state, organization, management and individual," that "cultural change has to come from good leadership," and that "beliefs are not sufficient; behavior is what counts." "Report on 2005 INMM/ESARDA Santa Fe Workshop: Changing the Safeguards Culture: Broader Perspectives and Challenges," *INMM Communicator*, Vol. 3, No. 3 (January 2006), p. 3.

234. IAEA, "The Evolution of IAEA Safeguards."

235. IAEA, "IAEA Safeguards: Staying Ahead of the Game," p. 27.

236. See R. Clark, "Setting the Culture from Day One"; K. Desson, "Organizational Culture—Why Does It Matter?"; and K. Dahlgren, "Having the Right Organizational Culture," all from Session 21, Enhancing the Organizational Culture to Prepare for Future Missions, IAEA Safeguards Symposium, November 1–5, 2010, http://www.iaea.org/safeguardssymposium2010.

237. Some, however, call for IAEA member states to examine and improve their own nuclear safeguards culture. See, for example, S.L. Frazar and S.V. Mladineo, "The Importance of Safeguards Culture," *Nonproliferation Review*, December 2013. One future challenge noted is to ensure that designers of the next generation of nuclear power plants imbibe a safeguards culture, so that they can appreciate the need for "safeguards by design." See for example, M. Stein, M. Morichi, L. Van den Durpel, T. Killeen, and B. Moran, "The Role of the Industry in Safeguards by Design," IAEA, No. IAEA-CN-184/190, p. 1.

238. Thomas E. Shea and David A. Thompson, "Human Factors Affecting IAEA Inspector Performance," Prepared for the U.S. Arms Control and Disarmament Agency (ACDA), AC2NC106 (unpublished), Washington, D.C., 1982/1983.

239. This paper also mentioned culture in the context of safety, security, and "quality management" throughout the Agency. IAEA Medium Term Strategy 2006–2011, Vienna, 2005, pp. 13 and 25, http://www.iaea.org/sites/default/files/mts2006_2011.pdf.

240. IAEA, Medium Term Strategy 2012–2017, https://www.iaea.org/sites/default /files/mts2012_2017.pdf; and IAEA, Medium Term Strategy 2018–2023, https://www .iaea.org/sites/default/files/16/11/mts2018_2013.pdf.

241. See "Human and Organizational Aspects of Assuring Nuclear Safety—Exploring 30 Years of Nuclear Safety," Proceedings of an International Conference, Vienna, Austria, February 22–26, 2016.

242. IAEA, A Harmonized Safety Culture Model, IAEA Working Document (last revised May 5, 2020), https://www.iaea.org/sites/default/files/20/05/harmonization _05_05_2020-final_002.pdf, accessed March 31, 2021.

243. IAEA, "Nuclear Security Culture: Implementing Guide," IAEA Nuclear Security Series No. 7, IAEA, Vienna, 2008, https://www-pub.iaea.org/MTCD/Publications/PDF /Pub1347_web.pdf.

244. Miklos Gaspar, "Security Culture: One for All, and All for One," *IAEA Bulletin*, December 2016, pp. 14–15.

245. James Casterton, "Engendering Safeguards Culture," Nuclear Safeguards Culture Workshop, Harvard University, Cambridge, Mass., May 9, 2014. Casterton personally had sympathy with and understanding of the cultural dimensions of change and sought to apply this where possible, but without using the word "culture."

246. Ibid.

247. Ibid.

248. In the case of safeguards, such stove-piping even took the form of individual inspectors and analysts "hoarding" information, a typical cultural work-around.

249. IAEA, Department of Safeguards, Long-Term Strategic Plan (2012–2023) Summary, 2011, www.iaea.org/safeguards/documents/LongTerm_Strategic_Plan_%2 820122023%29-Summary.pdf (hereafter Strategic Plan). Surprisingly, the Safeguards Department is the only one in the Agency to have undertaken such activity to date.

250. Ibid., p. 9.

251. Ibid., p. 5.

252. Eric Pujol, "Methodology for Long-Range Strategic Planning in the IAEA Department of Safeguards," IAEA Symposium on International Safeguards: Preparing for Future Verification Challenges, IAEA, Vienna, November 1, 2010.

253. Chaired by the Director of the Division of Concepts and Planning, the Strategic Planning Team comprised a staff member nominated by each of the six division heads, a member of the Section for Effectiveness Evaluation (SEE), and a representative of the DDG for Safeguards.

254. Cooley, Jill N., "Department of Safeguards Long-Term Strategic Plan, 2012–2023," IAEA Symposium on International Safeguards: Preparing for Future Verification Challenges, IAEA, Vienna, November 1, 2010, pp. 15 and 19.

255. Ibid., p. 19.

256. Shea and Thompson, p. 15.

257. Ibid.

258. IAEA, Strengthening the Effectiveness and Improving the Efficiency of Agency Safeguards: Report by the Director General, July 26, 2017, GC(61)/16, p. 10.

259. The initiative is described in Sriata Rao, "Organization Culture and Leadership in the Department of Safeguards," Topical Papers, *Journal of Nuclear Materials Management*, Volume 44, No. 2 (2016). Apparently there was already a "Leadership Blueprint" advocated by the Director General and senior management officials in guiding the behaviors and values of the organization, but this is not publicly available. See also Van Zyl de Villiers, Marguerite Leonardi, Carrie Mathews, Jenni Rissanen, and William Stanley, "Recent Developments in Performance Management in the IAEA Department of Safeguards," Topical Papers, *Journal of Nuclear Materials Management*, Volume 44, No. 2 (2016). The only reference to culture was in a box in a "Performance Map of the IAEA Department of Safeguards" that read "Foster a culture of integrity, collaboration and commitment to safety and security." (p. 10).

260. Ibid., p. 16.

261. The results of this assessment are not publicly available. Five years later, in May 2018, Varjoranta suddenly resigned.

262. Rao, pp. 16–17.

263. Rao, p. 16.

264. IAEA, "Managing Organizational Change in Nuclear Organizations," p. 21.

265. Email from Jim Casterton to the author, March 3, 2015.

266. Stephen Dahunsi, "External Factors that Shape Organizational Culture: A Nigerian Perspective," INMM/ESARDA Safeguards Culture Workshop, Texas A&M University, College Station, Texas, April 26–27, 2016.

Chapter Five

1.   For brief definitions of "artefacts," "espoused values," and "underlying assumptions," see Chapter 3 in this volume. For further elaboration of these concepts, see Edgar H. Schein, with Peter Schein, *Organizational Culture and Leadership* (Hoboken, N.J.: John Wiley and Sons, 2017), p. 6.

2.   Model Protocol Additional to the Agreement(s) between State(s) and the International Atomic Energy Agency for the Application of Safeguards, INFCIRC/540 (Corrected), September 1997 (hereafter Additional Protocol).

3.   IAEA, Department of Safeguards, Long-Term Strategic Plan (2012–2023) Summary, 2011, www.iaea.org/safeguards/documents/LongTerm_Strategic_Plan_%2 820122023%29-Summary.pdf (hereafter Strategic Plan).

4.   For details of all notable non-compliance cases to date, see Trevor Findlay, "Proliferation Alert! The IAEA and Non-Compliance Reporting," Report No. 2015-04, Project on Managing the Atom, Belfer Center for Science and International Affairs, Harvard University, Cambridge, Mass., October 2015.

5.   Herman Nackaerts, "Evolution of Safeguards Implementation," Presentation to Nuclear Safeguards Culture Workshop, Harvard University, Cambridge, Mass., May 9, 2014.

6.   Transcript of Panel Discussion with Current and Former SAGSI Representatives, "International Perspectives on Safeguards," 5.2, Foundations of International Safeguards video series, Pacific Northwest Nuclear Laboratories (PNNL), Tucson, Arizona, July 2007, p. 21.

7.   Additional Protocol, https://www.iaea.org/topics/additional-protocol/, accessed December 15, 2021.

8.   The Broader Conclusion is a determination by the Secretariat that "all nuclear material has remained in peaceful activities in a state." This is based on its finding that "there are no indications of diversion of declared nuclear material from peaceful nuclear activities and no indication of undeclared nuclear material or activities in the State as a whole." The Agency draws such a conclusion only when a member state has both a comprehensive safeguards agreement (CSA) and an Additional Protocol and the Secretariat has conducted the necessary evaluations. See IAEA, Safeguards Statement for 2020, B.1.1., https://www.iaea.org/sites/default/files/21/06/statement-sir-2020.pdf.

9.   IAEA, Safeguards Statement for 2019, B.1.1.1, https://www.iaea.org/newscenter/news/iaea-safeguards-now-applied-in-183-states-worldwide-safeguards-statement-2019. The state, Libya, was enmeshed in civil war and thus unable to fulfill its safeguards obligations.

10.   Statement by HE Mr Richard Sadleir, Resident Representative of Australia to the IAEA, Agenda Item 6(b): Safeguards Implementation Report for 2019, Australian Embassy and Permanent Mission, Vienna, June 16, 2020, https://austria.embassy.gov.au/vien/IAEABOG_Jun20_SIR.html.

11.   Strategic Plan, p. 4.

12.   Ibid., p. 9.

13.   Ibid., p. 4., The Nuclear Non-Proliferation (NPT) Review and Extension Conference in 1995 endorsed this, calling for IAEA safeguards to be "regularly assessed and evaluated"; see 1995 Review and Extension Conference of the Parties to the Treaty on the Non-Proliferation of Nuclear Weapons [NPT] Final Document, Decision 2: Principles and Objectives for Nuclear Non-Proliferation and Disarmament, para. 11, http://www.un.org/Depts/ddar/nptconf/2142.htm.

14.   Strategic Plan, p. 2.

15.   Ibid. This is not as ambitious as it sounds because there is only one other such agency at present, the Comprehensive Nuclear Test Ban Treaty Organization, and it is still in preparatory mode.

16.   Jack Boureston and Charles Ferguson claim that these reforms were "based on shoring up its existing operational technical needs rather than pushing for new authorities." Jack Boureston and Charles D. Ferguson, "Strengthening Nuclear Safeguards: Special Committee to the Rescue?" *Arms Control Today*, December 2005, www.armscontrol.org/print/1951.

17. Thomas E. Shea and David A. Thompson, "Human Factors Affecting IAEA Inspector Performance," Prepared for the U.S. Arms Control and Disarmament Agency (ACDA), AC2NC106 (unpublished), Washington, D.C., 1982/1983, p. 18.

18. Email from Robert Kelley to the author, April 12, 2015. Kelley says, "I learned a lot from [IAEA inspector] Maurizio Zifferero one cold March 1992 Sunday in Vienna when he told the Iraqi[s] that they had lied one time too many and that [we] were coming to Iraq to blow up the offending facilities with explosives, which we did. It was done coldly, professionally and correctly. They even had one more day to change his mind by telling the truth and the negotiators came in and said our story has not changed. They had one night to coordinate with Baghdad."

19. A similar reaction to UNSCOM's perceived excesses played out in the establishment of the Organization for the Prohibition of Chemical Weapons (OPCW), the preparatory work for which began in 1993. According to Maurizio Barbeschi, formerly in the Senior Policy Office of the OPCW's Verification Division, "one of the founding myths of the OPCW was that it would be completely unlike UNSCOM: it would work in a multilateral fashion and be 'above' the political games played during the Iraqi inspections." Maurizio Barbeschi, "Organizational Culture of the OPCW Secretariat," *Disarmament Forum*, No. 4 (2002), p. 47.

20. John Carlson, "Special Inspections Revisited," Paper Presented to Annual Meeting of the Institute of Nuclear Materials Management (INMM), Phoenix, Arizona, July 10–14, 2005, p. 1.

21. Ibid., p. 3.

22. Michael D. Rosenthal, et al., *Deterring Nuclear Proliferation: The Importance of IAEA Safeguards* (Upton, N.Y.: Brookhaven National Laboratory, 2019), p. 146. The authors note that the Board itself may also call for special inspections and did not foreclose its own options.

23. Laura Rockwood, Noah Mayhew, Artem Lazarev, and Mara Pfneisi, "IAEA Safeguards: Staying Ahead of the Game," Swedish Radiation Safety Authority, Report No. 2019:14, Stockholm, 2019, p. 24.

24. James Tape and Joseph Pilat, "Nuclear Safeguards and the Security of Nuclear Materials" in James Doyle, ed., *Nuclear Safeguards, Security, and Nonproliferation: Achieving Security with Technology and Policy* (Amsterdam: Elsevier for Los Alamos National Laboratory, 2008), p. 29.

25. Carlson, "Special Inspections Revisited," p. 4. This was confirmed by the author in interviews with Safeguards Department staff.

26. Rosenthal, et al., p. 146.

27. This is the case even though Iran had agreed in the JCPOA to act as if it had an Additional Protocol in force.

28. See, for example, Joint Statement by the Director General of the IAEA and the Vice-President of the Islamic Republic of Iran and Head of the AEOI [Atomic Energy Organization of Iran], August 20, 2020, https://www.iaea.org/newscenter/press releases/joint-statement-by-the-director-general-of-the-iaea-and-the-vice-president

-of-the-islamic-republic-of-iran-and-head-of-the-aeoi. In February 2021, Iran did just that, declaring that it would no longer implement "voluntary measures" agreed in the JCPOA, even though these were part of a legally-binding and Security Council–endorsed agreement. See Joint Statement by the Vice-President of the Islamic Republic of Iran and Head of the AEOI and the Director General of the IAEA, February 2021, https://www.iaea.org/newscenter/pressreleases/joint-statement-by-the-vice-presi dent-of-the-islamic-republic-of-iran-and-head-of-the-aeoi-and-the-director-general -of-the-iaea.

29. Rockwood, Mayhew, Lazarev, and Pfneisi, "IAEA Safeguards: Staying Ahead of the Game," pp. 24–26.

30. Ibid.

31. "Challenges in Verification," Presentation by IAEA Director General Yukiya Amano to the Center for Strategic and International Studies (CSIS), Washington, D.C., April 5, 2019, https://www.iaea.org/newscenter/statements/challenges-in-nu clear-verification.

32. *INMM Communicator*, Vol. 3, No. 3 (January 2006), p. 3.

33. Ibid.

34. John Patten, "The Role of Quality Management in Changing Safeguards Culture," INMM/ESARDA Joint Workshop on Changing Safeguards Culture: Broader Perspectives and Challenges, Santa Fe, New Mexico, October 30–November 1, 2005, p. 1.

35. To be fair, some organizational theorists share this view. See Paul Bate, *Strategies for Cultural Change* (Oxford: Butterworth Heinemann, 1994), especially pp. 8–24.

36. Unpublished presentations from Safeguards Culture Workshop, Texas A&M University, College Station, Texas, April 26–27, 2016. See http://www.inmm.org/Content /NavigationMenu/Events/UpcomingEvents/SafeguardsCultureWorkshop_Agenda _as_of_04-14-16.pdf.

37. Ibid.

38. Stephen V. Mladineo, "International Workshop on Safeguards Culture," *The Communicator*, INMM, June 2016, p. 3.

39. IAEA, "20/20 Vision for the Future: Background Report by the Director General for the Commission of Eminent Persons," Annex, "Report of the Commission of Eminent Persons on the Future of the Agency," GOV/2008/22-GC(52)/INF/4, IAEA, Vienna, 2008, p. 19.

40. For details, see S.B. Haber and M.T. Barriere, "Development of a Regulatory Organizational and Management Review Method," Research Report RSP-0060, Canadian Nuclear Safety Commission, Ottawa, Canada, 1998.

41. IAEA, Medium Term Strategy 2018–2023, 2017, https://www.iaea.org/sites /default/files/16/11/mts2018_2013.pdf. In theory this document is drafted by the Board, but in practice it is composed by the Secretariat.

42. Shea and Thompson, p. 13.

43. The new verification regime operated in cooperation with the UN Special Commission (UNSCOM) and the UN Monitoring, Verification and Inspection Commission (UNMOVIC).

44. See Mohamed ElBaradei, *The Age of Deception: Nuclear Diplomacy in Treacherous Times* (New York: Metropolitan Books, 2011), p. 62.

45. See remarks by Jill Cooley, Director, Concepts and Planning, Department of Safeguards, INMM/ESARDA Joint Workshop on Changing the Safeguards Culture: Broader Perspectives and Challenges, Santa Fe, New Mexico, October 30–November 2, 2005, reported in *INMM Communicator*, Vol. 3, No. 3 (January 2006).

46. Interview with David Waller by Elisabeth Roehrlich, IAEA Oral History Series, Washington, D.C., November 21, 2015.

47. Panel Discussion, "Retrospective of INFCIRCs 153 and 540," Foundations of International Safeguards video series, Pacific Northwest Nuclear Laboratories, October 2005, http://cgs.pnnl.gov/fois/videos/foundations/.

48. The OPCW is facing the same problem in trying to verify claims of chemical weapons use in Syria.

49. Jacques Baute is one of these experts. Discussion with Tariq Rauf by the author, Vienna, June 24, 2014.

50. See Jack Boureston and Charles D. Ferguson, "Strengthening Nuclear Safeguards: Special Committee to the Rescue?" Arms Control Association, 2005, http://www.arms control.org/act/2005_12DEC-Safeguards.

51. For details of safeguards implementation under the JCPOA, see Rosenthal, et al., pp. 184–196.

52. Nasser Karimi and Kiyoko Metzler, "Iran Officially Imposes Curbs on UN Nuclear Inspections," Associated Press, February 23, 2021.

53. "IAEA Secures 'Temporary Understanding' with Iran," World Nuclear News, February 22, 2021. In March 2021, the Iranians said they would impose further constraints on IAEA activities by denying access to surveillance footage of their nuclear facilities to the Agency.

54. The State-Level Concept (SLC) is the idea that all nuclear materials and facilities and nuclear-relevant activities in the entirety of a state's territory should be taken into account when considering the state's compliance with its safeguards obligations. This contrasts with the previous approach, which was only to consider materials and facilities that had been declared by the state. See IAEA, "The Conceptualization and Development of Safeguards Implementation at the State Level: Report by the Director General," GOV/2013/38, August 12, 2013; and "Supplementary Document to the Report on The Conceptualization and Development of Safeguards Implementation at the State Level (GOV/2013/38): Report by the Director General," GOV/2014/41, August 13, 2014.

55. John Barrett, who was Chair of the Board of Governors at the time, wrote a detailed account of this episode. See John Barrett, "The Travails of Governance: A View from the Board Chair," in Joseph F. Pilat, ed., *The International Atomic Energy*

Agency: Historical Reflections, Current Challenges, and Future Prospects (London: Routledge, 2022).

56. See Laura Rockwood, "The IAEA's State-level Concept and the Law of Unintended Consequences," *Arms Control Today*, Vol. 44, No. 7 (September 2014).

57. "The Future of IAEA Safeguards: Rebuilding the Vienna Spirit through Russian--U.S. Expert Dialogue," Nuclear Threat Initiative (NTI), Washington, D.C., and Center for Energy and Security Studies (CNESS), Moscow, November 2020, p. 34.

58. The downside of this approach, as Olli Heinonen points out, is that there is a reluctance to abandon such processes even when they are not working. See Olli Heinonen, "IAEA Safeguards: Evolving its 40-year old Obligations to Meet Today's Verification Undertakings," Presentation to Annual INMM Meeting, Palm Desert, California, July 14–18, 2013, p. 5.

59. Berhanykun Andemicael and John Mathiason, *Eliminating Weapons of Mass Destruction: Prospects for Effective International Verification* (London: Palgrave, 2005), p. 104.

60. See, for example, Henry Sokolski, ed., *Falling Behind: International Scrutiny of the Peaceful Atom* (Carlisle, Penn.: Nonproliferation Policy Education Center, Strategic Studies Institute, U.S. Army War College, 2008), especially Henry Sokolski, "Assessing the IAEA's Ability to Verify the NPT."

61. Heinonen, "IAEA Safeguards: Evolving its 40-year old Obligations to Meet Today's Verification Undertakings," p. 7.

62. According to David Fischer, the Agency has asked the nuclear weapon states to review the estimates—most recently, he then understood, in 1996. David Fischer, *History of the International Atomic Energy Agency: The First Forty Years* (Vienna: IAEA, 1997), p. 315.

63. John Carlson, "Future Directions in IAEA Safeguards," Discussion Paper, Project on Managing the Atom, Belfer Center for Science and International Affairs, Harvard University, Cambridge, Mass., November 2018, https://www.belfercenter.org/publication/future-directions-iaea-safeguards.

64. See Allan S. Krass, *Verification: How Much Is Enough?* (London: Taylor and Francis for the Stockholm International Peace Research Institute [SIPRI], 1985).

65. Hans Blix, IAEA Director General Statement, IAEA Seminar on "The Current and Future Safeguards Role of the IAEA: Challenges and Opportunities," Vienna, January 27, 1994, para. 13, https://www.iaea.org/newscenter/statements/seminar-current-and-future-safeguards-role-iaea-challenges-and-opportunities.

66. See Fischer, *History of the International Atomic Energy Agency: The First Forty Years*, pp. 285–286. Safeguards are not required to positively establish the physical diversion of nuclear material, let alone produce evidence of the manufacture of a nuclear weapon or explosive device. To establish that there has been a violation of an INFCIRC/153-type safeguards agreement, the Board has simply to conclude that "the Agency is *not able to verify* that there has been *no* diversion of nuclear material required to be safeguarded under the Agreement to nuclear weapons or other nuclear

explosive devices [emphases added]." IAEA, "The Evolution of Safeguards," Vienna, 1998, http://www-pub.iaea.org/MTCD/publications/PDF/NVS2_web.pdf.

67. Strategic Plan, p. 8.

68. Ibid., p. 5.

69. The auditor urged the Secretariat "to establish a comprehensive controlling system that could make such an important programme more transparent. This would enable the Secretariat to quantify the success it has realized in hard figures. Unfortunately, the Secretariat does not have such a controlling system in place. I recommend that the Member States request reports on the progress of integrated safeguards on a regular basis and in a transparent and comprehensible manner." Federal Court of Auditors, Germany, Report of the External Auditor on the audit of the financial statements of the International Atomic Energy Agency for the year ended 31 December 2011, Part V, IAEA, The Agency's Financial Statements for 2011, GC(56)/5, p. 149. The external auditor is always the national auditor of a member state and is appointed for a number of years.

70. Rockwood, Mayhew, Lazarev, and Pfneisi, "IAEA Safeguards: Staying Ahead of the Game," p. 26.

71. John Carlson, "Future Directions in IAEA Safeguards."

72. Ibid.

73. IAEA, Office of Internal Oversight Services, https://www.iaea.org/about/organizational-structure/offices-reporting-to-the-director-general/office-of-internal-oversight-services.

74. Audit Board of Indonesia, Audit Report on the International Atomic Energy Agency (IAEA), 2019, Part V, The Agency's Financial Statements for 2019, GC(64)/4, July 2020, pp. 203–204.

75. Ibid.

76. IAEA, The Agency's Program and Budget 2014–2015, GC(57)/2, August 2013, p. 21.

77. IAEA, *Atoms for Peace: A Pictorial History of the International Atomic Energy Agency* (Vienna: IAEA, 2007).

78. The first three programs were as follows—Program 1: Nuclear power, fuel cycle, and nuclear science; Program 2: Nuclear techniques for development and environmental protection; and Program 3: Nuclear safety and security. In 2012, the UN Joint Inspection Unit (JIU) noted, however, that most of the "development actors are not aware of the IAEA's dual mandate [sic] and the Agency's value in these realms continues to be underutilized." The JIU recommended that "the Agency should enhance its outreach efforts at every level to inform and remind partners and stakeholders that the Agency spends a significant portion of its budget on official development assistance, and should not, given its strong stake and mandate in developing [country] assistance, be reduced to the perception of being the 'nuclear watchdog.'" United Nations, Joint Inspection Unit (JIU), Review of Management and Administration in the International

Atomic Energy Agency (IAEA), JIU/REP/2012/Rev. 1, United Nations, Geneva, 2012, p. 31.

79. IAEA, Medium Term Strategy 2018–2023. It is notable, however, that the latest IAEA safeguards pamphlet for popular dissemination is subtitled "Serving Nuclear Non-Proliferation." IAEA, "IAEA Safeguards: Serving Nuclear Non-Proliferation," 2021, https://www.iaea.org/sites/default/files/18/09/sg-serving-nuclear-non-prolifera tion.pdf.

80. Undoubtedly influenced by the Agency's quiet defiance of U.S. views in Iraq, the Nobel Committee's citation focused almost exclusively on the IAEA's role in nuclear non-proliferation, with only passing references to safety and security and none to the other major IAEA programs involving the peaceful uses of nuclear energy: "In the nuclear nonproliferation regime, it is the IAEA which controls that [sic] nuclear energy is not misused for military purposes, and the Director General has stood out as an unafraid advocate of new measures to strengthen that regime." Transcript of 2005 Nobel Peace Prize Press Release, Oslo, October 7, 2005.

81. There is, however, an annual internal ceremony to award achievement medals, presided over by the Director General. The Iraq inspectors received several rewards for their work. Nic von Wielligh and Lydia von Wielligh-Steyn, *The Bomb: South Africa's Nuclear Weapons Programme* (Pretoria: Litera for the Institute for Security Studies, 2015), p. 238.

82. "The IAEA's Emerging Roles, Challenges, and Prospects for the Future: An Interview with the Director General," in Joseph F. Pilat, ed., *The International Atomic Energy Agency: Historical Reflections, Current Challenges, and Future Prospects* (London: Routledge, 2022), pp. 14–15.

83. For another example of the Agency's use of the term, see IAEA, *Atoms for Peace: A Pictorial History of the International Atomic Energy Agency*, p. 9. This author experienced the Agency's dim view of the term in its reaction to his *Unleashing the Nuclear Watchdog: Strengthening and Reform of the International Atomic Energy Agency*, Center for International Governance Innovation (CIGI), Waterloo, Ontario, 2013, https://www.cigionline.org/publications/2012/6/unleashing-nuclear-watchdog-strengthen ing-and-reform-of-iaea.

84. Interview with John Carlson by the author, Cambridge, Mass., June 5, 2014.

85. United Nations Sustainable Development Goals, https://www.un.org/sustainable development/sustainable-development-goals/. They were adopted in September 2015 to succeed the 2000 Millennium Development Goals.

86. Amano is quoted in Heinz Gärtner, "Yukiya Amano: One Year In," *Bulletin of the Atomic Scientists*, January 26, 2011, http://thebulletin.org/yukiya-amano-one-year.

87. "Amano Wants Development Recognition," World Nuclear News, January 23, 2015, http://www.world-nuclear-news.org/NP-Amano-wants-development-recognition -2301152.html.

88. Interview with senior IAEA official by the author, Vienna, June 24, 2014.

89. Strategic Plan, p. 2.

90. Interview with senior IAEA official by the author.

91. The Geneva Group comprises mostly Western countries that caucus on international organizations' budgetary matters, advocate for effectiveness and efficiency, and, for many years, have supported zero real growth. The current members are Australia, Belgium, Canada, France, Germany, Japan, Italy, Mexico, the Netherlands, Norway, Russia, South Korea, Spain, Sweden, Switzerland, Turkey, the United Kingdom, and the United States. The Vienna Group of Ten, established in 1980, consists of Australia, Austria, Canada, Denmark, Finland, Hungary, Ireland, the Netherlands, New Zealand, Norway, and Sweden. It encourages discussion and substantive progress at NPT Review Conferences on the so-called Vienna issues, including safeguards. See Statement by the Vienna Group of Ten, Ninth Review Conference of the Parties to the Treaty on the Non-Proliferation of Nuclear Weapons [NPT], April 30, 2015, https://www.un.org/en/conf/npt/2015/statements/pdf/VGroup10_en.pdf.

92. The former is clearly inaccurate because the Agency does not inspect nuclear weapons. The latter neglects, *inter alia*, the role of the UN Security Council, the Conference on Disarmament, and individual nations or groups of states.

93. See Gudrun Harrer, *Dismantling the Iraqi Nuclear Program: The Inspections of the International Atomic Energy Agency, 1991–1998* (Abingdon, Oxon: Routledge, 2014), p. 36.

94. For details, see Thomas E. Shea and Laura Rockwood, "Nuclear Disarmament: The Legacy of the Trilateral Initiative," Deep Cuts Working Paper No. 4, March 2015, University of Hamburg, Germany, https://deepcuts.org/images/PDF/DeepCuts_WP4_Shea_Rockwood_UK.pdf.

95. ElBaradei, p. 4.

96. Interview with Dieter Goethel by Elisabeth Roehrlich, IAEA Oral History Series, Vienna, August 18, 2016.

97. Mark Hibbs and Andreas Persbo, "The ElBaradei Legacy," *Bulletin of the Atomic Scientists*, September–October 2009, p. 20.

98. IAEA, Report of the Commission of Eminent Persons on the Future of the Agency: Note by the Director General, GOV/2008-GC(52)/INF/4, May 23, 2008.

99. Interview with John Carlson by the author, June 5, 2014. Tariq Rauf says that ElBaradei was disenchanted with SAGSI's domination by Western experts and tried to have the Safeguards Department redress the balance. It was his advisory board, however, so it is not clear why he could not have rebalanced it himself, as Hans Blix had done. Email from Tariq Rauf to the author, June 1, 2015.

100. Email from John Tilemann to the author, May 30, 2019.

101. Interview with national safeguards official by the author.

102. Andemicael and Mathiason, *Eliminating Weapons of Mass Destruction*, p. 214.

103. ElBaradei, p. 6.

104. ElBaradei is quoted in Roula Khalef and James Blitz, "IAEA Chief Embraces the Eye of the Storm," *Financial Times*, October 2, 2007.

105. ElBaradei, pp. 4–5.

106. John Carlson, "Future Directions in IAEA Safeguards."

107. This author had personal experience of this when ElBaradei was asked to rewrite his foreword to the Verification Research, Training and Information Center's *Verification Yearbook 2004*. We had to ask him to focus it less on advocating nuclear disarmament and more on the verification challenges. Trevor Findlay, ed., *Verification Yearbook 2004* (London: Verification Research, Training and Information Center [VERTIC], 2004).

108. While there is now an International Energy Agency, it is a subsidiary body of the Organization for Economic Cooperation and Development (OECD) and not a UN-style multilateral organization open to all states.

109. See Lawrence Scheinman, *The International Atomic Energy Agency and World Nuclear Order* (Washington, D.C.: Resources for the Future, 1987), pp. 288–292, on "depoliticization" of the IAEA; Yvonne Yew, "Diplomacy and Nuclear Non-Proliferation: Navigating the Non-Aligned Movement," Discussion Paper, Belfer Center for Science and International Affairs, Harvard University, Cambridge, Mass., June 13, 2011; and Findlay, "Unleashing the Nuclear Watchdog," pp. 16–17.

110. The Group of 77 (originally numbering 77, now 134), was formed in 1964 to coordinate its policies and press its demands regarding economic and social development at the United Nations and related bodies.

111. Fischer, *History of the International Atomic Energy Agency: The First Forty Years*, p. 93; and William Potter and Gaukhar Mukhatzhanova, "Nuclear Politics and the Non-Aligned Movement," Adelphi Paper No. 427, International Institute for Strategic Studies (IISS), London, 2012.

112. ElBaradei, p. 215.

113. "US Embassy Cables: New UN Chief is 'Director General of All States, But in Agreement with Us,'" *Guardian*, December 3, 2010, https://www.theguardian.com /world/us-embassy-cables-documents/230076.

114. Stephanie Liechtenstein, "Rafael Grossi Isn't America's—or Iran's, or North Korea's—Man," *Foreign Policy*, December 2019, p. 5, https://foreignpolicy.com/2019 /12/05rafael-grossi-iaea-america-iran-north-korea/.

115. IAEA Press Release, "New IAEA Laboratory Named after Late Director General Yukiya Amano," September 16, 2021, https://www.iaea.org/newscenter/news/new -iaea-laboratory-building-named-after-late-director-general-yukiya-amano.

116. See Findlay, "Unleashing the Nuclear Watchdog," pp. 5–6 and 21–22.

117. "US Embassy Cables: New UN Chief is 'Director General of All States, But in Agreement with Us.'"

118. See Peter F. Drucker, "What We Can Learn from Japanese Management," *Harvard Business Review*, March 1971, https://hbr.org/1971/03/what-we-can-learn-from-japan ese-management; and Charles Yang, "Demystifying Japanese Organizational Practices," *Harvard Business Review*, November 1984, https://hbr.org/1984/11/demystifying-jap anese-management-practices.

119. John Mathiason, *Invisible Governance: International Secretariats in Global Politics* (Bloomfield, Conn.: Kumarian Press, 2007), p. 71.

120. Interview with IAEA official by the author, Atlanta, July 23, 2014.

121. "Rafael Mariano Grossi," https://www.iaea.org/about/rafael-grossi, accessed March 29, 2021.

122. "Factbox: Five Facts about Next U.N. Nuclear Chief Rafael Grossi," Reuters, October 29, 2019.

123. Francois Murphy, "U.N. Nuclear Watchdog Picks Argentina's Grossi as Next Chief," Reuters, October 29, 2019.

124. Liechtenstein, "Rafael Grossi Isn't America's—or Iran's, or North Korea's—Man."

125. M. Campbell, et al., *At What Cost Success? Final Report of the External Review on the Management Processes of the International Atomic Energy Agency* (MANNET Report), Geneva, 2002 (hereafter MANNET Report), p. 1.

126. Ibid., p. 9.

127. Ibid., p. 10.

128. Ibid., p. 2.

129. Email from Tariq Rauf to the author, June 1, 2015.

130. Interview with IAEA staff member by the author, Vienna, June 25, 2014.

131. Pierre Goldschmidt, "Looking Beyond Iran and North Korea for Safeguarding the Foundations of Nuclear Nonproliferation," Nonproliferation Policy Education Center, Strategic Studies Institute, U.S. Army War College, November 15, 2011, http://www.npolicy.org/article.php?aid=1115&tid=4.

132. Email from Tariq Rauf to the author, June 1, 2015. Rauf dealt with verification cases involving Iran, Iraq, Libya, North Korea, South Korea, and Syria. Rauf says that ExPo did not question technical findings (to reject them), but asked for explanations to back up the technical findings of the Safeguards Department in order to demonstrate how the findings were reached, and to check their veracity. In some instances, ExPo asked the Safeguards Department to provide further explanation and to defer the case to the next meeting, usually a few weeks later. Marathon "conclusions" meetings (High-Level Committee for the Additional Protocol states, Low-Level Committee for CSA states) were held a couple of months prior to release of the SIR. By then, most if not all ExPo concerns had been resolved or dealt with. In a few cases, the conclusion was deferred to the next year. Email from Tariq Rauf to the author, March 24, 2021.

133. Interview with Olli Heinonen by the author, Cambridge, Mass., May 22, 2014.

134. Email from Tariq Rauf to the author, June 1, 2015.

135. Michael Barnett and Martha Finnemore, *Rules for the World: International Organizations in Global Politics* (Ithaca, N.Y.: Cornell University Press, 2004), p. 40.

136. There are parallels between the One House campaign and the UN's "Delivering as One" program. See Alisa Clarke, "Organizational Culture, System Evolution, and the United Nations of the 21st Century," *Journal of Organizational Studies*, Vol. 4, No. 1 (2013), p. 128.

137. Interview with John Carlson by the author, June 5, 2014.

138. Ibid.

139. This quote appears in Hibbs and Persbo, "The ElBaradei Legacy," pp. 18–19.

140. JIU, Review of Management and Administration in the International Atomic Energy Agency, p. 5.

141. Ibid.

142. Robert Kelley, "Creating a 'One House' Culture at the IAEA through Matrix Management," *SIPRI Policy Brief*, SIPRI, Stockholm, January 2014, p. 8.

143. JIU, Review of Management and Administration in the International Atomic Energy Agency, p. 5.

144. Ibid., p. 6.

145. Ibid., p. 6.

146. Ibid, p. 6.

147. IAEA Statute, Art. VII F, 1957.

148. The U.S. Central Intelligence Agency, for example, deals with highly secret information, yet they share it through established procedures and well-defined need-to-know.

149. See IAEA, "Overall Safeguards Management and Coordination," Project DDG-001, Development and Implementation Support Program for Nuclear Verification 2020–2021, STR-393, IAEA, Vienna, January 2020, pp. 16–17 and 43–48.

150. Ibid., p. 43.

151. Ibid., pp. 43, 45, and 48.

152. IAEA, Statement by IAEA Director General Yukiya Amano, October 2, 2018, https://www.iaea.org/newscenter/statements/statement-by-iaea-director-general-yukiya-amano-2-october-2018

153. "Challenges in Verification."

154. Ibid.

155. Rockwood, Mayhew, Lazarev, and Pfneisi, "IAEA Safeguards: Staying Ahead of the Game," p. 26.

156. Ibid., p. 57.

157. Ibid., p. 59.

158. Ibid., p. 27.

159. See IAEA, The Agency's Inspectorate: Memorandum by the Director General; and Annex, The Agency's Inspectors, GC(V) INF/39, Vienna, August 28, 1961 (the "Inspectors Document").

160. Treaty on the Non-Proliferation of Nuclear Weapons, also known as the Nuclear Non-Proliferation Treaty (NPT), http://disarmament.un.org/treaties/t/npt.

161. IAEA Statute, Art. VII.C provides that: "The staff shall include such qualified scientific and technical and other personnel as may be required to fulfil the objectives and functions of the Agency. The Agency shall be guided by the principle that its permanent staff shall be kept to a minimum."

162. IAEA Financial Statements 2011, p. 140, http://www.iaea.org/sites/default/files/mts2006_2011.pdf.

163. Interview with senior IAEA official by the author, Vienna, June 24, 2014.

164. IAEA, "Inviting Commitment: Working for the IAEA," IAEA Human Resources, Vienna, 2012, p. 36.

165. IAEA Financial Statements 2011, p. 140.

166. Ibid.

167. Email from Robert Kelley to the author, April 24, 2015. Kelley also worries about the costs of the system "associated with repatriation, travel, household goods movement (both ways), repatriation grant of many thousands and a new hire induction grant of thousands. So, you pay a very high premium, maybe $40,000 for non-European hires."

168. Ibid.

169. Interview with Robert Kelley by the author, Vienna, March 19, 2014.

170. IAEA, The Agency's Statement of Accounts for 2010, p. 36, http://www.un.org/en/auditors/panel/docs/IAEA_2010.pdf.

171. JIU, Review of Management and Administration in the International Atomic Energy Agency, p. 23.

172. Ibid., p. 1. There are several caveats about the validity of the findings. First, the staff survey was online and voluntary, with only a 37-percent response rate, potentially producing a selection bias of staff dissatisfied with the system. Second, the JIU conceded that of the twenty-two member states solicited for their views, only China, Egypt, India, South Africa, and South Korea responded to requests for meetings or telephone conferences.

173. Rockwood, Mayhew, Lazarev, and Pfneisi, "IAEA Safeguards: Staying Ahead of the Game," p. 31.

174. Ibid., p. 59.

175. Barnett and Finnemore, p. 40.

176. Ibid.

177. Andemicael and Mathiason point out that "despite the formal rotation policy, many of the key managers are long-termers. Staff who cannot demonstrate leadership qualities are less likely to reach management levels." Andemicael and Mathiason, *Eliminating Weapons of Mass Destruction*, p. 214.

178. MANNET Report, p. 23.

179. "The Future of IAEA Safeguards: Rebuilding the Vienna Spirit through Russian–U.S. Expert Dialogue," p. 39.

180. IAEA, Personnel: Staffing of the Agency's Secretariat: Report by the Director General, GOV/2017/38-GC(61)/18, p. 6.

181. IAEA, Personnel: Staffing of the Agency's Secretariat: Report by the Director General, GOV/2019/36-GC(63)/15, July 19, 2019, p. 6; and IAEA, Personnel: Staffing of the Agency's Secretariat: Report by the Director General, GOV/2017/38-GC(61)/18, p. 6.

182. IAEA, Personnel: Staffing of the Agency's Secretariat: Report by the Director General, GOV/2017/38-GC(61)/18, p. 6.

183. The Agency's Financial Statements for 2016, GC(61)/2, p. 26, https://www.iaea .org/About/Policy/GC/GC61/GC61Documents/English/gc61-2_en.pdf.

184. Office of the Comptroller and Auditor General of India, Report of the External Auditor on the Audit of the Financial Statements of the International Atomic Energy Agency for the Year Ended 31 December 2013, Part V, IAEA, The Agency's Financial Statements for 2013, GC(58)/5, pp. 138–140, https://www.iaea.org/About/Policy/GC /GC58/GC58Documents/English/gc58-5_en.pdf.

185. Email from Robert Kelley to the author, April 24, 2015.

186. Audit Board of Indonesia, Audit Report on the International Atomic Energy Agency (IAEA), 2016, Part V, IAEA, The Agency's Financial Statements for 2016, pp. 179–180.

187. Audit Board of Indonesia, Audit Report on the International Atomic Energy Agency (IAEA), 2019, Part V, The Agency's Financial Statements for 2019, GC(64)/4, July 2020, pp. 203–204.

188. IAEA, Personnel: Staffing of the Agency's Secretariat: Report by the Director General, GOV/2017/38-GC(61)/18, p. 5.

189. Ibid. There were similar figures for previous years.

190. IAEA, Personnel: Staffing of the Agency's Secretariat: Report by the Director General, GOV/2015/46-GC(59)/16, August 3, 2015, pp. 2–3, https://www.iaea.org /About/Policy/GC/GC59/GC59Documents/English/gc59-16_en.pdf.

191. IAEA, Personnel: Staffing of the Agency's Secretariat: Report by the Director General, GOV/2019/36-GC(63)/15, July 19, 2019, p. 2.

192. Ibid.

193. "Trial employment" was one of Shea and Thompson's recommendations in the early 1980s. Shea and Thompson, p. 66.

194. IAEA, Personnel: Staffing of the Agency's Secretariat: Report by the Director General, GOV/2017/38-GC(61)/16, p. 3; and IAEA, Personnel: Staffing of the Agency's Secretariat: Report by the Director General, GOV/2019/36-GC(63)/15, July 19, 2019, p. 4.

195. See IAEA, Strengthening the Effectiveness and Improving the Efficiency of Agency Safeguards, GC(61)/16, July 26, 2017, p. 4.

196. IAEA, Personnel: Women in the Secretariat: Report by the Director General, GOV/2015/47-GC(59)/17, August 3, 2015, p. 1.

197. For several years, the external auditor repeatedly found progress to be too slow and recommended increased efforts to boost the percentage of female employees in the Agency, especially in the professional category. IAEA, The Agency's Financial Statements for 2014, GC(59)/3, Recommendation 12, p. 165, http://www.un.org/en /auditors/panel/docs/IAEA%20Audit%20Report_2014.pdf.

198. IAEA Press Release, "IAEA Director General Yukiya Amano becomes a Gender Champion," June 17, 2017, https://www.iaea.org/newscenter/news/iaea-director-gen eral-yukiya-amano-becomes-a-gender-champion.

199. "IAEA Director General named to Board of International Gender Champions," December 8, 2020, https://www.iaea.org/newscenter/news/iaea-director-general -named-to-board-of-international-gender-champions.

200. IAEA, Personnel: Women in the Secretariat: Report by the Director General, GOV/2007/47-GC(51)/17, August 17, 2007, p. 3.

201. IAEA, Personnel: Women in the Secretariat: Report by the Director General, GOV/2019/37-GC(63)/16, July 19, 2019, pp. 5–6.

202. Ibid.

203. IAEA, Personnel: Women in the Secretariat: Report by the Director General, GOV/2015/47-GC(59)/17, p. 3.

204. IAEA, Personnel: Women in the Secretariat: Report by the Director General, GOV/2007/47-GC(51)/17, p. 1.

205. See World Nuclear University, http://www.world-nuclear-university.org/imis20 /wnu/.

206. "IAEA Safeguards: Staying Ahead of the Game," IAEA, Vienna, September 2007, p. 30.

207. IAEA, Personnel: Women in the Secretariat: Report by the Director General, GOV/2015/47-GC(59)/17, Annex, p. 2.

208. Interview with Shirley Johnson by the author, Atlanta, July 22, 2014.

209. IAEA, Strengthening the Effectiveness and Improving the Efficiency of Agency Safeguards: Report by the Director General, GC(64)/13, August 11, 2020, p. 14.

210. See Women in Nuclear (WiN), https://www.win-global.org/.

211. Panel Discussion, "Retrospective of INFCIRCs 153 and 540."

212. IAEA, Personnel: Women in the Secretariat: Report by the Director General, GOV/2017/39/GC(61)/19, p. 2.

213. Email from Robert Kelley to the author, April 25, 2014.

214. Email from Mark Schanfein to the author, May 14, 2016.

215. Ibid.

216. See, for example, Linda Klebe Treviño, Gary R. Weaver, David G. Gibson, and Barbara Ley Toffler, "Managing Ethics and Legal Compliance: What Works and What Hurts," *Harvard Business Review*, January 1999; and Mark S. Schwartz, "Developing and Sustaining an Ethical Corporate Culture: The Core Elements," *Harvard Business Review*, January 15, 2013.

217. Corey Hinderstein, "Inside Insights," *The Communicator*, INMM, June 2016, p. 3.

218. IAEA Core Ethical Values, https://www.iaea.org/about/employment/iaea-values, accessed April 1, 2021. The OPCW has gone further, with an 11-page "Guide to OPCW Core Values and Competencies." OPCW, "Guide to OPCW Core Values and Competencies," The Hague, 2020, https://www.opcw.org/sites/default/files/docu ments/2020/05/Guide%20to%20OPCW%20Core%20Values%20and%20Competen cies.pdf, accessed April 1, 2021.

219. IAEA, "Overall Safeguards Management and Coordination," p. 96.

220. Ibid.

221. See, for instance, IAEA, Strengthening the Effectiveness and Improving the Efficiency of Agency Safeguards: Report by the Director General, GC(64)/13, August 11, 2020, p. 14.

222. Audit Board of Indonesia, Audit Report on the International Atomic Energy Agency, 2019, Part V, The Agency's Financial Statements for 2019, GC(64)/4, 2020, p. 179.

223. Ibid., p. 181.

224. Ibid.

225. Ibid., p. 180.

226. IAEA, "Overall Safeguards Management and Coordination," pp. 90–96.

227. The Concepts and Planning Division in the Safeguards Department is responsible for, among other tasks, developing tools to support the Quality Management System, "root cause of identified conditions," training on Quality Management, and support for the Internal Quality Audit Program. See Major Program 4, Nuclear Verification, Subprogram 4.1.1, Concepts and Planning in IAEA, The Agency's Program and Budget 2020–2021, GC(63)/2, July 2019, pp. 130. In 2020, the Quality Management activities were merged with process design activities to form a combined integrated program called Process Design and Quality Management. Presumably, this will enhance both.

228. In the oil and construction industries, "accident-free day" campaigns with attendant rewards are an example of this approach.

229. In 2017, the Secretariat reported that "to ensure the health, safety and security of Agency staff in the field, particularly in light of elevated security levels in some locations, procedures specifying communication protocols during in-field emergencies were updated and information cards are now issued to staff. Training on radiation protection for staff in the Department of Safeguards is regularly delivered." IAEA, Strengthening the Effectiveness and Improving the Efficiency of Agency Safeguards: Report of the Director General, GC(61)/16, July 26, 2017, Vienna, p. 9.

230. Email from Jim Casterton to the author, March 2, 2015.

231. Interview with Olli Heinonen by the author, Harvard, May 22, 2014.

232. U.S. Government Accountability Office (GAO), "Nuclear Nonproliferation: IAEA Has Strengthened its Safeguards and Nuclear Security Programs, but Weaknesses Need to be Addressed," GAO-06–93, Washington D.C., October 2005, p. 19.

233. Ibid., p. 18.

234. JIU, Review of Management and Administration in the International Atomic Energy Agency, p. 26.

235. IAEA, "Training," Project SGCP-102, Development and Implementation Support Program for Nuclear Verification 2020–2021, STR-393, IAEA, Vienna, January 2020, p. 97.

236. Ibid.

237. Ibid.

238. "Training IAEA Inspectors: Newly Recruited Inspectors Take to the Field," https://www.iaea.org/newscenter/news/training-iaea-inspectors, accessed February 10, 2015.

239. IAEA, "Training," Project SGCP-102, Development and Implementation Support Program for Nuclear Verification 2020–2021, p. 100.

240. This training takes place with contributions from the U.S., UK, and Australian Support Programs.

241. "Training IAEA Inspectors: Newly Recruited Inspectors Take to the Field."

242. Ibid.

243. See Diana C. Pheysey, *Organizational Cultures: Types and Transformations* (London: Routledge, 1993), pp. 4–5.

244. Interview with IAEA official, Vienna, June 25, 2014.

245. IAEA, The Agency's Financial Statements for 2019, GC(64)/4, 2020, pp. 203–204.

246. See Sasha Henriques, "A Day in the Life of a Safeguards Inspector," July 27, 2016, https://www.iaea.org/newscenter/news/a-day-in-the-life-of-a-safeguards-inspect or#:~:text=The%20IAEA%20employs%20around%20385%20designated%20in spectors%20from%20around%2080%20countries.

247. James Casterton, Remarks at Seminar on IAEA Safeguards Culture, Carleton University, Ottawa, November 14, 2014.

248. IAEA, "Training," Project SGCP-102, Development and Implementation Support Program for Nuclear Verification 2020–2021, p. 104.

249. Ibid., p. 107.

250. Charles Duelfer, *Hide and Seek: The Search for Truth in Iraq* (New York: Public Affairs, 2009), p. 355.

251. Ibid., p. 166.

252. GAO, "Nuclear Nonproliferation: IAEA Has Strengthened its Safeguards and Nuclear Security Programs, but Weaknesses Need to be Addressed," p. 19.

253. IAEA, Strengthening the Effectiveness and Improving the Efficiency of the Safe-guards System including Implementation of Additional Protocols: Report by the Director General, GC(51)/8, Vienna, July 23, 2007, p. 9.

254. JIU, Review of Management and Administration in the International Atomic Energy Agency, p. 28.

255. IAEA Annual Report for 2013, IAEA, Vienna, 2014, p. 76, https://www.iaea.org/publications/reports/annual-report-2013-0.

256. IAEA, Strengthening the Effectiveness and Improving the Efficiency of Agency Safeguards: Report of the Director General, July 26, 2017, GC(61)/16, p. 8.

257. IAEA, "Training," Project SGCP-102, Development and Implementation Support Program for Nuclear Verification 2020–2021, p. 98.

258. Ibid., pp. 100–101.

259. Henriques, "A Day in the Life of a Safeguards Inspector."

260. Ibid.

261. Ibid.

262. Ibid.

263. Laura Rockwood, Noah Mayhew, Artem Lazarev, and Mara Pfneisi, "IAEA Safeguards: Staying Ahead of the Game," Swedish Radiation Safety Authority, Report No. 2019:14, Stockholm, 2019, p. 29.

264. OECD, *OECD Best Practice Principles for Regulatory Policy: Regulatory Enforcement and Inspections* (Paris: OECD, 2014), https://www.oecd.org/gov/regulatory-enforcement-and-inspections-9789264208117-en.htm.

265. Ibid., p. 63.

266. Ibid., p. 50. See also International Organization for Standardization (ISO), ISO/IEC/17020, "Conformity Assessment—Requirements for the operation of various types of bodies performing inspection" (Revised and confirmed in 2017), https://www.iso.org/obp/ui/#iso:std:iso-iec:17020:ed-2:v1:en.

267. Shea and Thompson, pp. 36–41.

268. Strategic Plan, p. 7.

269. Interview with senior IAEA official by the author, Vienna, June 24, 2014.

270. Shea and Thompson, p. 58.

271. OECD, *OECD Best Practice Principles for Regulatory Policy: Regulatory Enforcement and Inspections*, p. 48.

272. Ibid., p. 49.

273. Ian Ayres and John Braithwaite, *Responsive Regulation: Transcending the Deregulation Debate* (Oxford: Oxford University Press, 1992).

274. Kenneth W. Abbott and Duncan Snidal, "Taking Responsive Regulation Transnational: Strategies for International Organizations," *Regulation and Governance*, No. 7 (2013), https://onlinelibrary.wiley.com.doi.epdf, accessed April 20, 2021.

275. OECD, *OECD Best Practice Principles for Regulatory Policy: Regulatory Enforcement and Inspections*, p. 49.

276. Panel Discussion, "Retrospective of INFCIRCs 153 and 540."

277. IAEA, "Inviting Commitment: Working for the IAEA," IAEA Human Resources, Vienna, 2012.

278. Ibid.

279. Glassdoor, http://www.glassdoor.com/IAEA-Reviews-E474090.htm.

280. Ibid.

281. JIU, Review of Management and Administration in the International Atomic Energy Agency, p. 35.

282. Ibid.

Bibliography

Books

Ancona, Deborah G., Thomas A. Kochan, Maureen Scully, John Van Maanen, and Eleanor Westney, *Managing for the Future: Organizational Behavior and Processes*, 3rd ed. (Boston: South-Western College Publishing, 2004).

Andemicael, Berhanykun, and John Mathiason, *Eliminating Weapons of Mass Destruction: Prospects for Effective International Verification* (London: Palgrave, 2005).

Ayres, Ian, and Braithwaite, John, *Responsive Regulation: Transcending the Deregulation Debate* (Oxford: Oxford University Press, 1992).

Barnaby, Frank, *The Invisible Bomb: The Nuclear Arms Race in the Middle East* (London: I.B. Tauris, 1989).

Barnett, Michael, and Martha Finnemore, *Rules for the World: International Organizations in Global Politics* (Ithaca, N.Y.: Cornell University Press, 2004).

Bate, Paul, *Strategies for Cultural Change* (Oxford: Butterworth Heinemann, 1994).

Blix, Hans, *Disarming Iraq: The Search for Weapons of Mass Destruction* (London: Bloomsbury, 2005).

Bolton, John, *Surrender is Not an Option: Defending America at the United Nations and Abroad* (New York: Simon and Schuster, 2007).

Braut-Hegghammer, Målfrid, *Unclear Physics: Why Iraq and Libya Failed to Build Nuclear Weapons* (Ithaca, N.Y.: Cornell University Press, 2016).

Brown, Robert L., *Nuclear Authority: The IAEA and the Absolute Weapon* (Washington, D.C.: Georgetown University Press, 2015).

Burrows, William, and Robert Windrem, *Critical Mass: The Dangerous Race for Superpowers in a Fragmenting World* (New York: Simon and Schuster, 1994).

Butler, Richard, *Saddam Defiant: The Threat of Weapons of Mass Destruction, and the Crisis of Global Security* (London: Weidenfeld and Nicholson, 2000).

Chabris, Christopher, and Daniel Simons, *The Invisible Gorilla: How Our Intuitions Deceive Us* (New York: Broadway Paperbacks, 2009).

Cirincione, Joseph, Jon B. Wolfsthal, and Miriam Rajkumar, *Deadly Arsenals: Nuclear, Biological, and Chemical Threats* (Washington, D.C.: Carnegie Endowment for International Peace, 2005).

Conner, Daryl R., *Managing at the Speed of Change: How Resilient Managers Succeed and Prosper Where Others Fail* (New York: Villart, 1993).

Doyle, James, ed., *Nuclear Safeguards, Security, and Nonproliferation: Achieving Security with Technology and Policy* (Amsterdam: Elsevier for Los Alamos National Laboratory, 2008).

Duelfer, Charles, *Hide and Seek: The Search for Truth in Iraq* (New York: PublicAffairs, 2009).

ElBaradei, Mohamed, *The Age of Deception: Nuclear Diplomacy in Treacherous Times* (New York: Metropolitan Books, 2011).

Fischer, David, *History of the International Atomic Energy Agency: The First Forty Years* (Vienna: IAEA, 1997).

Fischer, David, and Paul Szasz, *Safeguarding the Atom: A Critical Appraisal* (London: Taylor and Francis for the Stockholm International Peace Research Institute [SIPRI], 1985).

Fuhrmann, Matthew, *Atomic Assistance: How "Atoms for Peace" Programs Cause Nuclear Insecurity* (Ithaca, N.Y.: Cornell University Press, 2012).

Garmonsway, G.N., *The Penguin English Dictionary* (Harmondsworth, UK: Penguin, 1965).

Geertz, Clifford, *The Interpretation of Cultures* (New York: Basic Books, 1973).

Gilovich, Thomas, Dale W. Griffin, and Daniel Kahneman, *Heuristics and Biases: The Psychology of Intuitive Judgment* (Cambridge: Cambridge University Press, 2002).

Goldblat, Jozef, ed., *Nonproliferation: The Why and the Wherefore* (London: Taylor and Francis for SIPRI, 1985).

Hagen, J.U., ed., *How Could This Happen? Managing Errors in Organizations* (Cham, Switzerland: Springer Nature Switzerland AG, 2018).

Hamza, Khidhir, with Jeff Stein, *Saddam's Bombmaker: The Terrifying Inside Story of the Iraqi Nuclear and Biological Weapons Agenda* (New York: Scribner, 2000).

Harrer, Gudrun, *Dismantling the Iraqi Nuclear Program: The Inspections of the International Atomic Energy Agency, 1991–1998* (Abingdon, Oxon: Routledge, 2014).

Hawkins, Darren G., David A. Lake, Daniel L. Nielson, and Michael J. Tierney, eds., *Delegation and Agency in International Organizations* (Cambridge: Cambridge University Press, 2006).

Hazzard, Shirley, *Defeat of an Ideal: The Self-Destruction of the United Nations* (London: Macmillan, 1973).

Hewlett, Richard G., and Jack M. Holl, *Atoms for Peace and War, 1953–1961: Eisenhower and the Atomic Energy Commission* (Berkeley: University of California Press, 1989).

Hurd, Ian, *International Organizations: Politics, Law, and Practice*, 2nd ed. (Cambridge: Cambridge University Press, 2014).

Hymans, Jacques C., *Achieving Nuclear Ambitions: Scientists, Politicians, and Proliferation* (Cambridge: Cambridge University Press, 2012).

IAEA, *Atoms for Peace: A Pictorial History of the International Atomic Energy Agency* (Vienna: IAEA, 2007).

IAEA, *International Atomic Energy Agency: Personal Reflections* (Vienna: IAEA, 1997).

Jacob, Philip E., and Alexine L. Atherton, *The Dynamics of International Organization: The Making of World Order* (Homewood, Ill.: Dorsey Press, 1965).

Kahneman, Daniel, *Thinking, Fast and Slow* (New York: Farrar, Strauss, and Giroux, 2011).

Kahneman, Daniel, and Amos Tversky, eds., *Choices, Values, and Frames* (Cambridge: Cambridge University Press, 2000).

Kahneman, Daniel, Paul Slovic, and Amos Tversky, eds., *Judgment Under Uncertainty: Heuristics and Bias* (Cambridge: Cambridge University Press, 1982).

Karns, Margaret P., and Karen A. Mingst, *International Organizations: The Politics and Processes of Global Governance*, 2nd ed. (Boulder, Colo.: Lynne Rienner, 2010).

Koremenos, Barbara, Charles Lipson, and Duncan Snidal, eds., *The Rational Design of International Institutions* (Cambridge: Cambridge University Press, 2004).

Krasno, Jean E., and James S. Sutterlin, *The United Nations and Iraq: Defanging the Viper* (New York: Praeger, 2003).

Krass, Allan S., *Verification: How Much Is Enough?* (London: Taylor and Francis for SIPRI, 1985).

Luck, Edward C., and Michael W. Doyle, eds., *International Law and Organization: Closing the Compliance Gap* (Lanham, Md.: Rowman and Littlefield, 2002).

Mathiason, John, *Invisible Governance: International Secretariats in Global Politics* (Bloomfield, Conn.: Kumarian Press, 2007).

McKnight, Allan, *Atomic Safeguards: A Study in International Verification* (New York: United Nations Institute for Training and Research [UNITAR], 1971).

Morgenthau, Hans J., *Politics Among Nations: The Struggle for Power and Peace* (New York: Alfred A. Knopf, 1962).

Nye, Joseph S., Jr., *The Future of Power* (New York: PublicAffairs, 2011).

Perrow, Charles, *Normal Accidents: Living with High-Risk Technologies* (Princeton, N.J.: Princeton University Press, 1999).

Pfeffer, Jeffrey, *Power: Why Some People Have It—And Others Don't* (New York: HarperCollins, 2010).

Pheysey, Diana C., *Organizational Cultures: Types and Transformations* (London: Routledge, 1993).

Pringle, Peter, and James Spigelman, *The Nuclear Barons* (New York: Avon Books, 1981).

Rittenberger, Volker, Bernhard Zangl, and Andreas Kruck, *International Organization*, 2nd ed. (Houndsmills, UK: Palgrave Macmillan, 2012).

Ritter, Scott, *Iraq Confidential: The Untold Story of the Intelligence Conspiracy to Undermine the UN and Overthrow Saddam Hussein* (New York: Nation Books, 2005).

Rosenthal, Michael D., et al., *Deterring Nuclear Proliferation: The Importance of IAEA Safeguards* (Upton, N.Y.: Brookhaven National Laboratory, 2019).

Schein, Edgar H., *The Corporate Culture Survival Guide* (San Francisco: Jossey-Bass, 1999).

Schein, Edgar H., *Organizational Culture and Leadership*, 3rd ed. (San Francisco: Jossey-Bass, 2004).

Schein, Edgar H., with Peter Schein, *Organizational Culture and Leadership*, 5th ed. (Hoboken, N.J.: John Wiley and Sons, 2017).

Scheinman, Lawrence, *The International Atomic Energy Agency and World Nuclear Order* (Washington, D.C.: Resources for the Future, 1987).

Schiff, Benjamin N., *International Nuclear Technology Transfer: Dilemmas of Dissemination and Control* (Lanham, Md.: Rowman and Littlefield, 1983).

Seaborg, Glenn T., with Benjamin T. Loeb, *Stemming the Tide: Arms Control in the Johnson Years* (Lexington, Mass: Lexington Books, 1971).

Snow, C.P., *The Two Cultures: A Second Look* (New York: New American Library, 1959).

Sokolski, Henry, ed., *Falling Behind: International Scrutiny of the Peaceful Atom* (Carlisle, Penn.: Nonproliferation Policy Education Center, Strategic Studies Institute, U.S. Army War College, 2008).

Spector, Leonard S., *The Undeclared Bomb* (Cambridge, Mass.: Ballinger, 1988).

Stockholm International Peace Research Institute (SIPRI), *Safeguards Against Nuclear Proliferation* (Stockholm: Almqvist and Wiksell for SIPRI, 1975).

Szasz, Paul C., *The Law and Practices of the International Atomic Energy Agency*, Legal Series No. 7 (Vienna: IAEA, 1970).

Trevan, Tim, *Saddam's Secrets: The Hunt for Iraq's Hidden Weapons* (London: Harper-Collins, 1999).

Vaughan, Diane, *The Challenger Launch Decision: Risky Technology, Culture, and Deviance at NASA* (Chicago: University of Chicago Press, 1991).

von Wielligh, Nic, and Lydia von Wielligh-Steyn, *The Bomb: South Africa's Nuclear Weapons Programme* (Pretoria: Litera for the Institute for Security Studies, 2015).

Wing, Christine, and Fiona Simpson, *Detect, Dismantle, and Disarm: IAEA Verification, 1992–2005* (Washington, D.C.: U.S. Institute of Peace, 2013).

Wood, Robert S., ed., *The Process of International Organization* (New York: Random House, 1971).

Book Chapters

Barrett, John, "The Travails of Governance: A View from the Board Chair," in Joseph F. Pilat, ed., *The International Atomic Energy Agency: Historical Reflections, Current Challenges, and Future Prospects* (London: Routledge, 2022).

Boureston, Jack, and Yana Feldman, "Verifying Libya's Nuclear Disarmament," in Trevor Findlay, ed., *Verification Yearbook 2004* (London: Verification Research, Training and Information Center [VERTIC], 2004).

Büchler, Carlos. L., "Safeguards: The Beginnings," in IAEA, *International Atomic Energy Agency: Personal Reflections* (Vienna: IAEA, 1997).

Boyer, Brian, and Mark Schanfein, "International Safeguards Inspection: An Inside Look at the Process," in James Doyle, ed., *Nuclear Safeguards, Security, and Nonproliferation: Achieving Security with Technology and Policy* (Amsterdam: Elsevier for Los Alamos National Laboratory, 2008).

Carroll, John S., "Understanding Safety Management Through Strategic Design, Political, and Cultural Approaches," in J.U. Hagen, ed., *How Could This Happen? Managing Errors in Organizations* (Cham, Switzerland: Springer Nature Switzerland AG, 2018).

Cochran, Thomas B., "Adequacy of IAEA's Safeguards for Achieving Timely Detection," in Henry Sokolski, ed., *Falling Behind: International Scrutiny of the Peaceful Atom* (Carlisle, Penn.: Nonproliferation Policy Education Center, Strategic Studies Institute, U.S. Army War College, 2008).

Everton, Craig, "In Defence of the Evolution of IAEA Safeguards," in *Verification and Implementation: A Biennial Collection of Analysis on International Agreements for Security and Development* (London: VERTIC, 2015).

Findlay, Trevor, "The Lessons of UNSCOM and UNMOVIC," in Trevor Findlay, ed., *Verification Yearbook 2004* (London: VERTIC, 2004).

Findlay, Trevor, "Lessons of UNSCOM and UNMOVIC for WMD Non-proliferation, Arms Control and Disarmament," in Waheguru Pal Singh Sidhu and Ramesh Thakur, eds., *Arms Control After Iraq: Normative and Operational Challenges* (Tokyo: United Nations University Press, 2006).

Heinonen, Olli, "International Atomic Energy Agency Inspections in Perspective," in Henry Sokolski, ed., *Moving Beyond Pretense: Nuclear Power and Nonproliferation* (Arlington, Va.: Strategic Studies Institute, U.S. Army War College, 2014).

Jolles, Paul, "From the East River to the Danube: Preparatory Commission and First Years in Vienna," in IAEA, *International Atomic Energy Agency: Personal Reflections* (Vienna: IAEA, 1997).

Keblušek, Emil, "The Spirit of Vienna," in IAEA, *International Atomic Energy Agency: Personal Reflections* (Vienna: IAEA, 1997).

Loosch, Reinhard, "The Emergence of the Group of 77 as a Major Player in the Board of Governors," in IAEA, *International Atomic Energy Agency: Personal Reflections* (Vienna: IAEA, 1997).

Mathiason, John, "Evolution of the International Public Service (1919–2006)," in John Mathiason, *Invisible Governance: International Secretariats in Global Politics* (Bloomfield, Conn.: Kumarian Press, 2007).

Quihillalt, Oscar A., and Carlos L. Büchler, "The Fifth General Conference of the IAEA (October 1961)," in IAEA, *International Atomic Energy Agency: Personal Reflections* (Vienna: IAEA, 1997).

Ramanna, Raja, "Positive Aspects of the Work of the International Atomic Energy Agency," in IAEA, *International Atomic Energy Agency: Personal Reflections* (Vienna: IAEA, 1997).

Rauf, Tariq, and Jan Lodding, "UNSCR 1540 and the IAEA," in Olivia Bosch and Peter van Ham, eds., *Global Non-Proliferation and Counter-Terrorism: The Impact of UNSCR 1540* (London: Royal Institute of International Affairs, 2007).

Sokolski, Henry, "Assessing the IAEA's Ability to Verify the NPT," in Henry Sokolski, ed., *Falling Behind: International Scrutiny of the Peaceful Atom* (Carlisle, Penn.: Nonproliferation Policy Education Center, Strategic Studies Institute, U.S. Army War College, 2008).

Tape, James, and Joseph Pilat, "Nuclear Safeguards and the Security of Nuclear Materials," in James Doyle, ed., *Nuclear Safeguards, Security, and Nonproliferation: Achieving Security with Technology and Policy* (Amsterdam: Elsevier for Los Alamos National Laboratory, 2008).

Wilson, Michael, "Safeguards and the IAEA Board of Governors: 1991–1993: Iraq, A Necessary Stimulus for Handling the DPRK," in IAEA, *International Atomic Energy Agency: Personal Reflections* (Vienna: IAEA, 1997).

Young, Oran R., "The United Nations and the International System," in Robert S. Wood, ed., *The Process of International Organization* (New York: Random House, 1971).

Journal Articles

Abbott, Kenneth W. and Snydal, Duncan, "Taking Responsive Regulation Transnational: Strategies for International Organizations," *Regulation & Governance*, No. 7 (2013), https://onlinelibrary.wiley.com.doi.epdf, accessed April 20, 2021.

Barnett, Michael, and Liv Coleman, "Designing Police: Interpol and the Study of Change in International Organizations," *International Studies Quarterly*, Vol. 49 (2005).

Barbeschi, Maurizio, "Organizational Culture of the OPCW Secretariat," *Disarmament Forum*, No. 4 (2002).

Boureston, Jack, and Charles D. Ferguson, "Strengthening Nuclear Safeguards: Special Committee to the Rescue?" *Arms Control Today*, December 2005, www.armscontrol.org/print/1951.

Brown, Robert L., and Jeffrey M. Kaplow, "Talking Peace, Making Weapons: IAEA Technical Cooperation and Nuclear Proliferation," *Journal of Conflict Resolution*, April 14, 2014.

Clarke, Alisa, "Organizational Culture, System Evolution, and the United Nations of the 21st Century," *Journal of Organizational Studies*, Vol. 4, No. 1 (2013).

Darrow, Mac, and Louise Arbour, "The Pillar of Glass: Human Rights in the Development Operations of the United Nations," *American Journal of International Law*, Vol. 103, No. 3.

Drucker, Peter F., "What We Can Learn from Japanese Management," *Harvard Business Review*, March 1971.

Fischer, David, "Safeguards Under the Non-Proliferation Treaty," *Disarmament*, Vol. 3, No. 2 (July 1980).

Frazar, S.L., and S.V. Mladineo, "The Importance of Safeguards Culture," *Nonproliferation Review*, December 2013.

Liechtenstein, Stephanie, "Rafael Grossi Isn't America's—or Iran's, or North Korea's—Man," *Foreign Policy*, December 2019, https://foreignpolicy.com/2019/12/05rafael-grossi-iaea-america-iran-north-korea/.

Meidinger, Errol, "Regulatory Culture: A Theoretical Outline," *Law and Policy*, October 1987.

Nackaerts, Herman, "The Future of Safeguards: Adapting to Change," *Journal of Nuclear Materials Management*, Fall 2011.

Nackaerts, Herman, "IAEA Safeguards: Cooperation as the Key to Change," *Journal of Nuclear Materials Management*, Fall 2011.

Pendley, Robert, Lawrence Scheinman, and Richard W. Butler, "International Safeguarding as Institutionalized Collective Behavior," *International Organization*, Vol. 29, No. 3 (June 1975).

Rao, Sriata, "Organization Culture and Leadership in the Department of Safeguards," *Journal of Nuclear Materials Management*, Volume 44, No. 2 (2016).

Rauf, Tariq, and Robert Kelley, "Nuclear Verification in Iraq," *Arms Control Today*, September 2014.

Rockwood, Laura, "The IAEA's State-Level Concept and the Law of Unintended Consequences," *Arms Control Today*, September 2014.

Schein, E.H., "The Three Cultures of Management: Implications for Organizational Learning," *Sloan Management Review*, No. 38 (Spring 1997).

Schwartz, Mark S., "Developing and Sustaining an Ethical Corporate Culture: The Core Elements," *Harvard Business Review*, January 2013.

Szasz, Paul C., "The Adequacy of International Nuclear Safeguards," *Journal of International Law and Economics*, Vol. 10 (1975).

Treviño, Linda Klebe, Gary R. Weaver, David G. Gibson, and Barbara Ley Toffler, "Managing Ethics and Legal Compliance: What Works and What Hurts," *Harvard Business Review*, January 1999.

Yang, Charles, "Demystifying Japanese Organizational Practices," *Harvard Business Review*, November 1984.

Zyl de Villiers, Van, Marguerite Leonardi, Carrie Mathews, Jenni Rissanen, and William Stanley, "Recent Developments in Performance Management in the IAEA Department of Safeguards," Topical Papers, *Journal of Nuclear Materials Management*, Volume 44, No. 2 (2016).

Newspaper, Newsletter, and Magazine Articles

"The 1970 Safeguards Committee," *IAEA Bulletin*, No. 1 (1990).

"An IAEA Inspection of the Iraqi Nuclear Complex 'Revealed No Non-Compliance,'" *Nucleonics Week*, November 26, 1981.

Cannon, Sandy, "U.S. at Odds over Need for Osirak Strike," *Nuclear Fuel*, June 22, 1981.

"Damage to Iraqi Reactor Still Unknown as Controversy Continues," *Nucleonics Week*, June 18, 1981.

Eklund, Sigvard, "Peaceful Nuclear Development Must Continue," *IAEA Bulletin*, No. 3 (1981).

"Factbox: Five Facts about Next U.N. Nuclear Chief Rafael Grossi," Reuters World News, October 29, 2019.

"France Sees Obligation to Supply Iraq with New Reactor," *Nucleonics Week*, June 25, 1981.

Frazar, S.L., and S.V. Mladineo, "Safeguards Culture: Lessons Learned," *ESARDA Bulletin*, No. 44 (June 2010).

Gaspar, Miklos, "Security Culture: One for All, and All for One," *IAEA Bulletin*, December 2016.

Grümm, Hans, "Safeguards and Tamuz: Setting the Record Straight," *IAEA Bulletin*, No. 4 (1992).

Hall, John, "The International Atomic Energy Agency: Origins and Early Years," *IAEA Bulletin*, No. 2 (1987).

"IAEA Secures 'Temporary Understanding' with Iran," *World Nuclear News*, February 22, 2021.

Jennekens, Jon, "IAEA Safeguards: A Look at 1970–1990 and Future Prospects," *IAEA Bulletin*, No. 1 (1990).

Karimi, Nasser, and Kiyoko Metzler, "Iran Officially Imposes Curbs on UN Nuclear Inspections," Associated Press, February 23, 2021.

Khalef, Roula, and James Blitz, "IAEA Chief Embraces the Eye of the Storm," *Financial Times*, October 2, 2007.

Kratzer, Myron B., "Prospective Trends in International Safeguards," *Nuclear News*, October 1980.

McGraw, Marsha, "No More Safeguards," *Christian Science Monitor*, July 14, 1981.

Miller, Judith, "Was Iraq Planning to Make the Bomb? Debate by Experts Seems Inconclusive," *New York Times*, June 19, 1981, http://www.nytimes.com/1981/06/19/world/was-iraq-planning-to-make-the-bomb-debate-by-experts-seems-inconclusive.html?pagewanted=all.

Muntzing, L. Manning, "Safeguards and Nuclear Safety: A Personal Perspective," *IAEA Bulletin*, No. 4 (1982).

Murphy, François, "U.N. Nuclear Watchdog Picks Argentina's Grossi as Next Chief," *Reuters World News*, October 29, 2019.

"Paying Tribute to 25 Years of Safeguards Leadership," *IAEA Bulletin*, No. 3 (1994).

"Secret Service Chief Says He's Working on Culture," *Washington Times*, March 17, 2015, http://www.washingtontimes.com/news/2015/mar/17/secret-service-director-says-hes-working-on-agency/?page=all.

Shane, Scott, "Agents Enjoy Status, But Intelligence Analysts Gain Attention," *New York Times*, March 27, 2015.

"Uber Report: Eric Holder's Recommendations for Change," *New York Times*, June 14, 2017.

"US Embassy Cables: New UN Chief is 'Director General of All States, But in Agreement with Us,'" *Guardian*, December 3, 2010, https://www.theguardian.com/world/us-embassy-cables-documents/230076.

Web-Only Articles

"Amano Wants Development Recognition," *World Nuclear News*, January 23, 2015, http://www.world-nuclear-news.org/NP-Amano-wants-development-recognition-2301152.html.

Boureston, Jack, and Charles D. Ferguson, "Strengthening Nuclear Safeguards: Special Committee to the Rescue?" Arms Control Association, 2005, http://www.armscontrol.org/act/2005_12DEC-Safeguards.

"From Aflaq to Tammuz: Iraq's Nuclear Weapons Program," http://nuclearweaponarchive.org/Iraq/IraqAtoZ.html, accessed January 4, 2018.

Gärtner, Heinz, "Yukiya Amano: One Year In," *Bulletin of the Atomic Scientists*, January 26, 2011, http://thebulletin.org/yukiya-amano-one-year.

Goldschmidt, Pierre, "Looking Beyond Iran and North Korea for Safeguarding the Foundations of Nuclear Nonproliferation," Nonproliferation Policy Education Center, Strategic Studies Institute, U.S. Army War College, November 15, 2011, http://www.npolicy.org/article.php?aid=1115&tid=4.

Heinonen, Olli, "IAEA Safeguards—Evolving its 40-Year old Obligations to Meet Today's Verification Undertakings," Belfer Center for Science and International Affairs, Harvard University, Cambridge, Mass., July 12, 2013, https://www.belfercenter.org/publication/iaea-safeguards-evolving-meet-todays-verification-undertakings.

Henriques, Sasha, "A Day in the Life of a Safeguards Inspector," July 27, 2016, https://www.iaea.org/newscenter/news/a-day-in-the-life-of-a-safeguards-inspector#:~:text

=The%20IAEA%20employs%20around%20385%20designated%20inspectors%20
from%20around%2080%20countries.

Hibbs, Mark, and Andreas Persbo, "The ElBaradei Legacy," *Bulletin of the Atomic Scientists*, September–October 2009.

Hinderstein, Corey, "Inside Insights," *The Communicator*, INMM, June 2016.

"Leaked Document Describes IAEA Transparency Proposal," FreedomInfo.org, April 24, 2015, http://www.freedominfo.org/2015/04/leaked-document-describes-iaea-transparency-proposal/.

Lewis, Jeffrey, "Replacing Olli," Arms Control Wonk, July 3, 2010, http://lewis.armscontrolwonk.com/archive/2792/replacing-olli.

Lowther, Adam, "A Year Later: Responding to Problems in the ICBM Force," *Bulletin of the Atomic Scientists*, February 20, 2015, http://thebulletin.org/year-later-responding-problems-icbm-force7984.

McIntosh, Toby, and William Burr, "IAEA Lacks Transparency, Observers, Researchers Say," FreedomInfo.org, http://www.freedominfo.org/2015/04/iaea-lacks-transparency-observers-researchers-say/.

Mladineo, Stephen V., "International Workshop on Safeguards Culture," *The Communicator*, INMM, June 2016.

"Rafael Mariano Grossi," https://www.iaea.org/about/rafael-grossi, accessed March 29, 2021.

"Saddam Hussein," Wikipedia, https://en.wikipedia.org/wiki/Saddam_Hussein, accessed February 2, 2021.

"Report on 2005 INMM/ESARDA Santa Fe Workshop: Changing the Safeguards Culture: Broader Perspectives and Challenges," *INMM Communicator*, Vol. 3, No. 3 (January 2006).

Schwartz, J.P., "Uranium Dioxide Caramel Fuel for an Alternative Fuel Cycle for Research and Test Reactors," Commissariat a l'Energie Atomique, Saclay, France, 1978, http://www.iaea.org/inis/collection/NCLCollectionStore/_Public/26/016/26016740.pdf.

Reports

Albright, David, Corey Gay, and Khidhir Hamza, "Development of the Al-Tuwaitha Site: What if the Public or the IAEA had Overhead Imagery?" Institute for Science and International Security (ISIS), April 26, 1999, http://isis-online.org/isis-reports/detail/development-of-the-al-tuwaitha-site-what-if-the-public-or-the-iaea-had-over/9.

Berdennikov, Grigory, John Carlson, Thomas Countryman, and Anton Khlopkov, "Principles and Recommendations for Implementation of the IAEA Safeguards System," in Nuclear Threat Initiative (NTI) and Center for Energy and Security Studies (CNESS), "The Future of IAEA Safeguards: Rebuilding the Vienna Spirit through Russian-U.S. Expert Dialogue," Washington, D.C., November 2020.

Berriman, Annette, Russel Leslie, and John Carlson, "The Role of 'Safeguards Criteria' in an Evolving Safeguards Environment," Australian Safeguards and Non-Proliferation Office, Canberra, 2016, https://www.dfat.gov.au/sites/default/files/inmm2006_safe guards_criteria.pdf.

Campbell, M., et al., *At What Cost Success? Final Report of the External Review on the Management Processes of the International Atomic Energy Agency* (MANNET Report), Geneva, 2002.

Carlson, John, "Future Directions in IAEA Safeguards," Discussion Paper, Project on Managing the Atom, Belfer Center for Science and International Affairs, Harvard University, Cambridge, Mass., November 2018, https://www.belfercenter.org/publication /future-directions-iaea-safeguards.

Carlson, John, "SAGSI: Its Role and Contribution to Safeguards Development," Australian Safeguards and Non-Proliferation Office, Canberra, 2006.

Chauvistré, Eric, "The Implications of IAEA Inspections under Security Council Resolution 687," Research Paper No. 11, UN Institute for Disarmament Research (UNIDIR), Geneva, 1992.

Findlay, Trevor, "A Standing United Nations Verification Body: Necessary and Feasible," *Compliance Chronicles*, No. 1, Canadian Center for Treaty Compliance, Ottawa, December 2005.

Findlay, Trevor, "Proliferation Alert! The IAEA and Non-Compliance Reporting," Report No. 2015–04, Project on Managing the Atom, Belfer Center for Science and International Affairs, Harvard University, Cambridge, Mass., October 2015.

Findlay, Trevor, "Unleashing the Nuclear Watchdog: Strengthening and Reform of the International Atomic Energy Agency," Center for International Governance Innovation (CIGI), Waterloo, Ontario, 2013, https://www.cigionline.org/publications/2012/6 /unleashing-nuclear-watchdog-strengthening-and-reform-of-iaea.

Haber, S.B., and M.T. Barriere, "Development of a Regulatory Organizational and Management Review Method," Research Report RSP-0060, Canadian Nuclear Safety Commission, Ottawa, 1998.

Institute of Nuclear Power Operations (INPO), "Principles for a Strong Nuclear Safety Culture," November 2004.

International Commission on Nuclear Non-Proliferation and Disarmament (ICNND), *Eliminating Nuclear Threats: A Practical Agenda for Global Policymakers* (Canberra and Tokyo: Paragon Press, 2009).

International Institute for Strategic Studies (IISS), "Iraq's Weapons of Mass Destruction: A Net Assessment," London, September 9, 2002.

Kelley, Robert, "Creating a 'One House' Culture at the IAEA through Matrix Management," *SIPRI Policy Brief*, SIPRI, Stockholm, January 2014.

Larrimore, Jim, Gothard Stein, and Grace Thompson, "Report on 2005 INMM/ESARDA Santa Fe Workshop: Changing the Safeguards Culture: Broader Perspectives and Challenges," *INMM Communicator*, Vol. 3, No. 3 (January 2006).

Mayhew, Noah, "A Lexical History of the State-Level Concept and Issues for Today," Vienna Center for Disarmament and Non-Proliferation (VCDNP), Vienna, December 2020, https://vcdnp.org/wp-content/uploads/2021/01/Lexical-History-of-the-State-level-Concept_Final.pdf.

Nuclear Threat Initiative (NTI) and Center for Energy and Security Studies (CNESS), "The Future of IAEA Safeguards: Rebuilding the Vienna Spirit through Russian-U.S. Expert Dialogue," Washington, D.C., November 2020.

Potter, William, and Gauhkar Mukhatzhanova, "Nuclear Politics and the Non-Aligned Movement," Adelphi Paper, No. 427, 2012.

Rockwood, Laura, Noah Mayhew, Artem Lazarev, and Mara Pfneisi, "IAEA Safeguards: Staying Ahead of the Game," Swedish Radiation Safety Authority, Report No. 2019:14, Stockholm, 2019.

Shea, Thomas E., and Laura Rockwood, "Nuclear Disarmament: The Legacy of the Trilateral Initiative," Deep Cuts Working Paper No. 4, University of Hamburg, Germany, March 2015, https://deepcuts.org/images/PDF/DeepCuts_WP4_Shea_Rockwood_UK.pdf.

World Institute for Nuclear Security (WINS), "Nuclear Security Culture," WINS International Best Practice Guide, Vienna, September 2011.

Conference and Workshop Papers and Presentations

Amano, Yukiya, IAEA Director, "Challenges in Verification," Presentation to the Center for Strategic and International Studies (CSIS), Washington, D.C., April 5, 2019, https://www.iaea.org/newscenter/statements/challenges-in-nuclear-verification.

Blix, Hans, IAEA Director General Statement, IAEA Seminar on "The Current and Future Safeguards Role of the IAEA: Challenges and Opportunities," Vienna, January 27, 1994, https://www.iaea.org/newscenter/statements/seminar-current-and-future-safeguards-role-iaea-challenges-and-opportunities.

Carlson, John, "Changing the Safeguards Culture: Views on and Expectations from the Workshop," INMM/ESARDA Workshop on Changing the Safeguards Culture: Broader Perspectives and Challenges, Santa Fe, New Mexico, October 30–November 2, 2005.

Carlson, John, "Safeguards in a Broader Policy Perspective: Verifying Treaty Compliance," INMM/ESARDA Workshop on Changing the Safeguards Culture: Broader Perspectives and Challenges, Santa Fe, New Mexico, October 30–November 2, 2005.

Carlson, John, "Special Inspections Revisited," Paper Presented to Annual Meeting of the Institute of Nuclear Materials Management (INMM), Phoenix, Arizona, July 10–14, 2005.

Casterton, James, "Engendering Safeguards Culture," Presentation at Nuclear Safeguards Culture Workshop, Harvard University, Cambridge, Mass., May 9, 2014.

Casterton, James, Remarks at Seminar on IAEA Safeguards Culture, Carleton University, Ottawa, November 14, 2014.

Clark, R., "Setting the Culture from Day One," Session 21, "Enhancing the Organizational Culture to Prepare for Future Missions," IAEA Safeguards Symposium, November 1–5, 2010, Vienna, http://www.iaea.org/safeguardssymposium2010.

Cooley, Jill, "Addressing Verification Challenges," International Safeguards Symposium, IAEA, Vienna, October 16–20, 2006, http://www.pub.iaea.org/MTCD/publications/PDF/P1298/P1298_Book.pdf.

Cooley, Jill N., "Department of Safeguards Long-Term Strategic Plan, 2012–2023," IAEA Symposium on International Safeguards: Preparing for Future Verification Challenges, IAEA, Vienna, November 1, 2010.

Cooley, Jill N., "Views on and Expectations from the Workshop," INMM/ESARDA Joint Workshop on Changing the Safeguards Culture: Broader Perspectives and Challenges, Santa Fe, New Mexico, October 30–November 2, 2005.

Dahlgren, K., "Having the Right Organizational Culture," Session 21, "Enhancing the Organizational Culture to Prepare for Future Missions," IAEA Safeguards Symposium, November 1–5, 2010, Vienna, http://www.iaea.org/safeguardssymposium2010.

Dahunsi, Stephen, "External Factors that Shape Organizational Culture: A Nigerian Perspective," INMM/ESARDA Safeguards Culture Workshop, Texas A&M University, College Station, Texas, April 26–27, 2016.

Desson, K., "Organizational Culture—Why Does It Matter?" Session 21, "Enhancing the Organizational Culture to Prepare for Future Missions," IAEA Safeguards Symposium, Vienna, November 1–5, 2010, http://www.iaea.org/safeguardssymposium2010.

"Disarming Iraq: Preparing for the Long-Term Monitoring of Iraq's Nuclear Weapons Capability," Washington Council on Non-Proliferation, Washington, D.C., October 26, 1992.

Ellis, Doris, and Sara Scott, "Views on and Expectations from the Workshop," INMM/ESARDA Workshop on Changing the Safeguards Culture: Broader Perspectives and Challenges, Santa Fe, New Mexico, October 30–November 2, 2005.

Findlay, Trevor, "The IAEA's Nuclear Safeguards Culture: 'Candy Concept' or Powerful Prism?" Presentation at Symposium on International Safeguards: "Linking Strategy, Implementation and People," IAEA, Vienna, October 21, 2014, published in IAEA, *Symposium on International Safeguards: Linking Strategy, Implementation and People*, Book of Abstracts, Presentations and Papers, IAEA-CN-220, https://www-legacy.iaea .org/safeguards/symposium/2014/home/eproceedings/sg2014_eproceedings_online .pdf.

Heinonen, Olli, "Five Decades of IAEA Safeguards," Paper Presented to INMM Meeting, Atlanta, Georgia, July 2014, www.belfercenter.org/publication/five-challenging -decades-iaea-safeguards.

Heinonen, Olli, "IAEA Safeguards—Evolving its 40-Year old Obligations to Meet Today's Verification Undertakings," Presentation to Annual INMM Meeting, Palm Desert, California, July 14–18, 2013.

Hosoya, M., E. Franklin Saburido, H. Nackaerts, G. Bernasconi, V. Wong, and K. Warthan, "Implementing an Integrated Safeguards Approach at Multi-Unit CANDU Stations: Potential Savings," IAEA-CN-148/60, "Addressing Verification Challenges," Proceedings of an International Safeguards Symposium, IAEA, Vienna, October 16–20, 2009, STI/PUB/1289.

IAEA, "Human and Organizational Aspects of Assuring Nuclear Safety—Exploring 30 Years of Nuclear Safety," Proceedings of an International Conference, Vienna, February 22–26, 2016.

Lockwood, D., M. Scholtz, L. Blair, and E. Wonder, "Next Generation Safeguards Initiative: Human Capital Development Programs," IAEA, IAEA-CN-184/108, November 2010.

Mladineo, Stephen V., Karyn R. Durbin, and Andrew Van Duzer, "Changing the Safeguards Culture: Broader Perspectives and Challenges," INMM/ESARDA Workshop on Changing the Safeguards Culture: Broader Perspectives and Challenges, Santa Fe, New Mexico, October 30–November 2, 2005.

Nackaerts, Herman, "Evolution of Safeguards Implementation," Presentation to Nuclear Safeguards Culture Workshop, Harvard University, Cambridge, Mass., May 9, 2014.

Patten, John, "The Role of Quality Management in Changing Safeguards Culture," INMM/ESARDA Workshop on Changing Safeguards Culture: Broader Perspectives and Challenges, Santa Fe, New Mexico, October 30–November 1, 2005.

Proceedings, INMM/ESARDA Workshop on Changing the Safeguards Culture: Broader Perspectives and Challenges, Santa Fe, New Mexico, October 30–November 2, 2005.

Pujol, Eric, "Methodology for Long-Range Strategic Planning in the IAEA Department of Safeguards," IAEA Symposium on International Safeguards: Preparing for Future Verification Challenges, IAEA, Vienna, November 1, 2010.

Rockwood, Laura, "Safeguards 101," Presentation at Belfer Center for Science and International Affairs, Harvard University, Cambridge, Mass., May 22, 2015.

"Safeguards Against Nuclear Proliferation," Remarks by Myron B. Kratzer, Director, Division of International Affairs, U.S. Atomic Energy Commission, at George Washington University—American Assembly, Airlie House, Warrenton, Va., May 11, 1967, reproduced in U.S. Congressional Research Service, *Nuclear Safeguards: A Reader*, Report Prepared by the U.S. Congressional Research Service for the Subcommittee on Energy Research and Production transmitted to the Committee on Science and Technology, U.S. House of Representatives, 98th Cong., 1st sess., December 1983, U.S. Government Printing Office, Washington, D.C., December 1983.

"The Politics of Safeguards," Transcript, Carnegie International Nuclear Policy Conference, Washington, D.C., March 24, 2015.

Unpublished presentations from Safeguards Culture Workshop, Texas A&M University, College Station, Texas, April 26–27, 2016, http://www.inmm.org/Content/Naviga tionMenu/Events/UpcomingEvents/SafeguardsCultureWorkshop_Agenda_as_of_04 -14-16.pdf.

Varjoranta, Tero, Keynote Address, IAEA International Safeguards Symposium, Vienna, October 20, 2014.

Documents

IAEA General Documents, including Board of Governors information documents (INF), General Conference information documents (INF) and Information Circulars (INFCIRC) (in chronological order)

Statute of the International Atomic Energy Agency, 1957.

The Agency's Safeguards System (1965, as provisionally extended in 1966 and 1968), INFCIRC/66/Rev. 2, September 16, 1968.

The Structure and Content of Agreements between the Agency and States Required in Connection with the Treaty on the Non-Proliferation of Nuclear Weapons, INF-CIRC/153 (Corrected), June 1972, https://www.iaea.org/sites/default/files/publications/documents/infcircs/1972/infcirc153.pdf.

Text of the Agreement between Iraq and the Agency for the Application of Safeguards in Connection with the Treaty on the Non-Proliferation of Nuclear Weapons, INF-CIRC/172, February 22, 1973, https://www.iaea.org/sites/default/files/publications/documents/infcircs/1973/infcirc172a1.pdf.

Code 3.1, Subsidiary Arrangement to the Agreement between the Government of [. . .] and the International Atomic Energy Agency for the Application of Safeguards in Connection with the Treaty on the Non-Proliferation of Nuclear Weapons, Fifth Revision, March 12, 1974, https://www.iaea.org/sites/default/files/sg-fm-1170-subsidiary-arrangement-code-1-9.pdf.

The Standard Text of Safeguards Agreements in Connection with the Treaty on the Non-Proliferation of Nuclear Weapons, Note by the Director General, GOV/INF/276, August 22, 1974.

Exchange of Letters Between the Government of the French Republic and the Government of the Republic of Iraq Supplementary to the Franco-Iraqi Co-operation Agreement for the Peaceful Utilization of Nuclear Energy Signed on 18 November 1975, INFCIRC/172/Add 1, September 1979.

IAEA Safeguards: An Introduction, IAEA/SG/INF/3, October 1981.

IAEA Safeguards, Aims, Limitations, and Achievements, IAEA/SG/INF.4, 1983.

Model Protocol Additional to the Agreement(s) between State(s) and the International Atomic Energy Agency for the Application of Safeguards, INFCIRC/540 (Corrected), September 1997.

"The Evolution of IAEA Safeguards," Vienna, 1998, www.pub.iaea.org/MTCD/publications/PDF/NVS2_web.pdf.

IAEA Safeguards Glossary, 2001 ed., International Nuclear Verification Series, No. 3, 2002.

Medium Term Strategy 2006–2011, 2005, http://www.iaea.org/sites/default/files/mts2006_2011.pdf.

IAEA Safety Glossary: Terminology Used in Nuclear Safety and Radiation Protection, 2007 Edition, http://www-pub.iaea.org/MTCD/publications/PDF/Pub1290_web.pdf.

"IAEA Safeguards: Staying Ahead of the Game," 2007.

Report of the Commission of Eminent Persons on the Future of the Agency: Note by the Director General, GOV/2008/22-GC (52)/INF/4, May 23, 2008.

"20/20 Vision for the Future: Background Report by the Director General for the Commission of Eminent Persons," Annex, "Report of the Commission of Eminent Persons on the Future of the Agency," GOV/2008/22-GC(52)/INF/4, IAEA, Vienna, 2008.

Medium Term Strategy 2012–2017, 2011, https://www.iaea.org/sites/default/files/mts 2012_2017.pdf.

"Nuclear Security Culture: Implementing Guide," IAEA Nuclear Security Series No. 7, IAEA, Vienna, 2008, https://www-pub.iaea.org/MTCD/Publications/PDF/Pub1347 _web.pdf.

Safeguards Department, Long-Term Strategic Plan (2012–2023) Summary, 2011, www .iaea.org/safeguards/documents/LongTerm_Strategic_Plan_%2820122023%29-Sum mary.pdf.

IAEA Department of Safeguards, Subsidiary Arrangement to the Agreement by the Government of [. . . . .] and the International Atomic Energy Agency for the Application of Safeguards in Connection with the Treaty on the Non-Proliferation of Nuclear Weapons, Fifth Revision, SG-FM-1170, November 2, 2011.

"Inviting Commitment: Working for the IAEA," IAEA Human Resources, Vienna, 2012.

"Managing Organizational Change in Nuclear Organizations," IAEA Nuclear Energy Series, No. NG-T-1.1, 2014.

Sections N to S, Joint Comprehensive Plan of Action (JCPOA), Vienna, 14 July 2015, Communication dated 24 July 2015 received from China, France, Germany, the Russian Federation, the United Kingdom, the United States of America (the E3/EU+3) and the Islamic Republic of Iran concerning the text of the Joint Comprehensive Plan of Action (JCPOA), INFCIRC/887, July 31, 2015.

IAEA Fact Sheet on DPRK safeguards, https://www.iaea.org/newscenter/focus/dprk /fact-sheet-on-dprk-nuclear-safeguards, accessed November 1, 2016.

Medium Term Strategy 2018–2023, 2017, https://www.iaea.org/sites/default/files/16 /11/mts2018_2013.pdf.

"Al Tuwaitha-Tamuz 2 Reactor," http://www-ns.iaea.org/projects/iraq/tuwaitha/tamuz -2.asp?s=8&l=66, accessed August 22, 2017.

"The IRT-5000," http://www-ns.iaea.org/projects/iraq/tuwaitha/irt5000.asp?l=66, accessed August 22, 2017.

Office of Legal Affairs, Country Fact Sheets, https://ola.iaea.org/ola/FactSheets/Coun tryDetails.asp?country=IQ, accessed August 22, 2017.

IAEA Mission Statement, http://www.iaea.org/About/mission.html, accessed November 2018.

"Development and Implementation Support Program for Nuclear Verification 2020–2021," STR-393, Vienna, January 2020.

A Harmonized Safety Culture Model, IAEA Working Document (last revised 05 May 2020), https://www.iaea.org/sites/default/files/20/05/harmonization_05_05_2020-final_002.pdf, accessed March 31, 2021.

Joint Statement by the Director General of the IAEA and the Vice-President of the Islamic Republic of Iran and Head of the AEOI [Atomic Energy Organization of Iran], August 20, 2020, https://www.iaea.org/newscenter/pressreleases/joint-statement-by-the-director-general-of-the-iaea-and-the-vice-president-of-the-islamic-republic-of-iran-and-head-of-the-aeoi.

IAEA Director General Rafael Grossi, Statement to Sixty-Fourth Regular Session of IAEA General Conference, September 21, 2020, https://www.iaea.org/newscenter/statements/statement-to-sixty-fourth-regular-session-of-iaea-general-conference.

Joint Statement by the Vice-President of the Islamic Republic of Iran and Head of the AEOI and the Director General of the IAEA, February 2021, https://www.iaea.org/newscenter/pressreleases/joint-statement-by-the-vice-president-of-the-islamic-republic-of-iran-and-head-of-the-aeoi-and-the-director-general-of-the-iaea.

IAEA Core Ethical Values, https://www.iaea.org/about/employment/iaea-values, accessed April 1, 2021.

Additional Protocol, https://www.iaea.org/topics/additional-protocol, accessed August 1, 2021.

IAEA Board of Governors Documents issued with GOV number only (in chronological order)

IAEA, Board of Governors, Official Records, GOV/OR.748, May 6, 1991.

Proposal for the Derestriction of Board Documents, GOV/2843, February 15, 1996.

Supplementary Document to the Report on The Conceptualization and Development of Safeguards Implementation at the State Level (GOV/2013/38): Report by the Director General, GOV/2014/41, August 13, 2014.

IAEA Documents with a General Conference (GC) number or with both a General Conference and Board of Governors (GOV) number (in chronological order)

The Agency's Inspectorate: Memorandum by the Director General, and Annex, "The Agency's Inspectors," GC(V) INF/39, Vienna, August 28, 1961, https://www-legacy.iaea.org/About/Policy/GC/GC05/GC05InfDocuments/English/gc05inf-39_en.pdf (the Inspectors' Document).

The Agency's Accounts for 1980, GC(XXV)/645, July 1981, http://www.iaea.org/inis /collection/NCLCollectionStore/_Public/40/086/40086035.pdf.

The Annual Report for 1983, GC(XXVIII)/713, July 1984, https://www.iaea.org/About /Policy/GC/GC28/GC28Documents/English/gc28-713_en.pdf.

The Annual Report for 1989, GC(XXXIV)/915, July 1990, https://www.iaea.org/About /Policy/GC/GC34/GC34Documents/English/gc34-915_en.pdf.

The Agency's Accounts for 1990, GC(XXXV)/954, August 1991, https://www.iaea.org /About/Policy/GC/GC35/GC35Documents/English/gc35-954_en.pdf.

The Annual Report for 1990, GC(XXXV)/953, October 1991, https://www.iaea.org /About/Policy/GC/GC35/GC35Documents/English/gc35-953_en.pdf.

Strengthening the Effectiveness and Improving the Efficiency of the Safeguards System: Report by the Director General, GC(39)/17, August 22, 1995, Annex 3.

Strengthening the Effectiveness and Improving the Efficiency of the Safeguards System including Implementation of Additional Protocols: Report by the Director General, GC(51)/8, July 23, 2007.

Personnel: Women in the Secretariat: Report by the Director General, GOV /2007/47-GC(51)/17, August 17, 2007.

The Agency's Accounts for 2010, GC(55)/4, 2011, http://www.un.org/en/auditors /panel/docs/IAEA_2010.pdf.

The Agency's Financial Statements for 2010, GC(55)/5, 2011, http://www.iaea.org /sites/default/files/mts2006_2011.pdf.

The Agency's Financial Statements for 2011, GC(56)/5, 2012, https://www-legacy.iaea .org/About/Policy/GC/GC56/GC56Documents/English/gc56-10_en.pdf.

Federal Court of Auditors, Germany, Report of the External Auditor on the audit of the financial statements of the International Atomic Energy Agency for the year ended 31 December 2011, Part V, The Agency's Financial Statements for 2011, GC(56)/5, 2012.

Personnel: Staffing of the Agency's Secretariat: Report by the Director General, GOV/2013/34/GC(57)/14, July 23, 2013.

The Conceptualization and Development of Safeguards Implementation at the State Level: Report by the Director General, GOV/2013/38, August 12, 2013,

The Agency's Program and Budget 2014–2015, GC(57)/2, August 2013, https:// www-legacy.iaea.org/About/Policy/GC/GC57/GC57Documents/English/gc57-2 _en.pdf.

The Agency's Financial Statements for 2013, GC(58)/5, 2014, https://www.iaea.org /About/Policy/GC/GC58/GC58Documents/English/gc58-5_en.pdf.

Office of the Comptroller and Auditor General of India, Report of the External Auditor on the Audit of the Financial Statements of the International Atomic Energy Agency for the Year Ended 31 December 2013, Part V, The Agency's Financial Statements for 2013, GC(58)/5, 2014, https://www.iaea.org/About/Policy/GC/GC58/GC58Docu ments/English/gc58-5_en.pdf.

The Annual Report for 2014, GC/GC59/7, 2015, https://www.iaea.org/About/Policy /GC/GC59/GC59Documents/English/gc59-7_en.pdf.

The Agency's Program and Budget 2016–2017, July 2015, GC(59)/2, https://www-leg acy.iaea.org/About/Policy/GC/GC59/GC59Documents/English/gc59-2_en.pdf.

Personnel: Staffing of the Agency's Secretariat: Report by the Director General, GOV /2015/46-GC(59)/16, August 3, 2015, https://www.iaea.org/About/Policy/GC/GC59 /GC59Documents/English/gc59-16_en.pdf.

Personnel: Women in the Secretariat: Report by the Director General, GOV /2015/47-GC(59)/17, August 3, 2015.

Strengthening the Effectiveness and Improving the Efficiency of Agency Safeguards: Report by the Director General, GC(60)/13, August 12, 2016, https://www-legacy.iaea .org/About/Policy/GC/GC60/GC60Documents/English/gc60-13_en.pdf.

The Agency's Financial Statements for 2016, GC(61)/2, 2017, https://www.iaea.org /About/Policy/GC/GC61/GC61Documents/English/gc61-2_en.pdf.

Strengthening the Effectiveness and Improving the Efficiency of Agency Safeguards: Report by the Director General, GC(61)/16, July 26, 2017, https://www.legacy.iaea .org/About/Policy/GC/GC61/GC61Documents/English/gc61-16_en.pdf.

Personnel: Staffing of the Agency's Secretariat: Report by the Director General, GOV /2017/38-GC(61)/18, August 2, 2017.

Personnel: Women in the Secretariat: Report by the Director General, GOV/2017 /39-GC(61)/19, August 2, 2017.

The Agency's Program and Budget 2020–2021, GC(63)/2, 2019, https://www.iaea.org /sites/default/files/gc/gc63-2.pdf.

Personnel: Women in the Secretariat: Report by the Director General, GOV/2019 /37-GC(63)/16, July 19, 2019.

The Agency's Financial Statements for 2019, GC(64)/4, Vienna, July 2020.

Strengthening the Effectiveness and Improving the Efficiency of Agency Safeguards, Report by the Director General, GC(64)/13, August 11, 2020.

The Annual Report for 2019, GC(64)/3, September 2020, https://www.iaea.org/sites/default/files/publications/reports/2019/gc64-3.pdf.

The Agency's Financial Statements for 2019, GC(64)/4, 2020.

Audit Board of Indonesia, Audit Report on the International Atomic Energy Agency, 2019, Part V, The Agency's Financial Statements for 2019, GC(64)/4, 2020.

Safeguards Statement for 2020, B.1.1., https://www.iaea.org/sites/default/files/21/06/statement-sir-2020.pdf.

United Nations Documents (in chronological order)

Fourth Review Conference of the Parties to the Treaty on the Non-Proliferation of Nuclear Weapons, Geneva, August 20–September 14, 1990, Report of Main Committee II, NPT/CONF.IV/MC.II/l, September 10, 1990.

Review and Extension Conference of the Parties to the Treaty on the Non-Proliferation of Nuclear Weapons, Final Document, May 1995, https://www.un.org/disarmament/wmd/nuclear/npt1995/official-documents.

United Nations, Report of the Secretary-General's High-Level Panel on Threats, Challenges and Change, New York, 2004.

United Nations, Joint Inspection Unit (JIU), Review of Management and Administration in the International Atomic Energy Agency (IAEA), JIU/REP/2012/Rev. 1, United Nations, Geneva, 2012.

United Nations, Sustainable Development Goals, 2015, https://sustainabledevelopment.un.org/?menu=1300.

United Nations Security Council Documents (in chronological order)

Note by the President of the Security Council, S/22746, June 28, 1991.

Consolidated Report on the First Two IAEA Inspections Under Security Council Resolution 687 (1991) of Iraqi Nuclear Capabilities, July 11, 1991, Security Council, Note by the Secretary-General, S/22788, July 15, 1991.

First Report on the Sixth IAEA On-Site Inspection in Iraq under Security Council Resolution 687 (1991), S/23122, October 8, 1991.

First semi-annual report (covering the period 17 June–17 December 1991) on the implementation by the IAEA of the plan for the destruction, removal or rendering harmless of items listed in paragraph 12 of UN Security Council Resolution 687 (1991), S/23295, December 17, 1991.

Eighth Report of the Director General of the International Atomic Energy Agency on the Implementation of the Agency's Plan for Future Ongoing Monitoring and Verification of Iraq's Compliance with Paragraph 12 of Resolution 687 (1991), Appendix, Note by Secretary-General, UN Security Council, S/995/844, October 6, 1995.

Report of the Twenty-Eighth IAEA On-Site Inspection in Iraq under Security Council Resolution 867 (1991), Attachment, Note by the Secretary-General, UN Security Council S/1995/1003, December 1, 1995.

Fourth Consolidated Report by the Director General of the International Atomic Energy Agency under Paragraph 16 of Security Council Resolution 1051 (1996), Appendix to Letter dated 6 October 1997 from the Director General of the International Atomic Energy Agency, Annex to UN Security Council, Note by Secretary General, S/1997/779, October 8, 1997, https://nsarchive2.gwu.edu/NSAEBB/NSAEBB80/wmd07.pdf.

Report of the First Panel Established pursuant to the Note by the President of the Security Council on 30 January 1999 (S/1999/100), concerning disarmament and current and future ongoing monitoring and verification issues, Annex 1, Letter dated 27 and 30 March 1999, respectively, from the Chairman of the Panels established Pursuant to the Note by the President of the Security Council of 30 January 1999 (S/1999/100) addressed to the President of the Security Council, S/1999/356, March 30, 1999 (the Amorim Report).

U.S. Government Documents (in chronological order)

Report on International Control of Atomic Energy, Prepared for the Secretary of State's Committee on Atomic Energy, Department of State Publication 2498, U.S. Government Printing Office, Washington, D.C., March 16, 1946 (the Acheson-Lilienthal Report).

Report of the Advisory Committee on U.S. Policy Toward the International Atomic Energy Agency, Hearing before the Joint Committee on Atomic Energy, Congress of the United States, Eighty-Seventh Congress, U.S. Government Printing Office, Washington, August 2, 1962 (the Smyth Report).

The International Atomic Energy Agency: An Appraisal with Recommendations for United States Policy: A Report to the Secretary of State of the Panel to Review the Activities of the International Atomic Energy Agency, U.S. State Department, Washington, D.C., 1972.

Donnelly, Warren, "Nuclear Weapons Proliferation and the International Atomic Energy Agency: An Analytic Report," prepared for the U.S. Senate Committee on Governmental Operations, Washington, D.C.: U.S. Government Printing Office, 1976.

Comptroller General of the United States, Assessment of U.S. and International Controls Over the Peaceful Uses of Nuclear Energy, Report to the Congress by the Comptroller General of the United States, ID-76–60, Appendix II, September 14, 1976.

U.S. Congressional Research Service, *Nuclear Safeguards: A Reader*, Report Prepared by the U.S. Congressional Research Service for the Subcommittee on Energy Research and Production transmitted to the Committee on Science and Technology, U.S. House of Representatives, 98th Cong., 1st sess., December 1983, U.S. Government Printing Office, Washington, D.C., December 1983.

_____, Boright, John P., "Historical Development and Current Trends in Nuclear Safeguards," Lecture at the International Training Course on Implementation of State Systems of Accounting for and Control of Nuclear Materials, March 15–30, 1982.

_____, Report Prepared by Warren Donnelly, Foreign Affairs and National Defense Division, Congressional Research Service, Library of Congress, June 16, 1981, U.S. Congress, Senate Committee on Foreign Relations, Hearings, The Israeli Airstrike, 97th Cong., 1st sess., June 18, 19, and 25, 1981.

_____, "IAEA Comments on the Testimony of Roger Richter on June 19, 1981, before the Senate Foreign Relations Committee," Appendix, U.S. Congress, Senate Committee on Foreign Relations, Hearings, The Israeli Airstrike, 97th Cong., 1st sess., June 18, 19, and 25, 1981.

_____, Nakićenović, Slobodan, "Comments on IAEA Safeguards," U.S. Congressional Research Service, Washington, D.C., August 1981.

_____, Pilat, Joseph, "The International Atomic Energy Agency (IAEA) and its Safeguards: A Fact Sheet," May 6, 1982.

_____, "Prepared Statement of Emanuel R. Morgan," IAEA Programs of Safeguards, Hearings before the Committee on Foreign Relations, U.S. Senate, 97th Cong., 1st Sess., December 2, 1981.

_____, Report to the Atomic Energy Commission by the *Ad Hoc* Advisory Panel on Safeguarding Special Nuclear Material, March 10, 1967.

_____, "Statement by Senator Alan Cranston," U.S. Congress, Senate Committee on Foreign Relations, Hearings, The Israeli Airstrike, 97th Cong., 1st sess., June 18, 19, and 25, 1981.

_____, "Testimony of Roger Richter," U.S. Congress, Senate Committee on Foreign Relations, Hearings, The Israeli Airstrike, 97th Cong., 1st sess., June 18, 19, and 25, 1981.

_____, "Testimony of Dr. Thomas E. Shea," IAEA Programs of Safeguards, Hearing before the Committee on Foreign Relations, U.S. Senate, 97th Cong., 1st sess., December 2, 1981.

U.S. Arms Control and Disarmament Agency (ACDA), "Review of the Negotiating History of the IAEA Safeguards Document INFCIRC/153," AC2NC103, Washington, D.C., July 30, 1984.

ACDA, "Review of the Negotiating History of IAEA Document INFCIRC/66/REV.2, 'The Agency's Safeguards System,'" Washington, D.C., September 5, 1984.

Paternoster, Richard R., "Nuclear Weapons Proliferation Indictors and Observables," LA-12430-MS, UC-700, Los Alamos National Laboratory, New Mexico, December 1992.

Office of Technical Assessment, *Nuclear Safeguards and the International Atomic Energy Agency*, OTA-ISS-615, Washington, D.C., April 1995.

Director of Central Intelligence, Comprehensive Report of the Special Advisor to the Director of Central Intelligence (DCI) on Iraq's WMD, Vol. II, "Nuclear," Washington, D.C., September 30, 2004.

Director of Central Intelligence, Addendums to the Comprehensive Report of the Special Advisor to the Director of Central Intelligence (DCI) on Iraq's WMD, Vol. II, "Nuclear," Washington, D.C., March 2005.

Government Accounting Office (GAO), "Nuclear Nonproliferation: IAEA Has Strengthened its Safeguards and Nuclear Security Programs, but Weaknesses Need to be Addressed," GAO-06–93, Washington, D.C., October 2005.

"Contributions to the IAEA," Office of Management and Budget, The White House, Washington, D.C., 2006; http://www.whitehouse.gov/omb/expectmore/summary/10 004639.2006.html (2006), cited in IAEA, "20/20 Vision for the Future: Background Report by the Director General for the Commission of Eminent Persons," Annex, "Report of the Commission of Eminent Persons on the Future of the Agency," GOV /2008/22-GC(52)/INF/4, IAEA, Vienna, 2008, p. 24.

Frazar, S.L., and S.V. Mladineo, "Safeguards Culture," PNNL-21555, Pacific Northwest National Laboratory, Seattle, Wash., July 2012.

Other Documents (in chronological order)

al-Mallah, Moyassar, Secretary-General of the IAEC, Letter to A.E. Cairo, IAEA, TA/ IRQ/3, IAEA Archives, Vienna, September 12, 1973.

International Energy Associates Ltd., "Review of the Negotiating History of the IAEA Safeguards Document INFCIRC/153," July 30, 1984.

Nuclear Threat Initiative (NTI), Iraq Nuclear Chronology, http://www.nti.org/media /pdfs/iraq_nuclear.pdf?_=1316466791.

Organization for Economic Cooperation and Development, *OECD Best Practice Principles for Regulatory Policy: Regulatory Enforcement and Inspections* (Paris: OECD, 2014), https://www.oecd.org/gov/regulatory-enforcement-and-inspections-9789264208117-en.htm.

Statement by HE Mr Richard Sadleir, Resident Representative of Australia to the IAEA, Agenda Item 6(b): Safeguards Implementation Report for 2019, Australian Embassy and Permanent Mission, Vienna, June 16, 2020, https://austria.embassy.gov.au/vien/IAEABOG_Jun20_SIR.html.

Unpublished documents (in chronological order)

Shea, Thomas E., and David A. Thompson, "Human Factors Affecting IAEA Inspector Performance," Prepared for the U.S. Arms Control and Disarmament Agency (ACDA), AC2NC106 (unpublished), Washington, D.C., 1982/1983.

Carroll, John S., "Introduction to Organizational Analysis: The Three Lenses," unpublished class paper, Sloan School of Management, MIT, revised June 2006.

Rockwood, Laura, "What the Agency has Missed in Iraq," Memorandum from the IAEA Office of Legal Affairs, March 17, 2003, Archive of Hans Blix, cited in Gudrun Harrer, *Dismantling the Iraqi Nuclear Program: The Inspections of the International Atomic Energy Agency, 1991–1998* (Abingdon, Oxon: Routledge, 2014).

Press Releases (in chronological order)

Statement by Dr. Sigvard Eklund, Director General of the International Atomic Energy Agency to the IAEA Board of Governors, IAEA Press Release, PR 81/6, Vienna, July 6, 1981.

IAEA Press Release, PR/91–24, July 18, 1991.

2005 Nobel Peace Prize Press Release, Oslo, October 7, 2005.

"Training IAEA Inspectors: Newly Recruited Inspectors Take to the Field," https://www.iaea.org/newscenter/news/training-iaea-inspectors, accessed February 10, 2015.

"IAEA Director General Yukiya Amano Becomes a Gender Champion," June 17, 2017, https://www.iaea.org/newscenter/news/iaea-director-general-yukiya-amano-becomes-a-gender-champion.

Statement by IAEA Director General Yukiya Amano, October 2, 2018, https://www.iaea.org/newscenter/statements/statement-by-iaea-director-general-yukiya-amano-2-october-2018.

"New IAEA Laboratory Named after Late Director General Yukiya Amano," September 16, 2021, https://www.iaea.org/newscenter/news/new-iaea-laboratory-building -named-after-late-director-general-yukiya-amano.

Web Sites

Glassdoor, http://www.glassdoor.com/IAEA-Reviews-E474090.htm.

"Culture Eats Strategy for Breakfast," Quote Investigator, https://quoteinvestigator .com/2017/05/23/culture-eats/.

World Nuclear University, http://www.world-nuclear-university.org/imis20/wnu/.

Interviews

By the Author (in chronological order)

Laura Rockwood, Vienna, July 3, 2013.

Robert Kelley, Vienna, March 19, 2014.

U.S. safeguards officials, State Department, Washington, D.C., May 19, 2014.

Olli Heinonen, Cambridge, Mass., May 22, 2014.

John Carlson, Cambridge, Mass., June 5, 2014.

Dimitri Perricos, Baden, Austria, June 20, 2014.

Jean-Maurice Crete, Vienna, June 25, 2014.

Tom Shea, Vienna, June 26, 2014.

Shirley Johnson, Atlanta, July 22, 2014.

IAEA officials, IAEA Safeguards Symposium, Vienna, October 20–24, 2014.

Olli Heinonen, Cambridge, Mass., June 5, 2015.

By Elisabeth Roehrlich, IAEA Oral History Series, University of Vienna, http://iaea-history.univic.at/oral-history-videos/ (in chronological order)

William Lichliter, Vienna, March 13, 2015.

Hans Blix, Stockholm, April 13, 2015.

Pier Roberto Danesi, Vienna, April 23, 2015.

Dimitri Perricos, Vienna, June 27, 2015.

Shirley Johnson, Vienna, August 20, 2015.

John Carlson, Vienna, November 13, 2015.

James Goodby, Washington, D.C., November 17, 2015.

David Waller, Washington, D.C., November 21, 2015.

Mohamed ElBaradei, Vienna, May 31, 2016.

Dieter Goethel, Vienna, August 18, 2016.

*By Yale-UN Oral History Project, https://archives.yale.edu/repositories
/12/resources/3915 (in chronological order)*

Rolf Ekéus, New Haven, Conn., February 3, 1998.

Scott Ritter, Larchmont, N.Y., October 27, 1998.

David Kay, McLean, Va., April 10, 2000.

Electronic Communication (in chronological order)

Jim Casterton, March 2, 2015.

Jim Casterton, March 3, 2015.

Robert Kelley, April 12, 2015.

Robert Kelley, April 24, 2015.

Former IAEA official, April 24, 2015.

Tariq Rauf, June 1, 2015.

Mark Schanfein, May 14, 2016.

John Tilemann, May 30, 2019.

Laura Rockwood, October 19, 2020.

Shirley Johnson, October 21, 2020.

Videos

Foundations of International Safeguards video series transcripts, Pacific Northwest Nuclear Laboratories (PNNL), http://cgs.pnnl.gov/fois/videos/foundations/ (in chronological order)

Panel Discussion with Myron Kratzer, Rich Hooper, and Ambassador Norman Wolf, "Retrospective of INFCIRCs 153 and 540," 4.0, Santa Fe, New Mexico, October 2005.

Discussion with Laura Rockwood, "Legal Foundations of Safeguards," 3.0, Richland, Wash., June 2006.

Panel Discussion with Current and Former SAGSI Representatives, "International Perspectives on Safeguards," 5.2, Tucson, Arizona, July 2007.

Panel Discussion with Jacques Baute, Rich Hooper, and Dimitri Perricos, "The Path Toward Strengthened Safeguards: Verification Challenges of the 1990s," 7.0, Vienna, December 2009.

Interview with Rich Hooper, "Concluding Remarks," 7.26, Vienna, December 2009.

About the Author

Trevor Findlay has had almost fifty years of experience in international relations as a scholar, practitioner, and author, working with governments, international organizations, non-governmental organizations, and academia. His lifelong specialization has been nuclear governance, including disarmament, arms control, non-proliferation, nuclear safeguards, the peaceful uses of nuclear energy, and nuclear security. He is considered one of the world's leading experts on the International Atomic Energy Agency.

Findlay is currently a Principal Fellow at the School of Social and Political Sciences at the University of Melbourne. From 2011 to 2014, he was a Senior Research Fellow with the Project on Managing the Atom at the Belfer Center for Science and International Affairs at the Harvard Kennedy School. From 2005 to 2015, he was a tenured professor at the Norman Paterson School of International Affairs at Carleton University in Ottawa, Canada, where he held the endowed William and Jeanie Barton Chair. He was previously Executive Director of the Verification Research, Training and Information Center (VERTIC) in London and a senior researcher at the Stockholm International Peace Research Institute (SIPRI) and at the Australian National University's Peace Research Center.

Findlay was a member of the UN Secretary-General's Advisory Board for Disarmament Matters and the Board of the UN Institute for Disarmament Research (UNIDIR) for several years and Chair of both for 2017. He was also a member of Japan's Eminent Persons Group for Substantive Advancement of Nuclear Disarmament and is currently a member of the Asia-Pacific Leadership Network (APLN).

Prior to his academic career, Findlay was in the Australian diplomatic service, with postings in Tokyo, Mexico City, and finally Geneva, where he was a member of Australian disarmament delegations to the Conference on Disarmament and the UN General Assembly and Disarmament Commission in New York.

Findlay is the author of multiple research reports, book chapters, and articles, as well as six books and monographs, including *Unleashing the Nuclear Watchdog: Strengthening and Reform of the International Atomic Energy Agency* (Waterloo, ON: CIGI, 2012); *Nuclear Energy and Global Governance: Ensuring Safety, Security and Non-proliferation* (London: Routledge, 2011) and *Nuclear Dynamite: The Peaceful Nuclear Explosions Fiasco* (Sydney: Brassey's Australia, 1990).

Index

Belfer Center Studies in
International Security

Published by The MIT Press
Steven E. Miller and Morgan L. Kaplan, series editors
Karen Motley, executive editor
Belfer Center for Science and International Affairs
Harvard Kennedy School, Harvard University

Acharya, Amitav, and Evelyn Goh, eds., *Reassessing Security Cooperation in the Asia-Pacific* (2007)

Agha, Hussein, Shai Feldman, Ahmad Khalidi, and Zeev Schiff, *Track-II Diplomacy: Lessons from the Middle East* (2003)

Allison, Graham, and Robert D. Blackwill, with Ali Wyne, *Lee Kuan Yew: The Grand Master's Insights on China, the United States, and the World* (2012)

Allison, Graham T., Owen R. Coté, Jr., Richard A. Falkenrath, and Steven E. Miller, *Avoiding Nuclear Anarchy: Containing the Threat of Loose Russian Nuclear Weapons and Fissile Material* (1996)

Allison, Graham T., and Kalypso Nicolaïdis, eds., *The Greek Paradox: Promise vs. Performance* (1996)

Arbatov, Alexei, Abram Chayes, Antonia Handler Chayes, and Lara Olson, eds., *Managing Conflict in the Former Soviet Union: Russian and American Perspectives* (1997)

Bennett, Andrew, *Condemned to Repetition? The Rise, Fall, and Reprise of Soviet-Russian Military Interventionism, 1973–1996* (1999)

Blackwill, Robert D., and Michael Stürmer, eds., *Allies Divided: Transatlantic Policies for the Greater Middle East* (1997)

Blackwill, Robert D., and Paul Dibb, eds., *America's Asian Alliances* (2000)

Blum, Gabriella, and Philip B. Heymann, *Laws, Outlaws, and Terrorists: Lessons from the War on Terrorism* (2010)

Brom, Shlomo, and Yiftah Shapir, eds., *The Middle East Military Balance 1999–2000* (1999)

Brom, Shlomo, and Yiftah Shapir, eds., *The Middle East Military Balance 2001–2002* (2002)

Brown, Michael E., ed., *The International Dimensions of Internal Conflict* (1996)

Brown, Michael E., and Šumit Ganguly, eds., *Fighting Words: Language Policy and Ethnic Relations in Asia* (2003)

Brown, Michael E., and Šumit Ganguly, eds., *Government Policies and Ethnic Relations in Asia and the Pacific* (1997)

Carter, Ashton B., and John P. White, eds., *Keeping the Edge: Managing Defense for the Future* (2001)

Chenoweth, Erica, and Adria Lawrence, eds., *Rethinking Violence: State and Non-State Actors in Conflict* (2010)

de Nevers, Renée, *Comrades No More: The Seeds of Political Change in Eastern Europe* (2003)

Elman, Colin, and Miriam Fendius Elman, eds., *Bridges and Boundaries: Historians, Political Scientists, and the Study of International Relations* (2001)

Elman, Colin, and Miriam Fendius Elman, eds., *Progress in International Relations Theory: Appraising the Field* (2003)

Elman, Miriam Fendius, ed., *Paths to Peace: Is Democracy the Answer?* (1997)

Falkenrath, Richard A., *Shaping Europe's Military Order: The Origins and Consequences of the CFE Treaty* (1994)

Falkenrath, Richard A., Robert D. Newman, and Bradley A. Thayer, *America's Achilles' Heel: Nuclear, Biological, and Chemical Terrorism and Covert Attack* (1998)

Feaver, Peter D., and Richard H. Kohn, eds., *Soldiers and Civilians: The Civil-Military Gap and American National Security* (2001)

Feldman, Shai, *Nuclear Weapons and Arms Control in the Middle East* (1996)

Feldman, Shai, and Yiftah Shapir, eds., *The Middle East Military Balance 2000–2001* (2001)

Findlay, Trevor, *Transforming Nuclear Safeguards Culture: The IAEA, Iraq, and the Future of Non-Proliferation* (2022)

Forsberg, Randall, ed., *The Arms Production Dilemma: Contraction and Restraint in the World Combat Aircraft Industry* (1994)

George, Alexander L., and Andrew Bennett, *Case Studies and Theory Development in the Social Sciences* (2005)

Gilroy, Curtis, and Cindy Williams, eds., *Service to Country: Personnel Policy and the Transformation of Western Militaries* (2007)

Hagerty, Devin T., *The Consequences of Nuclear Proliferation: Lessons from South Asia* (1998)

Heymann, Philip B., *Terrorism and America: A Commonsense Strategy for a Democratic Society* (1998)

Heymann, Philip B., *Terrorism, Freedom, and Security: Winning without War* (2003)

Heymann, Philip B., and Juliette N. Kayyem, *Protecting Liberty in an Age of Terror* (2005)

Howitt, Arnold M., and Robyn L. Pangi, eds., *Countering Terrorism: Dimensions of Preparedness* (2003)

Hudson, Valerie M., and Andrea M. den Boer, *Bare Branches: The Security Implications of Asia's Surplus Male Population* (2004)

Kayyem, Juliette N., and Robyn L. Pangi, eds., *First to Arrive: State and Local Responses to Terrorism* (2003)

Kokoshin, Andrei A., *Soviet Strategic Thought, 1917–91* (1998)

Lawson, Chappell, Alan Bersin, and Juliette Kayyem, eds., *Beyond 9/11: Homeland Security for the Twenty-First Century* (2020)

Lederberg, Joshua, ed., *Biological Weapons: Limiting the Threat* (1999)

Mansfield, Edward D., and Jack Snyder, *Electing to Fight: Why Emerging Democracies Go to War* (2005)

Martin, Lenore G., and Dimitris Keridis, eds., *The Future of Turkish Foreign Policy* (2004)

May, Ernest R., and Philip D. Zelikow, eds., *Dealing with Dictators: Dilemmas of U.S. Diplomacy and Intelligence Analysis, 1945–1990* (2007)

Phillips, David L., *Liberating Kosovo: Coercive Diplomacy and U.S. Intervention* (2012)

Poneman, Daniel B., *Double Jeopardy: Combating Nuclear Terror and Climate Change* (2019)

Rosecrance, Richard N., and Steven E. Miller, eds., *The Next Great War? The Roots of World War I and the Risk of U.S.-China Conflict* (2015)

Shaffer, Brenda, *Borders and Brethren: Iran and the Challenge of Azerbaijani Identity* (2002)

Shaffer, Brenda, ed., *The Limits of Culture: Islam and Foreign Policy* (2006)

Shields, John M., and William C. Potter, eds., *Dismantling the Cold War: U.S. and NIS Perspectives on the Nunn-Lugar Cooperative Threat Reduction Program* (1997)

Tucker, Jonathan B., ed., *Toxic Terror: Assessing Terrorist Use of Chemical and Biological Weapons* (2000)

Utgoff, Victor A., ed., *The Coming Crisis: Nuclear Proliferation, U.S. Interests, and World Order* (2000)

Weiner, Sharon K., *Our Own Worst Enemy? Institutional Interests and the Proliferation of Nuclear Weapons Expertise* (2011)

Williams, Cindy, ed., *Filling the Ranks: Transforming the U.S. Military Personnel System* (2004)

Williams, Cindy, ed., *Holding the Line: U.S. Defense Alternatives for the Early 21st Century* (2001)

Xu Qiyu, *Fragile Rise: Grand Strategy and the Fate of Imperial Germany, 1871–1914*, trans. Joshua Hill (2017)

Zoughbie, Daniel E., *Indecision Points: George W. Bush and the Israeli-Palestinian Conflict* (2014)

The Belfer Center for Science and International Affairs

Ashton B. Carter, Director
Eric Rosenbach, Co-Director
Harvard Kennedy School
79 JFK Street, Cambridge, MA 02138
Tel: (617) 495-1400 | Fax: (617) 495-8963
http://www.belfercenter.org | belfer_center@hks.harvard.edu

The Belfer Center is the hub of Harvard Kennedy School's research, teaching, and training in international security affairs, environmental and natural resource issues, and science and technology policy.

The Center has a dual mission: (1) to provide leadership in advancing policy-relevant knowledge about the most important challenges of international security and other critical issues where science, technology, environmental policy, and international affairs intersect; and (2) to prepare future generations of leaders for these arenas. Center researchers not only conduct scholarly research, but also develop prescriptions for policy reform. Faculty and fellows analyze global challenges from nuclear proliferation and terrorism to climate change and energy policy.

The Center's leadership begins with the recognition of science and technology as driving forces constantly transforming both the challenges we face and the opportunities for problem-solving. Building on the vision of founder Paul Doty, the Center addresses serious global concerns by integrating insights and research of social scientists, natural scientists, technologists, and practitioners in government, diplomacy, the military, and business.

The heart of the Belfer Center is its research community, including Harvard faculty, researchers, practitioners, and each year a new, international, interdisciplinary group of research fellows. Through publications and policy discussions, workshops, seminars, and conferences, the Center promotes innovative solutions to significant national and international challenges.

The Center's International Security Program, directed by Steven E. Miller, sponsors and edits the Belfer Center Studies in International Security and the quarterly journal *International Security*.

The Center is supported by an endowment established with funds from Robert and Reneé Belfer, the Ford Foundation, and Harvard University, by foundation grants, by individual gifts, and by occasional government contracts.